Lecture Notes in Applied and Computational Mechanics

Volume 18

Series Editors

Prof. Dr.-Ing. Friedrich Pfeiffer
Prof. Dr.-Ing. Peter Wriggers

Springer

Berlin
Heidelberg
New York
Hong Kong
London
Milan
Paris
Tokyo

Dynamics and Bifurcations of Non-Smooth Mechanical Systems

Remco I. Leine
Henk Nijmeijer

Springer

Dr. Remco I. Leine
ETH Zentrum
Center of Mechanics, CLA H21.2
Tannenstr. 3
8092 Zürich
Switzerland

Professor Dr. Henk Nijmeijer
Dept. of Mechanical Engineering
Eindhoven University of Technology
P.O. Box 513
5600 MB Eindhoven
The Netherlands

With 95 Figures

Library of Congress Control Number: 2004105237

ISBN 3-540-21987-0 Springer Verlag Berlin Heidelberg New York

Springer-Verlag is a part of Springer Science+Business Media

springeronline.com

Printed in Germany

Cover design: design & production GmbH, Heidelberg
Typesetting: Digital data supplied by authors

Printed on acid-free paper 62/3020Rw -5 4 3 2 1 0

Preface

The study of the dynamics of non-smooth systems is becoming a fashionable topic in the scientific community. A look at the programs of international conferences from the past few years reveals a rapidly increasing interest in non-smooth systems. Non-smooth models appear in many different disciplines. Mechanical engineers study stick-slip oscillations in systems with dry friction and the dynamics of impact phenomena. Electrical circuits contain diodes and transistors, which ideally behave in a non-smooth way. Control theorists have to deal with switching control laws. Similar problems of switching systems arise in air traffic management, economic models of markets and scheduling of automated railway systems. Although these examples come from very different fields, the mathematical structure and the related questions of interest are very similar. In particular, the time evolution is often described by non-smooth differential equations or (measure) differential inclusions with possible state re-initializations.

Our interest in non-smooth systems, which has led us to write a book on this topic, has originated in different ways. The first author conducted his MSc. research project at Shell Research (Rijswijk, The Netherlands) on torsional stick-slip vibrations in oilwell drillstrings, which initiated his interest in non-smooth mechanical systems, and went more deeply into the subject during his Ph.D. research under the supervision of Professor Dick van Campen. The second author became more interested in the dynamics of non-smooth systems during the time that the first author was working in his group. Three post-doctoral visits of the first author to Professor F. Pfeiffer (TU München), Dr. B. Brogliato (INRIA Grenoble) and Professor Ch. Glocker (ETH Zürich) drew our attention to the work of the Non-smooth Mechanics community. During the last decades, the Non-smooth Mechanics community has developed a formulation for non-smooth systems, the mathematical prerequisites (Convex Analysis), as well as dedicated numerical algorithms. Despite its use, the theory on non-smooth systems based on Convex Analysis is mostly unknown in the Nonlinear Dynamics community. The study of Nonlinear Dynamics is primarily concerned with bifurcations and chaos. A subdiscipline studies the nonlinear dynamical behaviour of low-dimensional non-smooth systems. We came to the conclusion, that it would be good to have a book which combines the knowledge of both the field of Nonlinear Dynamics and Non-smooth Mechanics. This book tries to fulfil that role and presents a framework for a class of non-smooth mechanical systems using techniques from both fields.

The book is written to address researchers in engineering and mathematics interested in the modelling, simulation and dynamics of non-smooth systems. Mechanical systems are taken as examples throughout the book, because of the practical relevance of and our every day experience with mechanical systems, and surely because of the background of the authors. The title of the book therefore combines the words 'dynamics and bifurcation' with the words 'non-smooth dynamical systems', with which the authors try to make this book a bridge between the two scientific communities. The same notation for the formulation of non-smooth systems is used as in the book of Glocker (2001).

Finally, we take the opportunity to thank the people and institutions which made this book possible. We especially like to acknowledge the encouragement of Professor Dick van Campen, being the initiator of the research on non-smooth systems at Eindhoven University of Technology. The first part of Chapter 7 was developed together with Dr. Nathan van de Wouw, to whom we are very grateful. The Royal Dutch Academy of Sciences as well as the Dutch Technology Foundation financed the research projects on which large parts of this book is based. Professor Christoph Glocker of the ETH Zürich has been a valuable source of information on Non-smooth Dynamics and the vivid discussions with him as well as his advise have greatly improved the book.

Zürich, January 2004 *Remco I. Leine*

Eindhoven, January 2004 *Henk Nijmeijer*

Contents

Notation

$\mathbb{R}$	set of all real numbers
$\mathbb{N}$	set of all natural numbers
inf	infinum
sup	supremum
$\forall$	for all
x	scalar
$\boldsymbol{x}$	column-vector in $\mathbb{R}^n$
$\boldsymbol{X}$	matrix in $\mathbb{R}^{n \times m}$
$\boldsymbol{I}$	identity matrix
$\det(\boldsymbol{A})$	determinant of $\boldsymbol{A}$
$\mathrm{adj}(\boldsymbol{A})$	adjoint matrix of $\boldsymbol{A}$
$\boldsymbol{x}^{\mathrm{T}}$	transpose of $\boldsymbol{x}$
$\boldsymbol{x} = \mathrm{col}\{x_i\}$	column-vector $\boldsymbol{x} = \begin{bmatrix} x_1 \; x_2 \; \cdots \; x_n \end{bmatrix}^{\mathrm{T}}$
$\boldsymbol{X} = \mathrm{mat}\{\boldsymbol{x}_i\}$	matrix of columns $\boldsymbol{X} = \begin{bmatrix} \boldsymbol{x}_1 \; \boldsymbol{x}_2 \; \cdots \; \boldsymbol{x}_n \end{bmatrix}$
$\boldsymbol{X} = \mathrm{diag}\{\boldsymbol{x}\}$	diagonal matrix with $X_{ii} = x_i$, $X_{ij} = 0$, $j \neq i$
x_i	i-th element of $\boldsymbol{x}$
$\lvert x \rvert$	absolute value of x
$\lVert \boldsymbol{x} \rVert$	Euclidian norm of $\boldsymbol{x}$
$\lVert \boldsymbol{x} \rVert_{\boldsymbol{M}}$	norm of $\boldsymbol{x}$ with respect to $\boldsymbol{M}$
$[\boldsymbol{x}]$	arbitrary scalar measure of $\boldsymbol{x}$
$\dot{x}(t)$	differentiation with respect to time, i.e. $\frac{dx(t)}{dt}$
$\boldsymbol{a} \perp \boldsymbol{b}$	$\boldsymbol{a}$ is orthogonal to $\boldsymbol{b}$
B_r	the n-dimensional open ball with radius r centred at the origin
$[a, b]$	the closed interval $\{x \in \mathbb{R} \mid a \leq x \leq b\}$
$\{a, b\}$	the set comprising the elements a and b
C^k	class of functions differentiable up to order k

$\emptyset$	empty set
$\bar{C}$	closure of C
$\overline{\text{co}}\, C$	closed convex hull of C
$\text{int}\, C$	interior of C
$\text{bdry}\, C$	boundary of C
$x \in C$	x is an element of C (inclusion)
$f(\boldsymbol{x})$	single-valued function $\mathbb{R}^n \to \mathbb{R}$
$\boldsymbol{f}(\boldsymbol{x})$	single-valued function $\mathbb{R}^n \to \mathbb{R}^n$
$\boldsymbol{F}(\boldsymbol{x})$	set-valued function $\mathbb{R}^n \to \mathbb{R}^n$
$f'(x)$	classical derivative of f with respect to x
$\nabla \boldsymbol{f}(\boldsymbol{x})$	classical gradient of $\boldsymbol{f}$ with respect to $\boldsymbol{x}$
$\partial f(\boldsymbol{x})$	generalized differential of f with respect to $\boldsymbol{x}$ (or subdifferential if f is a convex function)
$\boldsymbol{J}(\boldsymbol{x})$	(generalized) Jacobian matrix
$\text{graph}(f)$	graph of f
$\text{epi}(f)$	epigraph of f
$f^*(\boldsymbol{x}^*)$	conjugate function
$\Psi_C(\boldsymbol{x})$	indicator function of C at $\boldsymbol{x}$
$\Psi_C^*(\boldsymbol{x})$	support function of C at $\boldsymbol{x}$
$N_C(\boldsymbol{x})$	normal cone of C at $\boldsymbol{x}$
$K^\perp$	polar cone to the cone K
$\text{prox}_C(\boldsymbol{x})$	proximal point from $\boldsymbol{x}$ to the convex set C
$\text{dist}_C(\boldsymbol{x})$	distance from $\boldsymbol{x}$ to the convex set C
$\text{sign}(x)$	sign function with $\text{sign}(0) = 0$
$\text{Sign}(x)$	set-valued sign function
$\boldsymbol{\Phi}(t)$	(generalized) fundamental solution matrix
$\boldsymbol{S}$	saltation matrix
μ	friction coefficient; arbitrary bifurcation parameter
$(\boldsymbol{x}, \boldsymbol{y}) = LCP(\boldsymbol{A}, \boldsymbol{b})$	$\boldsymbol{y} = \boldsymbol{A}\boldsymbol{x} + \boldsymbol{b}$ and $\boldsymbol{0} \le \boldsymbol{y} \perp \boldsymbol{x} \ge \boldsymbol{0}$
$\boldsymbol{x}_{i+1} = \boldsymbol{P}(\boldsymbol{x}_i)$	discrete map

1 Introduction

Why should one bother to study non-smooth systems? After all, most models used in engineering practise are smooth and those which are non-smooth are often studied by smooth approximations. And why should one like to consider bifurcations of non-smooth systems? The number of textbooks on bifurcation theory is large and the field of bifurcations is becoming a classical subfield of dynamical systems. In this introductory chapter we explain why we *do* study non-smooth models and we will give a motivation for this monograph. After discussing different types of non-smooth systems and their characteristics and after giving the basic terminology, we present a literature survey and define the objective of the book. An outline of the succeeding chapters is given next.

1.1 Motivation

Physical phenomena such as impact, dry friction and backlash in mechanical systems are often studied by means of mathematical models with some kind of discontinuity or non-smoothness. Systems which can be described by a set of first-order ordinary differential equations with a discontinuous right-hand side form a subclass of discontinuous dynamical systems and are addressed as Filippov systems or non-smooth systems of Filippov-type. The focus of this book is on the dynamical behaviour of Filippov systems. Mechanical systems with dry friction constitute an important subclass of Filippov systems and are used throughout the book as examples to illustrate the theory and methods.

Dry friction is a nonlinearity which is abundant in nature, machines and other processes. Although friction may be a desirable property in brakes and violins, it is generally an impediment. The presence of dry friction can induce self-sustained vibrations like machine tool chattering, squealing noise of rail wheels, torsional vibration in oilwell drillstrings and many other vibrations which are highly detrimental to the performance of mechanical systems. Dry friction is therefore an important topic in scientific and engineering research.

During the last decades, tribologists have conducted many experiments on sliding contact and have explained the observed frictional phenomena with respect to surface chemistry and physics. The classical Amontons-Coulomb model has been validated but also more complex tribological models are now available which take some of the underlying mechanisms of friction into account. Mechanical engineers and applied mathematicians have been concerned with the dynamical behaviour of systems with friction. Studies on frictional dynamics are almost all based on the classical Amontons-Coulomb model for dry friction. The Amontons-Coulomb model has the advantage to have a limited amount of parameters and to account only for the essential features of dry friction, while more complex friction models are only validated for one type of material combination and are therefore not generally applicable. Furthermore, the Amontons-Coulomb model allows for an efficient numerical integration of the equations of motion of a system subjected to dry friction.

Application of the Amontons-Coulomb model to dynamical models of systems with dry friction results in differential equations (or differential inclusions) of Filippov-type. Filippov systems form a class of non-smooth systems described by differential equations with a discontinuous right-hand side. To the class of Filippov systems do not only belong mechanical systems with Amontons-Coulomb friction but also electrical circuits with diode elements, controlled systems with switching control laws, mechatronic systems with encoders and many other systems, being mechanical or non-mechanical, where a kind of switching is involved.

The undesired friction-induced vibrations can be prevented by changing the design of the system or can be combatted with the aid of control techniques. Knowledge of the dynamical behaviour of the system is therefore essential to improve the performance. Profound insight in the behaviour of dynamical systems can be gained from modern analysis techniques developed in the Nonlinear Dynamics community. The theoretical knowledge and available methods are, however, mostly applicable to smooth dynamical systems. This motivates a theoretical study on the nonlinear dynamics of Filippov systems.

The nonlinear dynamics of mechanical Filippov systems is explored in this monograph. Systems with dry friction constitute an important example of mechanical Filippov systems and are used as key applications of the theory throughout the book. Firstly, we try to provide a theoretical basis for practical investigations of systems of this class. Secondly, it may form a starting point for further mathematical investigations. Hence, in the book we attempt to establish a bridge between engineering-oriented and mathematics-oriented research in this field.

The next section reviews some basic terminology from the theory on nonlinear dynamics and bifurcations.

1.2 Nonlinear Dynamics and Bifurcations

Mathematical models are frequently used in many disciplines of science to study complex behaviour of systems. Systems that can be modelled by nonlinear differential equations are called *nonlinear dynamical systems*. Examples of such systems occur in economics (economical models of markets), biology (predator-prey models), chemistry (A-B-C reactions), electronics (electrical circuits with nonlinear elements) and mechanics (models of machines or moving bodies in general).

A set of differential equations or differential inclusions, describing a dynamical system, forms together with an initial condition an initial value problem. The solution to the initial value problem, provided that it exists, constitutes one or more solution curves in the state space through the initial point. Such a solution curve of an autonomous system is called a *trajectory* in the state space. If the solution is unique then there is a single trajectory through the initial point, whereas there are multiple trajectories if the solution is non-unique. The trajectory of a dynamical system starting from a particular initial state, can evolve towards a stationary state, usually referred to as an *equilibrium*. It may also evolve towards a periodic state, called *periodic solution* here. Equilibria and periodic solutions are both examples of *limit sets*. A *limit set* contains all possible limit points, i.e. points that can be approached in forward or backward (infinite) time, and therefore determines the asymptotic behaviour of the system. Limit sets include equilibria, periodic solutions, quasi-periodic solutions (a limit set that can be expressed as a countable sum of periodic functions with incommensurate frequencies) and chaotic attractors (bounded state behaviour that is not an equilibrium, not periodic and not quasi-periodic). Equilibria, periodic solutions and quasi-periodic solutions can be asymptotically stable, thereby attracting neighbouring trajectories. An isolated periodic solution, possessing a neighbourhood that contains no other periodic solutions, is called a *limit cycle*. The knowledge of the steady states of a system for a certain parameter set, together with their stability properties, is important as they partly determine the long term dynamical behaviour of the system.

It is often desirable to know how the equilibria and periodic solutions of a system change when a parameter of the system is changed. Such parameter studies are usually conducted by means of path-following (or continuation) techniques where a branch of equilibria or periodic solutions is followed while varying a parameter. The number and type of equilibria and periodic solutions (being stable or unstable) can change at a certain parameter value. This qualitative change in the structural behaviour of the system is loosely called *bifurcation*, an originally French word introduced by Poincaré [122]. A more strict definition of bifurcation will be given in Chapter 8.

Bifurcations occur in many different disciplines of science. Examples of bifurcations can be found in morphodynamics (the forming of meanders in rivers), structural mechanics (the buckling of a beam), flutter oscillation of suspension bridges, hunting motion of rail-way bogies and cardiac arrhythmias in malfunctioning hearts.

1.3 Non-smooth Systems versus Smooth Systems

Why should one bother to study non-smooth models if we can approximate the non-smoothness or discontinuity with a convenient smooth function? This question, which was already raised at the beginning of the chapter, is regularly turning up in discussions and has been a source of heated debate.

Physical systems can often operate in different modes, and the transition from one mode to another can sometimes be idealized as an instantaneous, discrete transition. Examples include mechanical systems with dry friction, impact and backlash or electrical circuits with diode elements. Since the time scale of the transition from one mode to another is often much smaller than the time scale of the dynamics of the individual modes, it may be very advantageous to model the transitions as being instantaneous. For instance, diode elements are often modelled as being ideal diodes and mathematical models of mechanical systems subjected to dry friction are considered to switch between a slip mode and a (pure) stick mode. The mathematical modelling of physical systems, therefore, may lead to non-smooth dynamical systems, which switch between different modes, where the dynamics in each mode is associated with a different set of smooth differential equations.

The reason why we do use non-smooth models is the desire to describe a system at a certain *abstraction level*, leaving part of the dynamics aside and describing only the global dynamics of interest. For instance, if we like to describe a bouncing tennis ball on a racket, then we may choose a simple restitution law relating post-impact conditions to pre-impact conditions resulting in a non-smooth model. The non-smooth 'restitution-type' model describes the global motion of the ball as a rigid body. However, if we are interested in the deformation of the ball and resulting stresses, then we need a model on a lower abstraction level and we may choose an elastic compliant model, for instance using a Hertz contact model or by using finite elements, which 'removes' the discontinuity of the velocity. A finite element model will give a much more accurate description of the transition phase during the impact at the cost of computational effort. A further step down in the abstraction level will arise when we consider the material to be plastic. Plasticity laws are often non-smooth, which will again result in a non-smooth model. It is therefore wrong to say that 'the physical reality is smooth'. *Models* of the physical reality are smooth or non-smooth depending on the abstraction level we choose to fit our needs.

The tennis ball example is somewhat inconclusive, because both a non-smooth rigid-body description and a smooth compliant model can be used to obtain a reasonable simulation of the fairly soft tennis ball without too much computational effort. However, if we would like to model the dynamics of a coin in a ticket machine of a car park, then there is no need to describe the deformation of the coin and various parts of the ticket machine. An engineer will naturally choose a rigid-body abstraction level which leads to a non-smooth model.

Dry friction in mechanical systems is often modelled by the set-valued Coulomb's law, resulting in a non-smooth model for the system. The complicated transition processes between 'slip' and 'stick' are generally unknown and are conveniently left aside by the non-smooth model, modelling the transition to be instantaneous. Very often, the non-smooth dry friction model is subsequently smoothened with an arctangent-type of approximation, in order to simplify the numerical simulation. The resulting set of differential equations is indeed smooth and standard integration techniques can be applied. However, the disadvantages are threefold:

1. Smoothening of discontinuous force laws results in stiff differential equations. The numerical integration of stiff differential equations requires much computational effort and an enormous step size reduction. Moreover, dedicated numerical integration techniques exist for non-smooth systems and will be discussed in Chapter 6.
2. Possible limit sets, which exist in the non-smooth model, might be lost due to the smoothening. For instance, the stationary state of a block on a slope can not be described with a smooth friction model (the block will always slide). The smoothening of a discontinuous (or set-valued) constitutive law does therefore not necessarily lead to a better description of the physical reality.
3. The 'event'-character of the non-smooth model is lost or blurred due to the smoothening, ruling out some analysis methods, such as the event map method, which rely on this.

Apparently, smoothening of an original non-smooth model destroys most of the advantages of the non-smooth modelling approach, which corresponds to the chosen abstraction level.

1.4 Terminology of Non-smooth Systems

In this section we try to clarify the terminology around non-smooth dynamical systems. The term 'non-smooth system' or 'discontinuous system' is often used in literature without stating explicitly which properties of the system are considered to be 'non-smooth'. Non-smooth dynamical systems can be divided in three types according to their degree of non-smoothness:

1. Non-smooth continuous systems with a discontinuous (or set-valued) Jacobian matrix, such as systems with purely elastic one-sided supports. Those systems are described by a continuous vector field but the vector field is non-smooth.

<table>
<tr><th>model</th><th>description</th><th>function</th></tr>
<tr>
<td>(1) elastic support
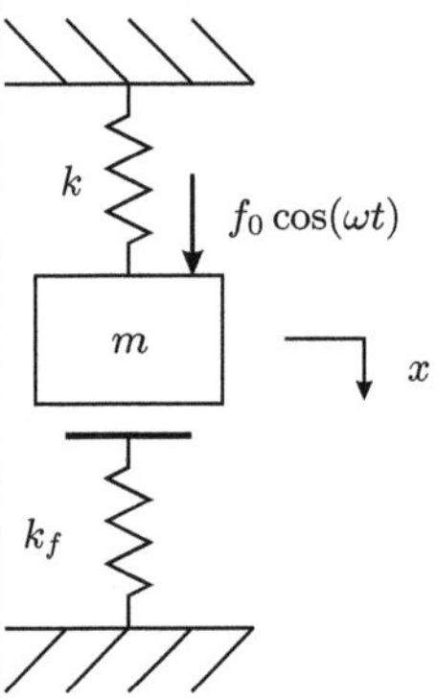
</td>
<td>$m\ddot{x} + kx = f_0 \cos(\omega t) - f(x)$
$f(x) = \begin{cases} 0, & x \leq 0 \\ k_f x, & x > 0 \end{cases}$</td>
<td>non-smooth continuous
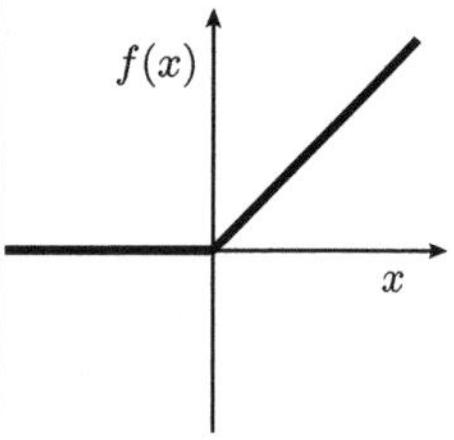
</td>
</tr>
<tr>
<td>(2) dry friction
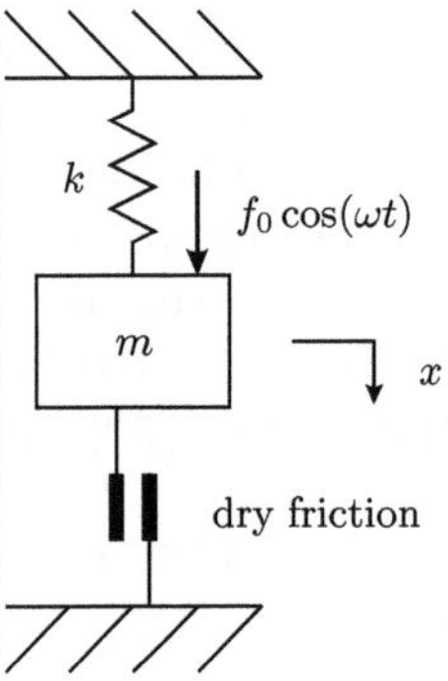
</td>
<td>$m\ddot{x} + kx = f_0 \cos(\omega t) - f(\dot{x})$
$f(\dot{x}) \in \begin{cases} -F_s, & \dot{x} < 0 \\ [-F_s, F_s], & \dot{x} = 0 \\ F_s, & \dot{x} > 0 \end{cases}$</td>
<td>discontinuous of Filippov-type
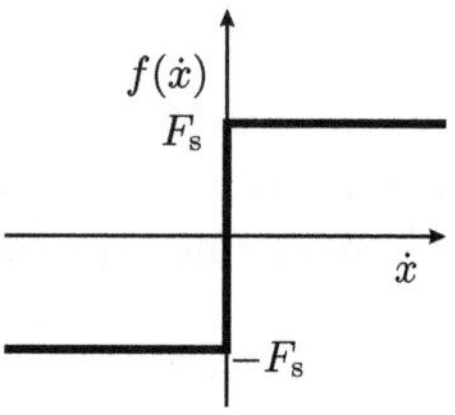
</td>
</tr>
<tr>
<td>(3) impact
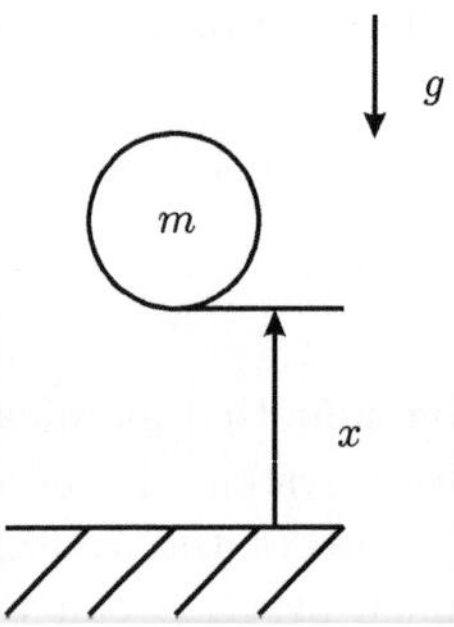
</td>
<td>$m\ddot{x} = -g + \lambda$
$x(t_k) = 0$
$\Lambda = \int_{t_k} \lambda \mathrm{d}t$
$\Lambda \geq 0$
$\dot{x}_+(t_k) + e\dot{x}_-(t_k) \geq 0$
$(\dot{x}_+(t_k) + e\dot{x}_-(t_k))\Lambda = 0$</td>
<td>impulsive
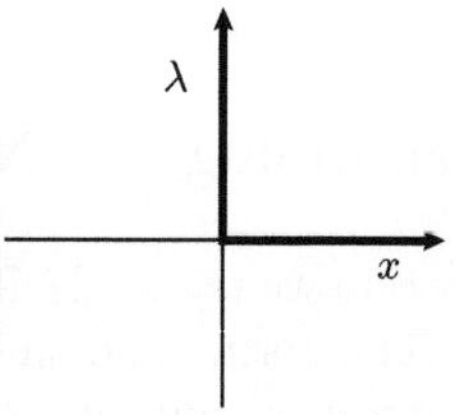
</td>
</tr>
</table>

Table 1.1. Non-smooth and discontinuous systems.

2. Systems described by differential equations with a discontinuous right-hand side, but with a time-continuous state. The vector field of those systems is discontinuous. Systems of this type can be transformed in differential inclusions with set-valued right-hand side and are called *Filippov systems* (Chapter 3). Examples are systems with visco-elastic supports and dry friction.
3. Systems which expose discontinuities (or jumps) in the state, such as impacting systems with velocity reversals.

Table 1.1 shows examples of each type of system together with its characteristic function for the non-smooth force. The first example is a mass on a one-sided elastic spring, where the spring force is a non-smooth continuous function of the position. The second example is a mass with a dry friction damper, where the dry friction force is a discontinuous (or set-valued) function of velocity. The third example is an impacting ball on a rigid floor, where the contact force exhibits a Dirac pulse at the instant of impact. More correct is to describe impulsive systems in terms of measure differential inclusions. In all three cases a kind of switching is involved and those systems are therefore often called 'switching systems' or 'differential equations with switching conditions' [27]. In the field of systems and control theory, the term *hybrid system* is frequently used for systems composed of continuous differential equations and discrete event parts [12]. Nowadays, the term 'hybrid system' is used for any system which exposes a mixed continuous and discrete nature [56]. Non-smooth (or switching/hybrid) systems can in some cases be considered as dynamical extensions of (non)linear complementarity problems, which gives another term: *complementarity systems* (see [56] and references therein).

1.5 Literature Survey

The scientific community interested in non-smooth mechanical systems seems to be divided in two subdisciplines. The first subdiscipline forms part of the Nonlinear Dynamics community and traditionally studies stick-slip oscillations *or* periodic motion with impacts. The emphasis lies more on the dynamics, bifurcations and chaos than on the (consistent) formulation of the non-smooth problem. Typically, the number of contact points considered does not exceed one or two and the systems are usually low-dimensional. The second subdiscipline is called the Non-smooth Mechanics community and typically uses methods from Non-smooth and Convex Analysis to formulate non-smooth constitutive laws. The number of contact points considered can reach over 10.000 (for instance in granular media). The Non-smooth Mechanics community also incorporates part of the Plasticity community, which formulates plastic flow rules using concepts of Convex Analysis, and has a strong link with the Non-smooth Optimization community. The primary goal of the Non-smooth Mechanics community is the proper formulation and efficient simulation of non-smooth mechanical systems. The nonlinear analysis of the systems considered by the Non-smooth Mechanics community is usually left aside, partly because of the complexity of the systems. The two subdisciplines hardly seem to have influenced each other during the last three decades, although this is slowly changing.

We will first briefly review the mathematical literature on Non-smooth Analysis and differential inclusions, which forms a prerequisite for the study of non-smooth mechanical systems. Subsequently, the contributions from Non-smooth Mechanics are discussed. Finally, the literature on non-smooth mechanical systems from the field of Nonlinear Dynamics is reviewed with an emphasis on bifurcations in non-smooth systems. Some publications contribute to both schools.

Non-smooth Analysis and Differential Inclusions

The book [128] by ROCKAFELLAR is regarded to be the classical reference on Convex Analysis. Meanwhile, many other books on Convex Analysis have been published, such as the books [60,61] by HIRIART-URRUTY and LEMARÉCHAL, the book [130] by ROCKAFELLAR and WETS and the short overview in the encyclopedia [44] of GAMKRELIDZE by THIKOMIROV.

The generalized gradient, which is used extensively here, is introduced in the book [14] by CLARKE and is also presented in the book [15] by CLARKE *et al.* The book [7] by AUBIN and CELLINA is an extensive reference on set-valued functions.

Theoretical results and numerical methods on linear complementarity problems can be found in the book [100] by MURTY and in the work of COTTLE and DANTZIG [17].

The fundamental work of FILIPPOV [36, 37, 131] extends a discontinuous differential equation to a differential inclusion (see Chapter 3). More results on differential inclusions can be found in the book [7] by AUBIN and CELLINA, the book [22] by DEIMLING and in the book [15] by CLARKE *et al.*. UTKIN introduced in [140–143] the concept of 'equivalent control', which is similar to Filippov's convex method, and is popular in control theory. Some recent literature on control theory focuses on hybrid systems and complementarity systems, which encompasses also Filippov systems. References on hybrid systems can be found in the work of HEEMELS [56] and VANDERSCHAFT and SCHUMACHER [147].

AIZERMAN and GANTMAKHER [1] derived jumping conditions of fundamental solution matrices (see Chapter 7) and their results have been extended to systems with a discontinuous state by MÜLLER [99].

Non-smooth Mechanics

The number of publications on non-smooth systems in the field of Non-smooth Mechanics is vast. The book [12] by BROGIATO gives an extensive review as well as the article of BROGLIATO, TEN DAM, PAOLI and ABADIE [13]. Only the main publications and those which are of relevance for the next chapters will be briefly reviewed here.

The work of MOREAU and JEAN [67, 96, 97] and of PANAGIOTOPUOLOS [112, 113] has fulfilled a pioneering role in the Non-smooth Mechanics community. MOREAU and JEAN developed a framework to describe frictional impact and unilateral contact using Convex Analysis to its full extent. The framework of MOREAU treats the dynamics of systems with impact on velocity–impulse level in terms of measure differential inclusions. This description leads in a natural way to a numerical integration method called the time-stepping method (see Chapter 6). The work of PANAGIOTOPOULOS focusses on variational formulations and on the formulation of contact laws in elastostatics. MOREAU, JEAN and PANAGIOTOPOULOS used iterative procedures, such as the Gauß–Seidel iteration method, to evaluate multi-contact problems. ALART and CURNIER [2] presented an augmented Lagrangian approach to the solution of static frictional contact problems (see also the work of SIMO and LAURSEN [76, 134] and KLARBRING [72]). The application of the augmented Lagrangian method to dynamic frictional contact problems is presented in LEINE and GLOCKER [80]. The augmented Lagrangian method was originally proposed by HESTENES [57] and POWELL [125] to solve nonlinear programming problems with equality constraints. ROCKAFELLAR [129] extended the method to inequality constraints (such as frictionless unilateral contact problems).

GLOCKER, PFEIFFER and co-workers applied the theory on linear complementarity problems and Convex Analysis on rigid multibody systems with impact and friction. In GLOCKER [46] and the book of PFEIFFER and GLOCKER [118], unilateral contact laws and Coulomb friction are formulated as linear complementarity problems on acceleration level and velocity level and simulation is performed using an event-driven integration method (see Chapter 6). The theory and methods were applied to a variety of engineering applications, see e.g. [121, 138, 151] and the overview in [118]. Later, spatial frictional contact problems were formulated as solvable nonlinear complementarity problems in GLOCKER [49, 51] and STIEGELMEYR [139]. The theory of set-valued force laws (i.e. set-valued constitutive laws) was put in a framework of non-smooth potential theory in the book [51] by GLOCKER. The book [51] discusses non-smooth extensions to the classical variational principles in dynamics, e.g. Gauß, Jourdain, d'Alembert/Lagrange (see also GLOCKER [47, 48]). A set-valued force law for spatial Coulomb-Contensou friction (combined sliding and drilling friction) can be found in LEINE and GLOCKER [80]. Time-stepping methods were found to be more convenient than event-driven integration methods for systems with many contact points and were studied by STIEGELMEYR [139]. Several other researchers studied the time-stepping method, such as STEWART and TRINKLE [137], ANITESCU and POTRA [5] and PAOLI and SCHATZMAN [114, 115] who also studied the convergence of the time-stepping method.

Non-existence and non-uniqueness of solutions of mechanical systems with frictional unilateral contact is discussed in GÉNOT and BROGLIATO [12, 41] and LEINE, BROGLIATO and NIJMEIJER [79].

Nonlinear Dynamics and Bifurcations

Publications on non-smooth mechanical systems in the field of Nonlinear Dynamics are mainly concerned with dry friction/stick-slip oscillations and impact. Some publications discuss the modelling and simulation of systems with impact and/or friction, while others are focussed on the bifurcations of periodic motion and chaos in non-smooth systems.

An extensive literature review on dry friction models can be found in review articles of ARMSTRONG *et al.* [6], FEENY *et al.* [29] and IBRAHIM [62, 63]. The book of JOHNSON [69] provides an introduction to contact mechanics from a more physical point of view.

The mathematical literature is mainly concerned with existence and uniqueness of solutions of discontinuous differential equations. In the work of DEIMLING [23], FEČKAN [34] and KUNZE [73, 74] dry friction problems are treated as differential inclusions and existence of periodic solutions is addressed.

During the last decades many textbooks about bifurcation theory for smooth systems appeared and bifurcations of periodic solutions in smooth vector fields are well understood, see for instance the books of GUCKENHEIMER and HOLMES [52], KUZNETSOV [75], SEYDEL [133] and WIGGINS [150]. However, much less research was done on bifurcations of periodic solutions in non-smooth vector fields.

ANDRONOV, VITT and KHAIKIN [4] treat periodic solutions of discontinuous systems. They reveal many aspects of discontinuous systems and address periodic solutions with sliding modes (Chapter 3) but do not treat periodic solutions in non-smooth systems with regard to Floquet theory underlying those solutions.

Many publications deal with bifurcations in non-smooth systems of Filippov-type. Published bifurcation diagrams are often constructed from data obtained by brute force techniques and only show stable branches of periodic solutions [8–10, 20, 42, 43, 58, 59, 74, 107, 123, 124, 135, 148] (this list is far from complete). Bifurcation diagrams calculated with path-following techniques show bifurcations to unstable periodic solutions but the bifurcations behave as conventional bifurcations in smooth systems [136, 145]. Experimental bifurcation diagrams of non-smooth systems compared with numerical results using the path-following technique can be found in [94] (dry friction) and in [144] (impact).

DANKOWICZ and NORDMARK [20] study bifurcations of stick-slip oscillations but the applied friction model, with internal states which allow for history and rate dependency, yields a non-smooth continuous system. A small number of publications show non-conventional bifurcations in Filippov systems which can not be understood with the classical bifurcation theory for smooth systems, see for instance [28,152]. YOSHITAKE and SUEOKA [152] address Floquet theory and remark that the Floquet multipliers 'jump' at the bifurcation point. The work of FEIGIN [30–33] and DI BERNARDO *et al.* [24–26] studies non-conventional bifurcations in Filippov systems by means of mappings and refers to those bifurcations as 'C-bifurcations'. Non-conventional bifurcations of non-smooth discrete mappings were also addressed by NUSSE and YORKE [106].

Non-conventional bifurcations of periodic solutions of Filippov systems are furthermore studied in the work of LEINE *et al.* [78,82–85,87,88], and those bifurcations are addressed as *discontinuous bifurcations*. The basic idea is that Floquet multipliers of Filippov systems can 'jump' when a parameter of the system is varied, as was noticed in [152]. If a Floquet multiplier jumps through the unit circle in the complex plane then a discontinuous bifurcation can be encountered. In the work of LEINE *et al.* it is explained how the discontinuous bifurcations come into being through jumps of the fundamental solution matrix and it is shown how discontinuous bifurcations are related to conventional bifurcations in smooth systems.

Another type of non-conventional bifurcation is the 'grazing bifurcation' which occurs in impacting systems. A grazing bifurcation can occur in a rigid multibody system with unilateral contacts if the pre-impact velocity becomes zero. Bifurcations in impacting systems are studied by [12,39,65,66,81,92,95, 105,117].

Numerical methods to calculate periodic solutions in discontinuous systems can be found in EICH and FÜHRER [27], LEINE *et al.* [86], MEIJAARD [93] and REITHMEIER [127].

1.6 Objective and Scope

Many practical problems in engineering are related to vibrations caused or influenced by transitions between modes (e.g. stick, slip, contact, no contact). Depending on the way of modelling, a mathematical model of the physical system may belong to one of the three classes of non-smooth dynamical systems mentioned in the Section 1.4. This urges for a description of the bifurcation behaviour of non-smooth dynamical systems. A theory for bifurcations in non-smooth systems is currently not completely developed. The theory on set-valued force laws and non-smooth potentials from Non-smooth Mechanics can provide a solid basis for the study of bifurcations in non-smooth systems, which might give an impulse for further research in the field. Furthermore, the analysis methods for non-smooth systems, developed by the Nonlinear Dynamics community, are useful tools for the Non-smooth Mechanics community to enlarge the insight in the dynamic behaviour of non-smooth systems. The main objective of the book is to present a framework for non-smooth mechanical systems using techniques from both fields.

The class of non-smooth systems has a too large spectrum of complicated phenomena to study in this book and we will therefore have to narrow our focus. Existence of solutions for systems with a discontinuous state is not *a priori* guaranteed, which complicates the study of bifurcations of mechanical systems with impact. In this sense, Filippov systems are less complex as existence of solutions is guaranteed (under some mild conditions, see Chapter 3). We will therefore confine our study to Filippov systems and non-smooth continuous systems (which can be regarded as a subclass of Filippov systems). Filippov systems embrace systems with dry friction and compliant impact but not systems with impact between rigid bodies. Filippov systems arise also in models of electrical circuits with (ideal) diode elements, controlled systems with encoders and in other scientific fields. In this book, however, we will focus on mechanical systems although the results apply to Filippov systems in general. As mentioned before, an important category is the class of systems with dry friction and we will address these types of systems in Sections 4.2, 6.5, 7.3, 9.8.2 and 9.9.1. We will especially pay much attention to periodic solutions of systems with dry friction. Bifurcations of periodic solutions of Filippov systems are closely related to bifurcations of fixed points in non-smooth continuous mappings. Fixed points of mappings can be regarded as equilibria of discrete-time systems. We will therefore also address bifurcations of equilibria in non-smooth continuous systems.

The aims of this book are:

- to provide the reader with mathematical background information on Non-smooth Analysis and Convex Analysis,
- to introduce the reader to differential inclusions,
- to present a framework to formulate rigid multibody systems with unilateral contact and friction,
- to give an overview of the existing numerical simulation methods for mechanical systems with friction,
- to give a thorough introduction to Floquet theory for non-smooth systems of Filippov-type.
- to study bifurcations of equilibria in non-smooth continuous systems and of periodic solutions in discontinuous systems of Filippov-type.

We do not aim to give a complete theory of bifurcations in non-smooth systems, but merely make clear why the bifurcation behaviour in non-smooth systems is so complicated and what the complications are when trying to formulate such a theory.

1.7 Outline

The book contains two parts. The first part (Chapters 2–6) provides the reader with the mathematical background and focuses on the modelling and simulation of mechanical systems with dry friction. The second part of the book (Chapters 7–9) proceeds with an investigation of periodic solutions and bifurcations in non-smooth dynamical systems.

In Chapter 2 some basic mathematical theory from non-smooth analysis is reviewed, including set-valued functions, some topics from Convex Analysis and the notion of the generalized differential for non-smooth functions. An introduction is given to linear complementarity problems, which will prove to be useful when analyzing non-smooth dynamical systems.

The 'convex method' proposed by Filippov extends a discontinuous differential equation to a differential inclusion. Chapter 3 briefly discusses the behaviour of trajectories of dynamical systems described by differential inclusions.

Dry friction plays an important role in the dynamic behaviour of mechanical systems. We briefly discuss the basic tribological phenomena related to dry friction in Chapter 4. Furthermore, a discussion is given of commonly used friction models which may lead or may not lead to differential inclusions.

Unilateral contact and friction are often modelled with set-valued force laws. Mechanical systems with set-valued force laws lead to differential inclusions and form an important class of non-smooth systems. In Chapter 5 we will discuss the mathematical formulation of mechanical multibody systems with unilateral contact and friction modelled with set-valued force laws.

The mathematical framework developed in Chapter 5 gives the basis for the numerical simulation methods presented in Chapter 6. In this chapter we present numerical techniques for the integration of differential inclusions with sliding modes. More sophisticated integration methods such as the event-driven integration method and time-stepping method for the simulation of mechanical systems with set-valued force laws are presented as well.

In Chapter 7, the discontinuous behaviour of fundamental solution matrices of Filippov systems is discussed. The definition of a fundamental solution matrix for smooth systems is extensively reviewed. The differences in the properties of the fundamental solution matrix between linear and nonlinear systems, as well as between autonomous and non-autonomous systems are discussed. Subsequently, the stability of periodic solutions is addressed by means of the concept of Floquet multipliers. The special behaviour of fundamental solution matrices of Filippov systems is treated and applied to examples.

Chapter 8 deals with bifurcations of equilibria of non-smooth continuous systems. The treatise is restricted to one-parameter bifurcations. The simplest discontinuous bifurcations are treated next to their smooth counterparts which enlarges the insight. A variety of examples are extensively treated, which show the complicated bifurcation behaviour in non-smooth systems. The results on bifurcations of equilibria of non-smooth continuous systems are used as a stepping stone to bifurcations of periodic solutions of Filippov systems.

In Chapter 9 we study bifurcations of periodic solutions in discontinuous systems of Filippov-type. It is explained how a discontinuous bifurcation of a periodic solution can be created when a periodic solution touches a non-smooth switching boundary. The relation with discontinuous bifurcations of equilibria in non-smooth continuous systems is discussed. Fundamental questions about discontinuous bifurcations of periodic solutions are discussed and a number of numerical examples which show discontinuous bifurcations are treated.

Finally, a short overview is given in Chapter 10 followed by concluding remarks on the state of the art of bifurcation theory for non-smooth dynamical systems.

2

Preliminaries on Non-smooth Analysis

In this chapter we review some basic mathematical theory from non-smooth analysis [7, 15, 44, 60, 61, 128]. We begin with a brief introduction to sets and set-valued functions (Section 2.1). The notion of continuity of functions is relaxed in Section 2.2 to semi-continuity for set-valued functions and the notion of the classical derivative of smooth functions is extended in Section 2.3 to generalized differentials for non-smooth functions. Some definitions from convex analysis are reviewed in Section 2.4. Finally, an introduction is given to linear complementarity problems, which will prove to be useful when analyzing non-smooth dynamical systems.

2.1 Sets and Set-valued Functions

A number of properties of sets and set-valued functions will be briefly reviewed. Let C be a subset of the normed space $\mathbb{R}^n$, equipped with the Euclidean norm $\|\cdot\|$.

Definition 2.1 (Closed Set). *A set $C \subset \mathbb{R}^n$ is* closed *if it contains all its limit points. Every limit point of a set C is the limit of some sequence $\{\boldsymbol{x}_k\}$ with $\boldsymbol{x}_k \in C$ for all $k \in \mathbb{N}$.*

We define the *closure* of a set C as the smallest closed set containing C, i.e. $\bar{C} = C \cup \{$ limit points of $C\}$.

Definition 2.2 (Bounded Set). *A set $C \subset \mathbb{R}^n$ is* bounded *if there exists a number c such that $\|\boldsymbol{x}\| \le c$ for all $\boldsymbol{x} \in C$.*

An important property in Non-smooth Analysis is the convexity of sets.

Definition 2.3 (Convex Set). *A set $C \subset \mathbb{R}^n$ is* convex *if for each $\boldsymbol{x} \in C$ and $\boldsymbol{y} \in C$ also $(1-q)\boldsymbol{x} + q\boldsymbol{y} \in C$ for arbitrary q with $0 \le q \le 1$.*

It follows that a convex set contains all line segments between any two points in the set. We define the closed convex hull of $\boldsymbol{x}, \boldsymbol{y} \in \mathbb{R}^n$ as the line segment between $\boldsymbol{x}$ and $\boldsymbol{y}$, i.e. the smallest closed convex set containing $\boldsymbol{x}$ and $\boldsymbol{y}$

$$\overline{\text{co}}\{\boldsymbol{x}, \boldsymbol{y}\} = \{(1-q)\boldsymbol{x} + q\boldsymbol{y}, \forall\, q \in [0,1]\}. \tag{2.1}$$

A function $f : \mathbb{R} \to \mathbb{R}$ associates to any element x of its domain a single function value $f(x)$. A function is therefore single-valued which excludes vertical lines, loops and surfaces on its graph, i.e.

$$\text{graph}(f) = \{(x, f(x)) \mid x \in \mathbb{R}\}. \tag{2.2}$$

We define the epigraph of f to be the set above the graph of f

$$\text{epi}(f) = \{(x, y) \mid y \geq f(x); x \in \mathbb{R}\}. \tag{2.3}$$

Definition 2.4 (Convex Function). *A function f is convex when* epi(f) *is a convex set.*

For a smooth function $f : \mathbb{R}^n \to \mathbb{R}$, it follows that $f(\boldsymbol{x})$ is convex if

$$f(\boldsymbol{x}^*) \geq f(\boldsymbol{x}) + \nabla f(\boldsymbol{x})^{\mathrm{T}}(\boldsymbol{x}^* - \boldsymbol{x}), \quad \forall \boldsymbol{x}^*. \tag{2.4}$$

Similarly, we define a smooth vector-valued function $\boldsymbol{f} : \mathbb{R}^n \to \mathbb{R}^n$ to be convex if

$$\boldsymbol{f}(\boldsymbol{x}^*) \geq \boldsymbol{f}(\boldsymbol{x}) + \nabla \boldsymbol{f}(\boldsymbol{x})^{\mathrm{T}}(\boldsymbol{x}^* - \boldsymbol{x}), \quad \forall \boldsymbol{x}^*, \tag{2.5}$$

where $\nabla \boldsymbol{f}$ is the transpose of the Jacobian of $\boldsymbol{f}$.

A *set-valued function* $F(x)$, $x \in \mathbb{R}$, is a function almost everywhere, except at a finite number of isolated points where $F(x)$ forms a subset of $\mathbb{R}$. A set-valued function can therefore contain vertical segments on its graph. The image of $F(x)$, being a set, is in general not closed or convex. Convexity of a function $F(x)$ should not be confused with convexity of the image of $F(x)$.

Definition 2.5 (Convex Image). *A function F has a* convex image *on $X \subset \mathbb{R}$ if $F(x)$ is a convex set for all fixed values $x \in X$.*

We will reserve lowercase letters for single-valued functions, $\boldsymbol{f}(\boldsymbol{x}) \in \mathbb{R}^n$, and uppercase letter for set-valued functions, $\boldsymbol{F}(\boldsymbol{x}) \subset \mathbb{R}^n$. The term multi-valued function or multifunction is sometimes used instead of set-valued function [60].

2.2 Continuity and Semi-continuity

A function $\boldsymbol{f}(\boldsymbol{x})$ is *continuous* at $\boldsymbol{x} \in X$, $X \subset \mathbb{R}^n$, provided that for all $\varepsilon > 0$, there exists a $\delta > 0$ so that $\boldsymbol{y} \in \boldsymbol{x} + B_\delta \subset X$ implies $\|\boldsymbol{f}(\boldsymbol{x}) - \boldsymbol{f}(\boldsymbol{y})\| \leq \varepsilon$. For single-valued functions this means that we can draw the graph of the function without taking the pencil of the paper. A function $\boldsymbol{f}(\boldsymbol{x})$ is *smooth* if it is continuously differentiable up to any order in $\boldsymbol{x}$.

The notion of continuity can be relaxed for set-valued functions to semi-continuity [15].

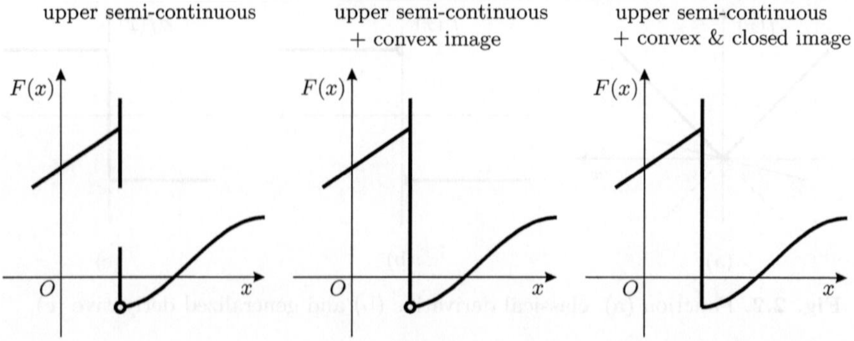

Fig. 2.1. Illustration of upper semi-continuity, convexity and closedness of a set-valued function.

Definition 2.6 (Upper Semi-continuity of Set-valued Functions). *A set-valued function $\boldsymbol{F}(\boldsymbol{x})$ is upper semi-continuous in $\boldsymbol{x}$ if*

$$\lim_{\boldsymbol{y}\to\boldsymbol{x}}\left(\sup_{\boldsymbol{a}\in\boldsymbol{F}(\boldsymbol{y})}\inf_{\boldsymbol{b}\in\boldsymbol{F}(\boldsymbol{x})}\|\boldsymbol{a}-\boldsymbol{b}\|\right)\to 0.$$

This condition is equivalent to the condition that for all $\varepsilon > 0$ there exists a $\delta > 0$ such that $\|\boldsymbol{x}-\boldsymbol{y}\| < \delta \Rightarrow \boldsymbol{F}(\boldsymbol{y}) \subset \boldsymbol{F}(\boldsymbol{x}) + B_\varepsilon$. If a function $\boldsymbol{F}$ is set-valued at a distinct $\boldsymbol{x}$, then the graph of the function in a neighbourhood around $\boldsymbol{x}$ is connected to the set $\boldsymbol{F}(\boldsymbol{x})$. Upper semi-continuity does not imply convexity.

Definition 2.7 (Lower Semi-continuity of Set-valued Functions). *A set-valued function $\boldsymbol{F}(\boldsymbol{x})$ is lower semi-continuous in $\boldsymbol{x}$ if*

$$\lim_{\boldsymbol{y}\to\boldsymbol{x}}\left(\inf_{\boldsymbol{a}\in\boldsymbol{F}(\boldsymbol{y})}\sup_{\boldsymbol{b}\in\boldsymbol{F}(\boldsymbol{x})}\|\boldsymbol{a}-\boldsymbol{b}\|\right)\to 0.$$

This condition is equivalent to the condition that for all $\varepsilon > 0$ there exists a $\delta > 0$ such that $\|\boldsymbol{x}-\boldsymbol{y}\| < \delta \Rightarrow \boldsymbol{F}(\boldsymbol{x}) \subset \boldsymbol{F}(\boldsymbol{y}) + B_\varepsilon$. A finite valued function $\boldsymbol{F}(\boldsymbol{x})$, that is both upper and lower semi-continuous is also continuous. Upper and lower semi-continuity of set-valued functions are sometimes called outer and inner semi-continuity [60].

The notions of upper semi-continuity, convexity and closedness of (the images of) set-valued functions are illustrated in Figure 2.1.

2.3 Generalized Differential of Continuous Functions

The classical derivative of smooth continuous functions will be extended in this section to the generalized derivative (and differential) of Clarke for non-smooth continuous functions.

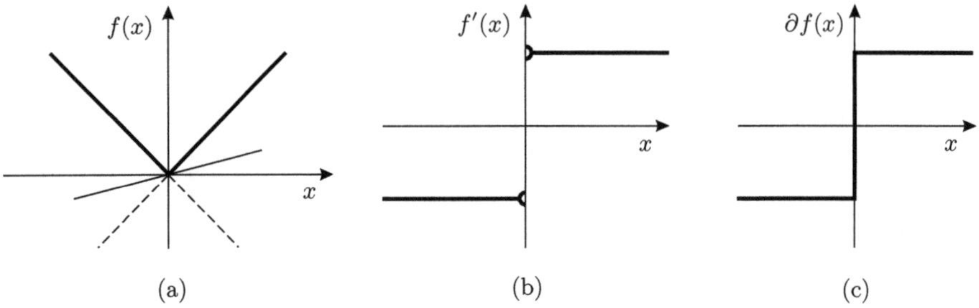

Fig. 2.2. Function (a), classical derivative (b) and generalized derivative (c).

Consider a scalar continuous piecewise differentiable function $f(x)$ with a kink (i.e. non-smooth point) at one value of x, such as $f(x) = |x|$ (Figure 2.2). The derivative $f'(x)$ is defined by the tangent line to the graph of f when the graph is smooth at x

$$f'(x) = \frac{\partial f}{\partial x}(x) = \lim_{y \to x} \frac{f(y) - f(x)}{y - x}. \tag{2.6}$$

Although the function is not absolutely differentiable at every point x, it possesses at each x a left and right derivative defined as

$$f'_{-}(x) = \lim_{y \uparrow x} \frac{f(y) - f(x)}{y - x}, \qquad f'_{+}(x) = \lim_{y \downarrow x} \frac{f(y) - f(x)}{y - x}. \tag{2.7}$$

The *generalized derivative* of f at x is declared as *any* value $f'_q(x)$ included between its left and right derivatives [15, 19]. Such an intermediate value can be expressed as a convex combination of the left and right derivatives.

$$f'_q(x) = (1-q) f'_{-}(x) + q f'_{+}(x) \ , 0 \leq q \leq 1. \tag{2.8}$$

Geometrically, a generalized derivative is the slope of *any* line drawn through the point $(x, f(x))$ and between the left and right tangent lines (drawn by dashed lines in Figure 2.2a). The set of all the generalized derivatives of f at x, more generally the convex hull of the derivative extremes, is called the *generalized differential* of f at x

$$\begin{aligned} \partial f(x) &= \overline{\text{co}}\{f'_{-}(x), f'_{+}(x)\} \\ &= \{f'_q(x) \mid f'_q(x) = (1-q) f'_{-}(x) + q f'_{+}(x), \forall q \mid 0 \leq q \leq 1\}. \end{aligned} \tag{2.9}$$

The generalized differential of Clarke at x is the set of the slopes of all the lines included in the cone bounded by the left and right tangent lines and is a closed convex set (Figure 2.2b,c). In non-smooth analysis, the generalized differential is for instance used to define a local extremum of f at x by $0 \in \partial f$, which is the generalized form of $f'(x) = 0$ in smooth analysis [15, 44].

Infinitely many directional derivatives exist for functions in $\mathbb{R}^n$, whereas only two directional derivatives exist for scalar functions (the left and right derivative). For $\boldsymbol{f} : \mathbb{R}^n \to \mathbb{R}^m$, differentiable almost everywhere, we define the *generalized differential of Clarke* as [44]

$$\partial \boldsymbol{f}(\boldsymbol{x}) = \bigcap_{\delta>0} \overline{\text{co}}\{\nabla \boldsymbol{f}(\boldsymbol{y}) \mid \boldsymbol{y} \in \boldsymbol{x} + B_\delta\} \quad \subset \mathbb{R}^{n\times m}, \tag{2.10}$$

with the gradient

$$\nabla \boldsymbol{f}(\boldsymbol{x}) = \left(\frac{\partial \boldsymbol{f}(\boldsymbol{x})}{\partial \boldsymbol{x}}\right)^{\mathrm{T}} \quad \subset \mathbb{R}^{n\times m}. \tag{2.11}$$

The generalized differential (2.10) simplifies to (2.9) for the scalar case. We define the *generalized Jacobian of Clarke* as the transpose of the generalized differential

$$\boldsymbol{J}(\boldsymbol{x}) = (\partial \boldsymbol{f}(\boldsymbol{x}))^{\mathrm{T}} \quad \subset \mathbb{R}^{m\times n}, \tag{2.12}$$

which reduced for a smooth function $\boldsymbol{f}$ to $\boldsymbol{J}(\boldsymbol{x}) = \nabla \boldsymbol{f}(\boldsymbol{x})^{\mathrm{T}}$. Note that $\boldsymbol{f}(\boldsymbol{x})$ can be convex or non-convex in the above definitions. The image of the generalized differential $\partial \boldsymbol{f}(\boldsymbol{x})$ is for continuous functions always a closed convex set. If $f(\boldsymbol{x})$ is a scalar convex function, then the following relation holds

$$\partial f(\boldsymbol{x}) = \{\boldsymbol{y} \mid f(\boldsymbol{x}^*) \geq f(\boldsymbol{x}) + \boldsymbol{y}^{\mathrm{T}}(\boldsymbol{x}^* - \boldsymbol{x}); \forall \boldsymbol{x}^*\} \quad \subset \mathbb{R}^n. \tag{2.13}$$

The generalized differential of a scalar convex function, defined by (2.13), is called the *subdifferential*.

2.4 Definitions from Convex Analysis

In this section, the basic concepts of *normal cone*, *indicator function*, *support function*, *proximal point* and *distance* are introduced. We will use the notion of the subdifferential to show the relation between the indicator function and the normal cone.

A cone is a subset of $\mathbb{R}^n$ consisting of rays (half lines emanating from the origin).

Definition 2.8 (Cone). *A subset $C \subset \mathbb{R}^n$ is called a* cone *if for any $\boldsymbol{x} \in C$ and $\lambda > 0$ also $\lambda \boldsymbol{x} \in C$ holds. A cone is convex if the subset $C \subset \mathbb{R}^n$ is convex.*

A vector $\boldsymbol{y}$ is *normal* to $\boldsymbol{x}$ with respect to C if $\boldsymbol{y}$ makes an obtuse angle with every line segment in C starting from $\boldsymbol{x}$.

Definition 2.9 (Normal Vector). *Let C be a convex set and $\boldsymbol{x} \in C$. A vector $\boldsymbol{y}$ is* normal *to $\boldsymbol{x}$ with respect to C if*

$$\boldsymbol{y}^{\mathrm{T}}(\boldsymbol{x}^* - \boldsymbol{x}) \leq 0, \quad \boldsymbol{x} \in C, \forall \boldsymbol{x}^* \in C.$$

The normal cone of a set C at $\boldsymbol{x}$ is the set of rays that are normal to $\boldsymbol{x}$.

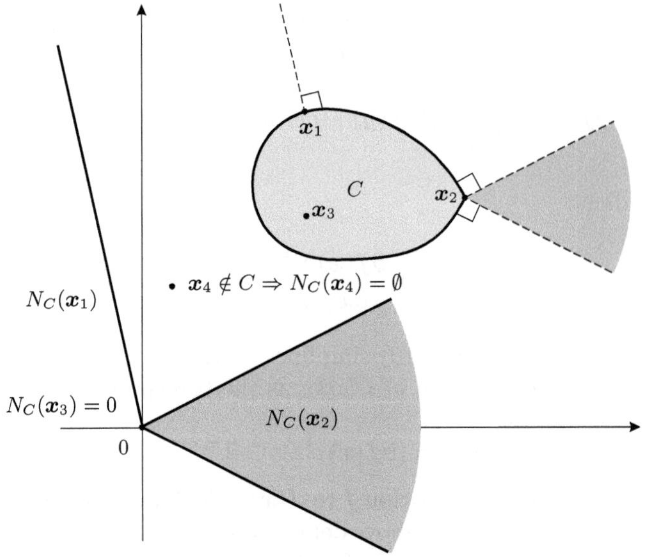

Fig. 2.3. Normal cone at different points of a convex set [46].

Definition 2.10 (Normal Cone). *Let C be a convex set and $\boldsymbol{x} \in C$. The set of vectors $\boldsymbol{y}$ that are normal to $\boldsymbol{x} \in C$ form the* normal cone *of C in $\boldsymbol{x}$*

$$N_C(\boldsymbol{x}) = \{\boldsymbol{y} \mid \boldsymbol{y}^{\mathrm{T}}(\boldsymbol{x}^* - \boldsymbol{x}) \leq 0, \quad \boldsymbol{x} \in C, \forall \boldsymbol{x}^* \in C\}.$$

If $\boldsymbol{x}$ is in the interior of C then $N_C(\boldsymbol{x}) = 0$. If $\boldsymbol{x} \notin C$ then $N_C(\boldsymbol{x}) = \emptyset$.

Definition 2.11 (Polar Cone). *Let K be a convex cone. The set of normal vectors $\boldsymbol{y}$ to $\boldsymbol{0}$ with respect to K form the* polar cone *of K*

$$K^{\perp} = \{\boldsymbol{y} \mid \boldsymbol{x}^{\mathrm{T}}\boldsymbol{y} \leq 0, \ \forall \boldsymbol{x} \in K\}.$$

It therefore holds that $K^{\perp} = N_K(\boldsymbol{0})$.

Definition 2.12 (Indicator Function). *Let C be a convex set. The* indicator function *of C is defined as*

$$\Psi_C(\boldsymbol{x}) = \begin{cases} 0, & \boldsymbol{x} \in C, \\ +\infty, & \boldsymbol{x} \notin C. \end{cases}$$

The indicator function is a convex function. With the definition of the sub-differential (2.13) and the indicator function it follows that

$$\begin{aligned} \partial\Psi_C(\boldsymbol{x}) &= \{\boldsymbol{y} \mid \Psi_C(\boldsymbol{x}^*) \geq \Psi_C(\boldsymbol{x}) + \boldsymbol{y}^{\mathrm{T}}(\boldsymbol{x}^* - \boldsymbol{x}), \quad \boldsymbol{x} \in C, \forall \boldsymbol{x}^* \in C\} \\ &= \{\boldsymbol{y} \mid 0 \geq \boldsymbol{y}^{\mathrm{T}}(\boldsymbol{x}^* - \boldsymbol{x}), \quad \boldsymbol{x} \in C, \forall \boldsymbol{x}^* \in C\}. \end{aligned} \tag{2.14}$$

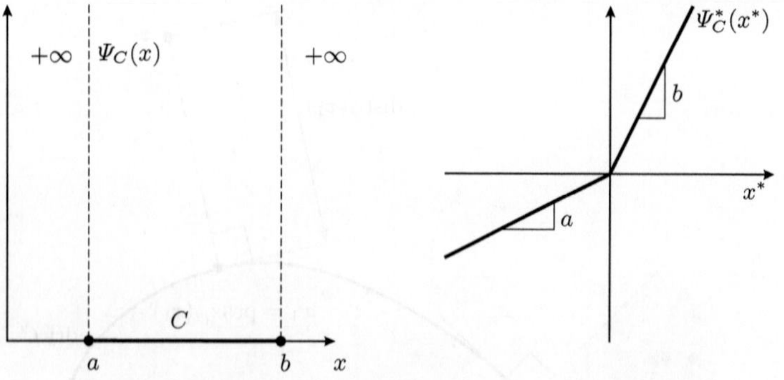

Fig. 2.4. Indicator function and support function [46].

This is exactly the definition of the normal cone at C. The subdifferential of the indicator function at $\boldsymbol{x} \in C$ is therefore the normal cone of C at $\boldsymbol{x}$,

$$\partial\Psi_C(\boldsymbol{x}) = N_C(\boldsymbol{x}). \tag{2.15}$$

Definition 2.13 (Conjugate Function). *Let f be a convex function. The function f^* is called the* conjugate function *of f and is defined as*

$$f^*(\boldsymbol{x}^*) = \sup_{\boldsymbol{x}}\{\boldsymbol{x}^{\mathrm{T}}\boldsymbol{x}^* - f(\boldsymbol{x})\}.$$

Definition 2.14 (Support Function). *Let C be a convex set. The conjugate function of the indicator function Ψ_C is called* support function,

$$\begin{aligned}\Psi_C^*(\boldsymbol{x}^*) &= \sup_{\boldsymbol{x}}\{\boldsymbol{x}^{\mathrm{T}}\boldsymbol{x}^* - \Psi_C(\boldsymbol{x})\}\\ &= \sup_{\boldsymbol{x}}\{\boldsymbol{x}^{\mathrm{T}}\boldsymbol{x}^* \mid \boldsymbol{x} \in C\}.\end{aligned}$$

Definition 2.15 (Proximal point). *The proximal point of a convex set C to a point $\boldsymbol{z}$ is the closest point in C to $\boldsymbol{z}$*

$$\mathrm{prox}_C(\boldsymbol{z}) = \operatorname*{argmin}_{\boldsymbol{x}^* \in C} \|\boldsymbol{z} - \boldsymbol{x}^*\|, \quad \boldsymbol{z} \in \mathbb{R}^n. \tag{2.16}$$

Let $\boldsymbol{x} = \mathrm{prox}_C(\boldsymbol{z})$ then it holds that $\boldsymbol{x} \in C$ and

$$\begin{cases} \boldsymbol{x} = \boldsymbol{z}, & \text{if } \boldsymbol{z} \in C, \\ \boldsymbol{x} \in \mathrm{bdry}\, C, & \text{if } \boldsymbol{z} \notin C. \end{cases} \tag{2.17}$$

where $\mathrm{bdry}\, C$ denotes the boundary of C. The vector $\boldsymbol{z} - \mathrm{prox}_C(\boldsymbol{z})$ is an element of the normal cone of C at the proximal point:

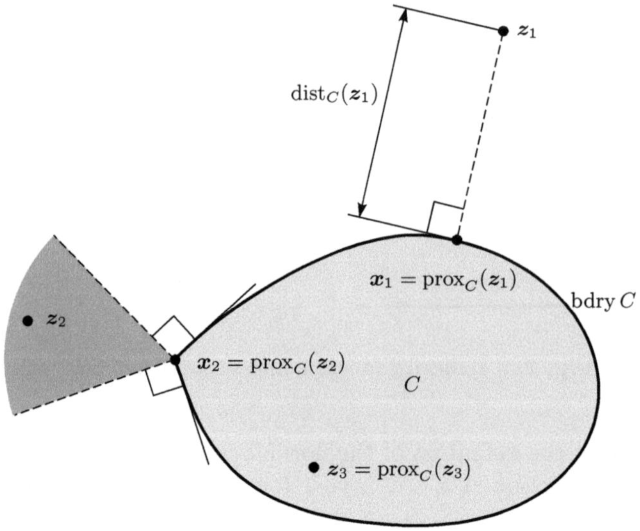

Fig. 2.5. Normal cone, proximal point and distance to a convex set, $z_1, z_2 \notin C$, $z_3 \in C$.

$$\begin{aligned} x = \text{prox}_C(z) &\Leftrightarrow x = \underset{x^* \in C}{\text{argmin}} \frac{1}{2}\|z - x^*\|^2 \\ &\Leftrightarrow x = \underset{x^*}{\text{argmin}} \frac{1}{2}\|z - x^*\|^2 + \Psi_C(x^*) \\ &\Leftrightarrow z - x \in \partial\Psi_C(x) \\ &\Leftrightarrow z - x \in N_C(x) \end{aligned} \tag{2.18}$$

Furthermore, by substituting $z = x - ry$, and using $-y \in N_C(x) \Leftrightarrow -ry \in N_C(x)$ for $r > 0$ we obtain

$$x = \text{prox}_C(x - ry),\ r > 0 \quad \Longleftrightarrow \quad -y \in N_C(x). \tag{2.19}$$

Definition 2.16 (Distance). *The distance from a point $z \in \mathbb{R}^n$ to a convex set C is the Euclidian distance from the point to its proximal point in C*

$$\text{dist}_C(z) = \|z - \text{prox}_C(z)\|. \tag{2.20}$$

It therefore holds that $\text{prox}_C(x) = x$ and $\text{dist}_C(x) = 0$ for all $x \in C$.

Lemma 2.17. *Let C be a convex set. It holds that*

$$\nabla \frac{1}{2} \text{dist}_C^2(z) = z - \text{prox}_C(z). \tag{2.21}$$

Proof. Consider the convex function $f(z) = \frac{1}{2}\text{dist}_C^2(z)$. The subdifferential $\partial f(z) = \nabla f(z)$ is a continuous function, because $f(z)$ is differentiable. The definition of the subdifferential (2.13) applied on $f(z)$ yields the condition

$$\frac{1}{2}\operatorname{dist}_C^2(\boldsymbol{z}^*) \geq \frac{1}{2}\operatorname{dist}_C^2(\boldsymbol{z}) + \partial f(\boldsymbol{z})^{\mathrm{T}}(\boldsymbol{z}^* - \boldsymbol{z}), \quad \forall \boldsymbol{z}^*. \tag{2.22}$$

Take $\boldsymbol{z}^*$ to be within the translate of the normal cone $N_C(\boldsymbol{z})$, i.e. $\boldsymbol{z}^* = \operatorname{prox}_C(\boldsymbol{z}) + \alpha(\boldsymbol{z} - \operatorname{prox}_C(\boldsymbol{z}))$ for $\alpha \geq 0$. For this choice of $\boldsymbol{z}^*$ it holds that $\operatorname{dist}_C(\boldsymbol{z}^*) = \alpha \operatorname{dist}_C(\boldsymbol{z})$ and (2.22) becomes

$$\frac{\alpha^2 - 1}{2}\operatorname{dist}_C^2(\boldsymbol{z}) \geq (\alpha - 1)\partial f(\boldsymbol{z})^{\mathrm{T}}\left(\boldsymbol{z} - \operatorname{prox}_C(\boldsymbol{z})\right), \quad \alpha \geq 0, \tag{2.23}$$

or, dividing by $\alpha - 1$,

$$\begin{aligned} \frac{\alpha+1}{2}\operatorname{dist}_C^2(\boldsymbol{z}) &\geq \partial f(\boldsymbol{z})^{\mathrm{T}}\left(\boldsymbol{z} - \operatorname{prox}_C(\boldsymbol{z})\right), \quad \alpha > 1, \\ \frac{\alpha+1}{2}\operatorname{dist}_C^2(\boldsymbol{z}) &\leq \partial f(\boldsymbol{z})^{\mathrm{T}}\left(\boldsymbol{z} - \operatorname{prox}_C(\boldsymbol{z})\right), \quad 0 < \alpha < 1. \end{aligned} \tag{2.24}$$

We conclude from (2.24), considering $\alpha = 1 \pm \varepsilon$ for infinitely small values of $\varepsilon > 0$ and using the continuity of $\partial f(\boldsymbol{z})$, that

$$\operatorname{dist}_C^2(\boldsymbol{z}) = \partial f(\boldsymbol{z})^{\mathrm{T}}(\boldsymbol{z} - \operatorname{prox}_C(\boldsymbol{z})), \tag{2.25}$$

from which follows $\partial f(\boldsymbol{z}) = \boldsymbol{z} - \operatorname{prox}_C(\boldsymbol{z})$. □

The following property is useful when studying the augmented Lagrangian in Section 5.5:

$$\begin{aligned} &f(\boldsymbol{x}) = \boldsymbol{x}^{\mathrm{T}}\boldsymbol{y} + \tfrac{1}{2}\|\boldsymbol{x}\|^2 - \tfrac{1}{2}\operatorname{dist}_C^2(\boldsymbol{y} + \boldsymbol{x}) \\ &\Longrightarrow \nabla f(\boldsymbol{x}) = \boldsymbol{y} + \boldsymbol{x} - \left(\boldsymbol{y} + \boldsymbol{x} - \operatorname{prox}_C(\boldsymbol{y} + \boldsymbol{x})\right) = \operatorname{prox}_C(\boldsymbol{y} + \boldsymbol{x}). \end{aligned} \tag{2.26}$$

The concepts of proximal point, distance and normal cone are illustrated in Figure 2.5.

2.5 The Linear Complementarity Problem

A linear complementarity problem (LCP) [100] is a set of linear equations

$$\boldsymbol{y} = \boldsymbol{A}\boldsymbol{x} + \boldsymbol{b} \tag{2.27}$$

subjected to the complementarity conditions

$$\boldsymbol{y} \geq \boldsymbol{0}, \qquad \boldsymbol{x} \geq \boldsymbol{0}, \qquad \boldsymbol{y}^{\mathrm{T}}\boldsymbol{x} = 0, \tag{2.28}$$

for which the vectors $\boldsymbol{x} \in \mathbb{R}^n$ and $\boldsymbol{y} \in \mathbb{R}^n$ have to be solved for given $\boldsymbol{A} \in \mathbb{R}^{n \times n}$ and $\boldsymbol{b} \in \mathbb{R}^n$. The complementarity conditions (2.28) are often written more briefly as

$$\boldsymbol{0} \leq \boldsymbol{y} \perp \boldsymbol{x} \geq \boldsymbol{0}, \tag{2.29}$$

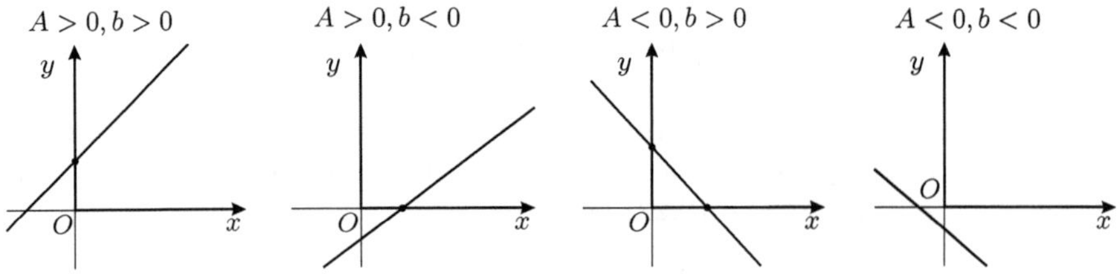

Fig. 2.6. Four different modes of a scalar LCP.

where $\boldsymbol{a} \perp \boldsymbol{b}$ denotes $\boldsymbol{a}^{\mathrm{T}}\boldsymbol{b} = 0$. With the aid of the notions of the generalized differential and indicator function we can express the complementarity conditions in 7 equivalent forms [46]

$$\begin{aligned} \boldsymbol{y} \geq \boldsymbol{0}, \quad \boldsymbol{x} \geq \boldsymbol{0}, \quad \boldsymbol{y}^{\mathrm{T}}\boldsymbol{x} = 0, \\ \boldsymbol{y}^{\mathrm{T}}(\boldsymbol{x}^* - \boldsymbol{x}) \geq 0, \quad \boldsymbol{x} \geq \boldsymbol{0}, \quad \forall \boldsymbol{x}^* \geq 0, \\ (\boldsymbol{y}^* - \boldsymbol{y})^{\mathrm{T}}\boldsymbol{x} \geq 0, \quad \boldsymbol{y} \geq \boldsymbol{0}, \quad \forall \boldsymbol{y}^* \geq 0, \\ -\boldsymbol{y} \in \partial\Psi_{\mathbb{R}^+}(\boldsymbol{x}), \\ \boldsymbol{x} \in \partial\Psi_{\mathbb{R}^-}(-\boldsymbol{y}), \\ -\boldsymbol{y} \in N_{\mathbb{R}^+}(\boldsymbol{x}), \\ \boldsymbol{x} \in N_{\mathbb{R}^-}(-\boldsymbol{y}). \end{aligned} \tag{2.30}$$

An LCP can have a unique solution, multiple solutions or no solution at all. Consider a scalar LCP $y = Ax + b$ with $y, x \geq 0, yx = 0$. The complementarity conditions express that if $y > 0$, then it must hold that $x = 0$ and vice versa. The graph of the complementarity conditions forms a corner in the x-y plane, consisting of the positive x-axis, positive y-axis and the origin. A solution of the LCP is a crossing of the line $y = Ax + b$ with the corner defined by the complementarity conditions. The scalar LCP has a unique solution for $A > 0$, two solutions for $A < 0 \wedge b > 0$, no solution for $A < 0 \wedge b < 0$ or infinitely many solutions for $A = b = 0$. The solutions of the LCP are indicated in Table 2.1.

Table 2.1. LCP solutions of a scalar LCP.

A	b	number of solutions	LCP solutions
$A > 0$	$b > 0$	unique solution	$x = 0 \wedge y = b$
$A > 0$	$b < 0$	unique solution	$x = -\frac{b}{A} \wedge y = 0$
$A < 0$	$b > 0$	two solutions	$x = -\frac{b}{A} \wedge y = 0$ or $x = 0 \wedge y = b$
$A < 0$	$b < 0$	no solution	$\emptyset$

All existing solutions of an n dimensional LCP can be found with enumerative methods, which treat the problem by a combinatorial evolution of the complementarity condition $x_i y_i = 0$. From the complementarity condition it follows that when $x_i > 0$, then $y_i = 0$, and vice versa. An LCP of dimension n therefore provides 2^n different combinations of n variables which are allowed to be larger than zero at the same time. For large dimensions, enumerative methods become numerically expensive since 2^n grows very rapidly. A more efficient algorithm is the complementary pivot algorithm, usually referred to as Lemke's algorithm. A drawback of Lemke's algorithm is, that it is not guaranteed to find a solution for arbitrary $\boldsymbol{A}$ [17].

A generalization of the linear complementarity problem is the *nonlinear* complementarity problem (NCP)

$$\mathbf{0} \leq \boldsymbol{f}(\boldsymbol{x}) \perp \boldsymbol{x} \geq \mathbf{0}, \tag{2.31}$$

where $\boldsymbol{f}$ is a nonlinear function in $\boldsymbol{x}$.

2.6 Summary

In this chapter we have introduced some basic concepts of Non-smooth and Convex Analysis. Section 2.1 discusses different properties of sets. If the image of a function is a set, then the function is said to be set-valued. The notion of continuity has been relaxed in Section 2.2 to semi-continuity for set-valued functions. The generalized differential of Clarke has been introduced in Section 2.3 as a generalization of the derivative of non-smooth continuous functions. The generalized differential agrees with the subdifferential for scalar convex functions. Dynamical systems for which the time-derivative of the state is an element of a set-valued function are described by differential inclusions. In the next chapter we will discuss existence and uniqueness of differential inclusions. Section 2.4 on Convex Analysis has introduced the normal cone and the indicator function of convex sets. The normal cone of a convex set appears to be identical to the subdifferential of the indicator function. The indicator function and its conjugate function can serve as a non-smooth potential for a set-valued force law. The normal cone and indicator function will be of use when defining set-valued contact laws in Chapter 5. The proximal point will turn out to be quintessential for the augmented Lagrangian method (Section 5.5).

3

Differential Inclusions

This chapter briefly discusses the behaviour of trajectories of dynamical systems described by differential inclusions. Historically, the concept of a differential inclusion has arisen by considering differential equations with a discontinuous right-hand side. Filippov has introduced in the 1960s a solution concept for discontinuous differential equations by a convexification the discontinuous right-hand side [36]. The 'convex method' proposed by Filippov extends a discontinuous differential equation to a differential inclusion. Filippov's convex method is somewhat historical as models of physical systems are nowadays directly described by differential inclusions. Still, differential inclusions which can be regarded as a convexification of a discontinuous differential equation obey certain existence properties and will be named Filippov systems.

The chapter starts with an existence and uniqueness result for continuous dynamical systems. After discussing an example which illustrates the basic problem for a discontinuous system, a concise review will be given in Section 3.2 of Filippov's convex method. Existence and uniqueness problems will be addressed and much attention will be paid to the sliding mode behaviour of trajectories of differential inclusions in Section 3.3.

3.1 Dynamical Systems

We consider a dynamical system described by a set of ordinary differential equations

$$\dot{\boldsymbol{x}}(t) = \boldsymbol{f}(t, \boldsymbol{x}(t)), \qquad \boldsymbol{x}(t) \in \mathbb{R}^n, \tag{3.1}$$

where $\boldsymbol{x}$ is the state vector and $\boldsymbol{f}(t, \boldsymbol{x}(t))$ is the right-hand side vector describing the time derivative of the state vector. We will assume that $\boldsymbol{f}(t, \boldsymbol{x})$ is linearly bounded [15], i.e. there exist positive constants γ and c such that

$$\|\boldsymbol{f}(t, \boldsymbol{x})\| \leq \gamma \|\boldsymbol{x}\| + c, \quad \forall\, (t, \boldsymbol{x}). \tag{3.2}$$

If the vector field is smooth then a solution $\boldsymbol{x}(t)$ of the system (3.1) exists for any given initial condition and is unique for all $t \in \mathbb{R}$, i.e. solutions of (3.1) are globally (in time) defined. In fact, smoothness of the vector field is not a necessary condition for existence and uniqueness of the solution, as can be concluded from the following theorem (see [15], theorem 1.1, page 178):

Theorem 3.1 (Existence and uniqueness of continuous systems). *Suppose that* $\boldsymbol{f}(t, \boldsymbol{x})$ *is continuous, and let* $(t_0, \boldsymbol{x}_0) \in \mathbb{R} \times \mathbb{R}^n$ *be given. Then the following holds:*

1. *There exists a solution of system (3.1) on an open interval* $(t_0 - \delta, t_0 + \delta)$, *for* $\delta > 0$, *satisfying* $\boldsymbol{x}(t_0) = \boldsymbol{x}_0$.
2. *If in addition we assume that* $\boldsymbol{f}(t, \boldsymbol{x})$ *is linearly bounded, so that (3.2) holds, then there exists a solution of system (3.1) on* $(-\infty, \infty)$ *such that* $\boldsymbol{x}(t_0) = \boldsymbol{x}_0$.
3. *We now add the hypothesis that* $\boldsymbol{f}(t, \boldsymbol{x})$ *is locally Lipschitz at* $\boldsymbol{x}$, *i.e. there exists a constant* $L(\boldsymbol{x}) > 0$ *and* $r > 0$ *such that*

$$\|\boldsymbol{f}(t, \boldsymbol{x}_1) - \boldsymbol{f}(t, \boldsymbol{x}_2)\| \leq L(\boldsymbol{x}) \|\boldsymbol{x}_1 - \boldsymbol{x}_2\|, \ \forall \boldsymbol{x}_1, \boldsymbol{x}_2 \in B_r + \boldsymbol{x}.$$

Then there exists a unique solution of (3.1) on $(-\infty, \infty)$ *such that* $\boldsymbol{x}(t_0) = \boldsymbol{x}_0$.

Note that above theorem deals with systems that are continuous but are allowed to be *non-smooth.*

3.2 Filippov's Solution Concept

Existence and uniqueness of solutions of continuous dynamical systems have been discussed in Section 3.1. However, differential equations stemming from systems with switches may be discontinuous, i.e. the right-hand side $\boldsymbol{f}$ can be discontinuous in $\boldsymbol{x}$. The theory of Filippov [36, 37, 131] gives a generalized[1] definition of the solution of differential equations which incorporates systems with a discontinuous right-hand side. The solution $\boldsymbol{x}(t)$ in the sense of Filippov to a differential equation with a discontinuous right-hand side (also called Filippov systems) is continuous in time. Systems with a discontinuous solution, i.e. 'jumps' in $\boldsymbol{x}(t)$ at certain time instances t (occurring for systems with impact between rigid bodies), are not described by the theory of Filippov. Filippov's theory will be briefly outlined in this chapter.

In order to make things as clear as possible, we first look at a very simple one-dimensional example (see [74]). Consider the following differential equation with a discontinuous right-hand side

[1] Note : 'generalized' in the sense that the definition holds for a larger class of differential equations.

$$\dot{x} = f(x) = 1 - 2\,\text{sign}(x) = \begin{cases} 3, & x < 0, \\ 1, & x = 0, \\ -1, & x > 0, \end{cases} \tag{3.3}$$

with sign(0) = 0. For a given initial condition $x(0) \neq 0$ we can obtain a solution of the IVP (initial value problem)

$$x(t) = \begin{cases} 3t + C_1, & x < 0, \\ -t + C_2, & x > 0, \end{cases} \tag{3.4}$$

with constants C_1 and C_2 being determined by the initial condition. Each solution reaches $x = 0$ in finite time. If the solution arrives at $x = 0$, it can not leave $x = 0$, because $\dot{x} > 0$ for $x < 0$ and $\dot{x} < 0$ for $x > 0$. The solution will therefore stay at $x = 0$, which implies $\dot{x}(t) = 0$. Note that $x(t) = 0$ with $\dot{x}(t) = 0$ is not a solution in the classical sense since $0 \neq 1 - 2\,\text{sign}(0)$. The natural idea to extend the notion of solution is to replace the right-hand side $f(x)$ by a *set-valued* function $F(x)$ such that $F(x) = f(x)$ for all x for which f is continuous in x. At the points for which f is discontinuous in x a suitable choice of $F(x)$ is required. The differential equation is then replaced by the differential inclusion [36,37]

$$\dot{x} \in F(x). \tag{3.5}$$

Define the set-valued Sign function as the generalized differential of $|x|$

$$\text{Sign}(x) := \partial|x| = \begin{cases} \{-1\}, & x < 0, \\ [-1,1], & x = 0, \\ \{1\}, & x > 0, \end{cases} \tag{3.6}$$

which is set-valued at $x = 0$. With this definition $x(t) = 0$ is a unique solution in *forward time* of the differential inclusion

$$\dot{x} \in 1 - 2\,\text{Sign}(x), \tag{3.7}$$

with initial condition $x(0) = 0$. The solution of (3.7) is non-unique in *backward time*. For instance, the solutions of (3.7) with initial condition $x(-1) = 1$ and initial condition $x(-1) = -3$ evolve both to $x(0) = 0$. In fact, there exist infinitely many initial conditions at $t = -1$ which evolve to $x(0) = 0$. Knowledge about the solution for $t \in (-\infty, 0)$ is therefore lost by the non-uniqueness in backward time.

We also give a one-dimensional example of non-uniqueness in forward time. Consider the one-dimensional differential inclusion

$$\dot{x}(t) \in \text{Sign}(x(t)), \quad x(0) = 0. \tag{3.8}$$

This IVP has three solutions

$$x(t) = \begin{cases} -t, & x < 0, \\ 0, & x = 0, \\ t, & x > 0. \end{cases} \tag{3.9}$$

The solution of (3.8) is non-unique in forward time but unique in backward time, i.e. $x(t) = 0$ for $t \in (-\infty, 0]$.

The above examples are one-dimensional. We now need to define a differential equation with a discontinuous right-hand side in a more general sense for any dimension n. We restrict ourselves to differential equations with a right-hand side that is discontinuous at a number of hyper-surfaces. We start with a single hyper-surface. The state space $\mathbb{R}^n$ is split into two subspaces $\mathcal{V}_-$ and $\mathcal{V}_+$ by a hyper-surface Σ such that $\mathbb{R}^n = \mathcal{V}_- \cup \Sigma \cup \mathcal{V}_+$. The hyper-surface Σ is called the *switching boundary* and is defined by a scalar switching boundary function $h(\boldsymbol{x})$ [2]. The state $\boldsymbol{x}$ is in Σ when

$$h(\boldsymbol{x}) = 0 \Longleftrightarrow \boldsymbol{x} \in \Sigma. \tag{3.10}$$

The subspaces $\mathcal{V}_-$ and $\mathcal{V}_+$ and switching boundary Σ can be formulated as

$$\begin{aligned} \mathcal{V}_- &= \{\boldsymbol{x} \in \mathbb{R}^n \mid h(\boldsymbol{x}) < 0\}, \\ \Sigma &= \{\boldsymbol{x} \in \mathbb{R}^n \mid h(\boldsymbol{x}) = 0\}, \\ \mathcal{V}_+ &= \{\boldsymbol{x} \in \mathbb{R}^n \mid h(\boldsymbol{x}) > 0\}. \end{aligned} \tag{3.11}$$

The switching boundary is assumed to be continuous but can be allowed to be non-smooth. The normal $\boldsymbol{n}(\boldsymbol{x})$ perpendicular to a locally smooth switching boundary Σ is given by

$$\boldsymbol{n}(\boldsymbol{x}) = \nabla h(\boldsymbol{x}), \tag{3.12}$$

or, if $h(\boldsymbol{x})$ is non-smooth, by making use of the generalized differential

$$\boldsymbol{n}(\boldsymbol{x}) \in \partial h(\boldsymbol{x}), \tag{3.13}$$

where we assume $\partial h(\boldsymbol{x})$ to be bounded. Note that, if $\mathcal{V}_-$ is locally convex in the neighbourhood of $\boldsymbol{x}$, then $\boldsymbol{n}(\boldsymbol{x}) \in N_{\mathcal{V}_-}(\boldsymbol{x})$. If $\mathcal{V}_-$ is locally concave then $\mathcal{V}_+$ is locally convex in the neighbourhood of $\boldsymbol{x}$, and $-\boldsymbol{n}(\boldsymbol{x}) \in N_{\mathcal{V}_+}(\boldsymbol{x})$. The normal $\boldsymbol{n}(\boldsymbol{x})$ therefore belongs to a convex cone.

A switching boundary function $h(\boldsymbol{x})$ that defines a certain switching boundary Σ is not unique. Different switching boundary functions can define the same Σ. We assume that the switching boundary function $h(\boldsymbol{x})$ is chosen such that it always holds that

$$\partial h(\boldsymbol{x}) \not\ni \boldsymbol{0}. \tag{3.14}$$

[2] The switching boundary function $h(\boldsymbol{x})$ is considered to be autonomous. Non-autonomous systems can give rise to non-autonomous switching boundary functions. However, non-autonomous time-periodic systems can be transformed into autonomous systems having autonomous switching boundary functions.

The normal $\boldsymbol{n}(\boldsymbol{x})$ on a locally smooth part of Σ is therefore always properly defined.

The right-hand side of the dynamics $\dot{\boldsymbol{x}} = \boldsymbol{f}(t, \boldsymbol{x})$ is assumed to be locally continuous, smooth and linearly bounded for all $\boldsymbol{x} \notin \Sigma$. From this assumption it follows that the solution $\boldsymbol{x}(t)$ within each subspace $\mathcal{V}_-$ and $\mathcal{V}_+$ exists and is unique (cf. Theorem 3.1).

The set-valued extension of $\boldsymbol{f}(t, \boldsymbol{x})$ of (3.1) for $\boldsymbol{x} \in \Sigma$ is given by the closed convex hull of all the limits

$$\boldsymbol{F}(t, \boldsymbol{x}) = \overline{\text{co}}\{\boldsymbol{y} \in \mathbb{R}^n \mid \boldsymbol{y} = \lim_{\tilde{\boldsymbol{x}} \to \boldsymbol{x}} \boldsymbol{f}(t, \tilde{\boldsymbol{x}}), \tilde{\boldsymbol{x}} \in \mathbb{R}^n \backslash \Sigma\}, \tag{3.15}$$

where the notation (2.1) has been used. All the limits exist because $\boldsymbol{f}(t, \boldsymbol{x})$ is assumed to be locally continuous, smooth and linearly bounded for all $\boldsymbol{x} \notin \Sigma$.

We are now able to consider the following n-dimensional nonlinear system with discontinuous right-hand side

$$\dot{\boldsymbol{x}}(t) = \boldsymbol{f}(t, \boldsymbol{x}(t)) = \begin{cases} \boldsymbol{f}_-(t, \boldsymbol{x}(t)), & \boldsymbol{x} \in \mathcal{V}_-, \\ \boldsymbol{f}_+(t, \boldsymbol{x}(t)), & \boldsymbol{x} \in \mathcal{V}_+, \end{cases} \tag{3.16}$$

with the initial condition $\boldsymbol{x}(0) = \boldsymbol{x}_0$. As mentioned before, the right-hand side $\boldsymbol{f}(t, \boldsymbol{x})$ is assumed to be discontinuous but such that it is piecewise continuous and smooth on $\mathcal{V}_-$ and $\mathcal{V}_+$ and discontinuous on Σ. The function $\boldsymbol{f}_-(t, \boldsymbol{x})$ is therefore assumed to be C^1 on $\mathcal{V}_- \cup \Sigma$ and $\boldsymbol{f}_+(t, \boldsymbol{x})$ is assumed to be C^1 on $\mathcal{V}_+ \cup \Sigma$. It is not required that $\boldsymbol{f}_-(t, \boldsymbol{x})$ and $\boldsymbol{f}_+(t, \boldsymbol{x})$ agree on Σ. The system described by (3.16) does not define $\boldsymbol{f}(t, \boldsymbol{x}(t))$ if $\boldsymbol{x}(t)$ is on Σ. We can overcome this problem with the following set-valued extension $\boldsymbol{F}(t, \boldsymbol{x})$

$$\dot{\boldsymbol{x}}(t) \in \boldsymbol{F}(t, \boldsymbol{x}(t)) = \begin{cases} \boldsymbol{f}_-(t, \boldsymbol{x}(t)), & \boldsymbol{x} \in \mathcal{V}_-, \\ \overline{\text{co}}\{\boldsymbol{f}_-(t, \boldsymbol{x}(t)), \boldsymbol{f}_+(t, \boldsymbol{x}(t))\}, & \boldsymbol{x} \in \Sigma, \\ \boldsymbol{f}_+(t, \boldsymbol{x}(t)), & \boldsymbol{x} \in \mathcal{V}_+, \end{cases} \tag{3.17}$$

where the convex set with two right-hand sides $\boldsymbol{f}_-$ and $\boldsymbol{f}_+$ can be cast in

$$\overline{\text{co}}\{\boldsymbol{f}_-, \boldsymbol{f}_+\} = \{(1-q)\boldsymbol{f}_- + q\boldsymbol{f}_+, \forall q \in [0,1]\}. \tag{3.18}$$

The extension (or convexification) of a discontinuous system (3.16) into a convex differential inclusion (3.17) is known as *Filippov's convex method.*

It was stated that the set-valued extension $\boldsymbol{F}$ of $\boldsymbol{f}$ should be *suitable*. If the discontinuous system (3.16) is a mathematical model of a physical system, then we are interested in a solution concept that guarantees existence of solutions. For practical reasons therefore, we demand that the choice for $\boldsymbol{F}$ guarantees existence of solutions. Existence can be guaranteed with the notion of upper semi-continuity of set-valued functions. The following existence theorem is proven in [7] (theorem 3, page 98):

Theorem 3.2 (Existence of solution of a differential inclusion). *Let $\boldsymbol{F}$ be a set-valued function. We assume that $\boldsymbol{F}$ is upper semi-continuous and that $\boldsymbol{F}(t, \boldsymbol{x})$ is closed, convex and bounded for all $t \in \mathbb{R}$ and $\boldsymbol{x} \in \mathbb{R}^n$. Then, for each $\boldsymbol{x}_0 \in \mathbb{R}^n$ there exists a $\tau > 0$ and an absolutely continuous function $\boldsymbol{x}(t)$ defined on $[0, \tau]$, which is a solution of the initial value problem*

$$\dot{\boldsymbol{x}} \in \boldsymbol{F}(t, \boldsymbol{x}(t)), \quad \boldsymbol{x}(0) = \boldsymbol{x}_0.$$

The theorem holds at values of $\boldsymbol{x}$ for which $\boldsymbol{F}(t, \boldsymbol{x})$ is locally single-valued and continuous, because of the boundedness restriction (see Theorem 3.1). To illustrate the theorem for set-valued $\boldsymbol{F}(t, \boldsymbol{x})$ we once more look at the example of (3.7)

$$\dot{x} \in 1 - a \operatorname{Sign}(x), \text{with } |a| > 1. \tag{3.19}$$

To allow for the solution $x(t) = 0$ we must demand that $0 \in 1 - a \operatorname{Sign}(0)$. In order to guarantee the existence of a solution we therefore have to define $\operatorname{Sign}(0)$ to be the set $[-1, 1]$. This set is upper semi-continuous with the values of $\operatorname{sign}(x) = \pm 1$ for $x \neq 0$, and the set is closed, convex and bounded.

Filippov's convex method together with the above existence theorem defines the *solution in the sense of Filippov* for a discontinuous differential equation.

Definition 3.3 (Solution in the sense of Filippov). *An absolute continuous function $\boldsymbol{x}(t) : [0, \tau] \to \mathbb{R}^n$ is said to be a solution of $\dot{\boldsymbol{x}}(t) = \boldsymbol{f}(t, \boldsymbol{x})$ (3.16) in the sense of Filippov if for almost all*[3] *$t \in [0, \tau]$ it holds that*

$$\dot{\boldsymbol{x}}(t) \in \boldsymbol{F}(t, \boldsymbol{x}(t)),$$

where $\boldsymbol{F}(t, \boldsymbol{x}(t))$ is the closed convex hull of all the limits of $\boldsymbol{f}(t, \boldsymbol{x}(t))$ as in (3.15).

Remark: If $\boldsymbol{x}(t)$ is in a region where the vector field is continuous, $\boldsymbol{x}(t) \in \mathcal{V}$, then of course $\boldsymbol{F}(t, \boldsymbol{x}(t)) = \boldsymbol{f}(t, \boldsymbol{x}(t))$ must hold. If the solution $\boldsymbol{x}(t)$ slides along a switching boundary, $\boldsymbol{x}(t) \in \Sigma$, then $\dot{\boldsymbol{x}}(t) \in \boldsymbol{F}(t, \boldsymbol{x}(t))$. However, $\dot{\boldsymbol{x}}(t)$ is not defined at time instances t_Σ where the solution $\boldsymbol{x}(t)$ arrives at a switching boundary Σ or leaves from Σ (the solution $x(t)$ arrives at or leaves from Σ if there exists an arbitrary small $\varepsilon > 0$ and $t^* \in t_\Sigma + B_\varepsilon \backslash 0$ such that $\boldsymbol{x}(t^*) \notin \Sigma$ and $\boldsymbol{x}(t_\Sigma) \in \Sigma$). The set of t for which the solution $\boldsymbol{x}(t)$ arrives at or leaves from Σ is of Lebesgue measure zero.

It was assumed that $\boldsymbol{f}(t, \boldsymbol{x})$ is linearly bounded for $\boldsymbol{x} \notin \Sigma$. In addition, $\boldsymbol{F}(t, \boldsymbol{x}(t))$ is assumed to be bounded at values $(t, \boldsymbol{x})$ for which $\boldsymbol{F}$ is set-valued. Consequently, $\boldsymbol{F}(t, \boldsymbol{x}(t))$ is linearly bounded, i.e. there exist positive constants γ and c such that for all $t \in [0, \infty)$ and $\boldsymbol{x} \in \mathbb{R}^n$ holds:

$$\|\boldsymbol{F}(t, \boldsymbol{x})\| \leq \gamma \|\boldsymbol{x}\| + c.$$

Solutions $\boldsymbol{x}(t)$ to (3.17) therefore exist on $[0, \infty)$ (see [7, 15]) but uniqueness is not guaranteed.

---

[3] *for almost all* t means except for a set t of Lebesgue measure 0.

3.3 Sliding Modes

Solutions of differential inclusions do not have to be unique. Obviously, the solution of the IVP where $\boldsymbol{x}_0 \notin \Sigma$ is locally unique, because $\boldsymbol{f}_-(t, \boldsymbol{x})$ and $\boldsymbol{f}_+(t, \boldsymbol{x})$ are smooth. Uniqueness problems of IVPs for initial conditions on Σ will be illustrated in the following examples which show three basic ways in which the vector field around Σ can behave.

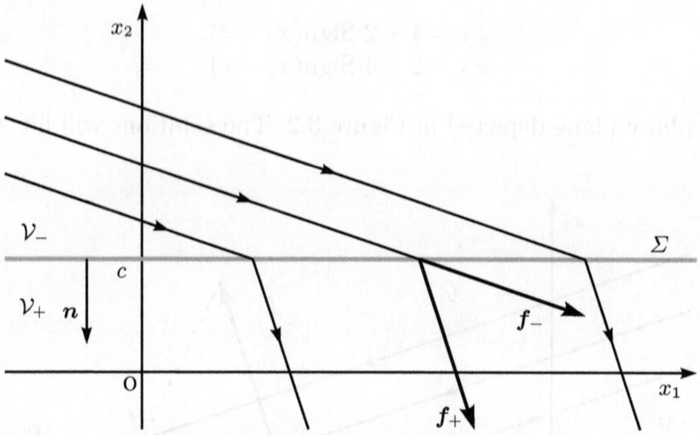

Fig. 3.1. Transversal intersection.

Example 3.1
Consider the differential inclusion

$$\begin{aligned} \dot{x}_1 &\in 4 + 2 \operatorname{Sign}(x_2 - c), \\ \dot{x}_2 &\in -4 + 2 \operatorname{Sign}(x_2 - c), \end{aligned} \tag{3.20}$$

which is set-valued at $\Sigma = \{x_2 = c\}$. We take $h(\boldsymbol{x}) = c - x_2$ as switching boundary function which defines the subspaces $\mathcal{V}_-$ and $\mathcal{V}_+$ by (3.11) with $\boldsymbol{x} = [x_1\ x_2]^{\mathrm{T}}$. The normal $\boldsymbol{n}(\boldsymbol{x})$ to Σ is given by $\boldsymbol{n}(\boldsymbol{x}) = [0, -1]^{\mathrm{T}}$. The phase plane of the system is depicted in Figure 3.1. The vector field is pushing the solution to Σ in the space $\mathcal{V}_- = \{x_2 > c\}$ and pushing from Σ in the space $\mathcal{V}_+ = \{x_2 < c\}$. A solution of (3.20) with an initial condition in $\mathcal{V}_-$ will after some time hit Σ, cross it transversally and proceed in $\mathcal{V}_+$. This is called a transversal intersection. Note that the word 'transversal' refers to the *solution* which is transversal to Σ and does not refer to the vector field $\boldsymbol{f}$. Any solution of (3.20) with an initial condition in $\mathcal{V}_-$, exposing a transversal intersection, therefore exists and is unique. A necessary condition for a transversal intersection at Σ is that there exists a normal $\boldsymbol{n}(\boldsymbol{x}) \in \partial h(\boldsymbol{x})$ such that

$$\boldsymbol{n}(\boldsymbol{x})^{\mathrm{T}} \boldsymbol{f}_-(t, \boldsymbol{x}(t)) \cdot \boldsymbol{n}(\boldsymbol{x})^{\mathrm{T}} \boldsymbol{f}_+(t, \boldsymbol{x}(t)) > 0, \qquad \boldsymbol{x}(t) \in \Sigma, \tag{3.21}$$

where $\boldsymbol{n}^{\mathrm{T}}\boldsymbol{f}_-$ and $\boldsymbol{n}^{\mathrm{T}}\boldsymbol{f}_+$ are the projections of $\boldsymbol{f}_-$ and $\boldsymbol{f}_+$ on the normal $\boldsymbol{n}$ to the switching boundary Σ.

The vector field can also push the solution to Σ in both $\mathcal{V}_-$ and $\mathcal{V}_+$. This will be demonstrated in the following example.

Example 3.2
Consider the differential inclusion

$$\begin{aligned} \dot{x}_1 &= 4 + 2\ \mathrm{Sign}(x_2 - c), \\ \dot{x}_2 &= 2 - 4\ \mathrm{Sign}(x_2 - c), \end{aligned} \tag{3.22}$$

with the phase plane depicted in Figure 3.2. The solutions will hit Σ but can

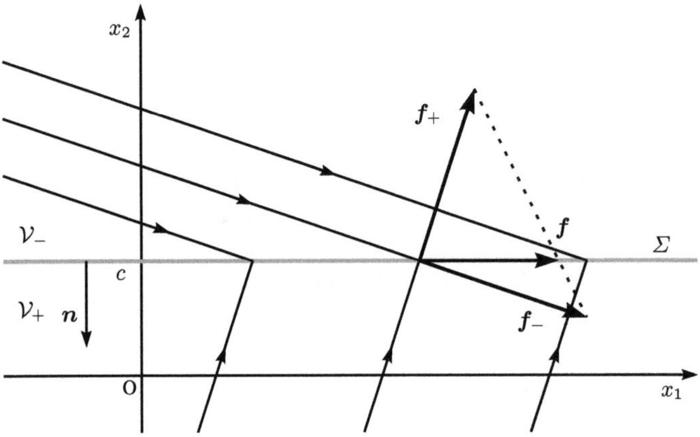

Fig. 3.2. Attractive sliding mode.

not leave it and will therefore move along the switching boundary Σ. This is often called a *sliding mode.* Because the switching boundary attracts the solution, we call this an *attractive sliding mode.* During the sliding mode the solution will continue along Σ with time derivative $\boldsymbol{f}$ given by

$$\boldsymbol{f} = \alpha \boldsymbol{f}_+ + (1 - \alpha)\boldsymbol{f}_-, \tag{3.23}$$

with

$$\alpha = \frac{\boldsymbol{n}(\boldsymbol{x})^{\mathrm{T}}\boldsymbol{f}_-}{\boldsymbol{n}(\boldsymbol{x})^{\mathrm{T}}(\boldsymbol{f}_- - \boldsymbol{f}_+)}. \tag{3.24}$$

For the system (3.22) we obtain

$$\boldsymbol{f}_+ = \begin{bmatrix} -2 \\ 6 \end{bmatrix}, \quad \boldsymbol{f}_- = \begin{bmatrix} 6 \\ -2 \end{bmatrix}, \quad \boldsymbol{n}(\boldsymbol{x}) = \begin{bmatrix} 0 \\ -1 \end{bmatrix}, \quad \alpha = \frac{1}{4}, \tag{3.25}$$

with $\boldsymbol{x} = [x_1\ x_2]^{\mathrm{T}}$ and $h(\boldsymbol{x}) = c - x_2$. The scalar α can be regarded as the value for q in (3.18) that chooses one $\boldsymbol{f} \in \boldsymbol{F}$ such that it lies along Σ. The solution of (3.22), being an attractive sliding mode, exists and is unique in forward time. An attractive sliding mode at Σ occurs if

$$\boldsymbol{n}(\boldsymbol{x})^{\mathrm{T}} \boldsymbol{f}_{-}(t, \boldsymbol{x}(t)) > 0 \quad \text{and} \quad \boldsymbol{n}(\boldsymbol{x})^{\mathrm{T}} \boldsymbol{f}_{+}(t, \boldsymbol{x}(t)) < 0, \tag{3.26}$$

with $\boldsymbol{x}(t) \in \Sigma$ and $\forall \boldsymbol{n}(\boldsymbol{x}) \in \partial h(\boldsymbol{x})$. The inequality signs in (3.26) depend of course on the choice of $h(\boldsymbol{x})$ (in this case $h(\boldsymbol{x}) > 0$ has been chosen in the subspace $\mathcal{V}_+$ and $h(\boldsymbol{x}) < 0$ in $\mathcal{V}_-$).

The third possible case is depicted in Figure 3.3 which is the vector field of the following example.

Example 3.3
Consider the differential inclusion

$$\begin{aligned} \dot{x}_1 &= -4 - 2\ \mathrm{Sign}(x_2 - c), \\ \dot{x}_2 &= -2 + 4\ \mathrm{Sign}(x_2 - c). \end{aligned} \tag{3.27}$$

Note that this vector field is the vector field of (3.22) in reverse time. Here

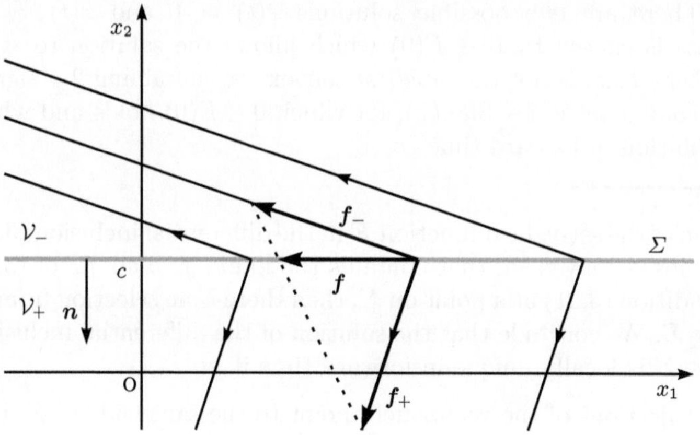

Fig. 3.3. Repulsive sliding mode.

the solutions are diverging from Σ. A solution which starts close to Σ will move away from it. But a solution emanating from Σ can stay on Σ, obeying Filippov's solution, or leave Σ by entering either $\mathcal{V}_-$ or $\mathcal{V}_+$. This type of vector field around the switching boundary is addressed as *repulsive sliding mode* as the vector field is repulsing from Σ. The IVP with initial condition on Σ has three possible solutions. The solution still exists but is not unique in forward time. A repulsive sliding mode at Σ occurs if there exists a normal $\boldsymbol{n}(\boldsymbol{x}) \in \partial h(\boldsymbol{x})$ such that

$$\boldsymbol{n}(\boldsymbol{x})^{\mathrm{T}} \boldsymbol{f}_{-}(t, \boldsymbol{x}(t)) < 0 \quad \text{and} \quad \boldsymbol{n}(\boldsymbol{x})^{\mathrm{T}} \boldsymbol{f}_{+}(t, \boldsymbol{x}(t)) > 0, \qquad \boldsymbol{x}(t) \in \Sigma. \quad (3.28)$$

In the preceding examples a locally unique solution in forward time has been found for the transversal intersection and for the attractive sliding mode, whereas non-uniqueness in forward time occurs for the repulsive sliding mode. One can come to the false conclusion that (3.21) is a sufficient condition for local uniqueness in forward time. As is shown by the following counter example this is not the case.

Example 3.4
Consider the one-dimensional differential inclusion

$$\dot{x} \in F(x) = \begin{cases} 2 + \operatorname{sign}(x), & x \neq 0, \\ [-1, 4], & x = 0, \end{cases}$$

with initial condition $x(0) = 0$. Existence of the solutions is guaranteed because F is upper semi-continuous, non-empty, closed, convex and bounded (Theorem 3.2). Equation (3.21) is also fulfilled. However, the solution is not unique. There are two possible solutions $x(t) = 0$ and $x(t) = 3t$. Non-uniqueness is caused by $0 \in F(0)$ which allows the solution to stay on Σ. Remark that $F(x)$ is not the *smallest* convex set containing $2 + \operatorname{sign}(x)$. The smallest convex set is $2 + \operatorname{Sign}(x)$, for which $0 \notin F(0)$ holds and which has a unique solution in forward time.

The image of the set-valued function $\boldsymbol{F}$ in the differential inclusion (3.17) is the smallest closed convex set that contains the limits $\boldsymbol{f}_{-}$ and $\boldsymbol{f}_{+}$ of (3.16). If $\boldsymbol{F}$ obeys condition (3.21) at a point on Σ, then there is no selection from $\boldsymbol{F}$ which lies along Σ. We conclude that the solution of the differential inclusion (3.17) with $\boldsymbol{x}_0 \in \Sigma$ is locally unique in forward time if

1. the projections of the vector field point to the same side of Σ, i.e.

$$\boldsymbol{n}(\boldsymbol{x}_0)^{\mathrm{T}} \boldsymbol{f}_{-}(t_0, \boldsymbol{x}_0) \cdot \boldsymbol{n}(\boldsymbol{x}_0)^{\mathrm{T}} \boldsymbol{f}_{+}(t_0, \boldsymbol{x}_0) > 0, \quad \forall \boldsymbol{n}(\boldsymbol{x}) \in \partial h(\boldsymbol{x}),$$

or if

2. the projections point to Σ, i.e.

$$\boldsymbol{n}(\boldsymbol{x}_0)^{\mathrm{T}} \boldsymbol{f}_{-}(t_0, \boldsymbol{x}_0) > 0 \text{ and } \boldsymbol{n}(\boldsymbol{x}_0)^{\mathrm{T}} \boldsymbol{f}_{+}(t_0, \boldsymbol{x}_0) < 0, \quad \forall \boldsymbol{n}(\boldsymbol{x}) \in \partial h(\boldsymbol{x}).$$

A complication of Filippov systems is the possible occurence of 'accumulation points' [37, 56, 142]. At an accumulation point, an infinite number of mode switches occur in a finite time as is illustrated in the following example.

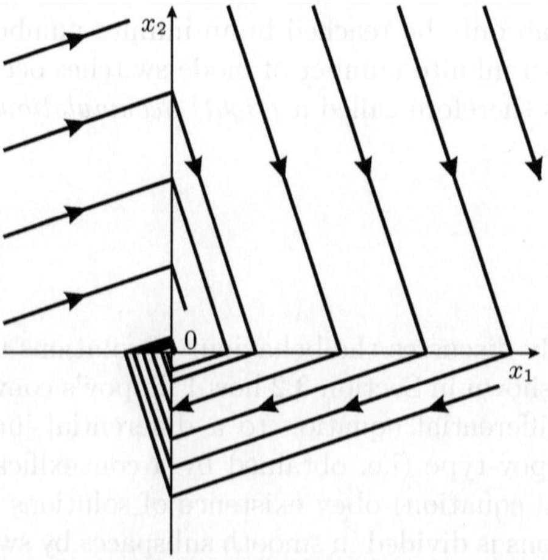

Fig. 3.4. Trajectories towards an accumulation point.

Example 3.5
Consider the differential inclusion

$$\dot{\boldsymbol{x}} \in \boldsymbol{F}(\boldsymbol{x}) = \begin{bmatrix} -\operatorname{Sign}(x_1) + 2\operatorname{Sign}(x_2) \\ -2\operatorname{Sign}(x_1) - \operatorname{Sign}(x_2) \end{bmatrix}, \tag{3.29}$$

with $\boldsymbol{x} = \begin{bmatrix} x_1 \; x_2 \end{bmatrix}^{\mathrm{T}}$. The system is piecewise constant and characterized by two switching boundaries $\Sigma_1 = \{x_1 = 0\}$ and $\Sigma_2 = \{x_2 = 0\}$ with normals $\boldsymbol{n}_1 = \begin{bmatrix} 1 \; 0 \end{bmatrix}^{\mathrm{T}}$ and $\boldsymbol{n}_2 = \begin{bmatrix} 0 \; 1 \end{bmatrix}^{\mathrm{T}}$. The origin $\boldsymbol{x} = \boldsymbol{0}$ is the only equilibrium of the system, $\boldsymbol{0} \in \boldsymbol{F}(\boldsymbol{0})$, and is located at the intersection of the two switching boundaries. If a solution hits a switching boundary, then a transversal intersection occurs as can be inferred by checking condition (3.21) at $\Sigma_{1,2} \backslash \boldsymbol{0}$. We take a positive definite function $v = |x_1| + |x_2|$ as Lyapunov function. Its time-derivative becomes set-valued when any of the arguments x_1 or x_2 vanishes,

$$\begin{aligned} \dot{v} &= \partial_{x_1} v \, \dot{x}_1 + \partial_{x_2} v \, \dot{x}_2 \\ &= \operatorname{Sign}(x_1)(-\operatorname{Sign}(x_1) + 2\operatorname{Sign}(x_2)) + \operatorname{Sign}(x_2)(-2\operatorname{Sign}(x_1) - \operatorname{Sign}(x_2)) \\ &= -\operatorname{Sign}(x_1)^2 - \operatorname{Sign}(x_2)^2. \end{aligned} \tag{3.30}$$

It therefore holds that $\dot{v} = -2$ for $\boldsymbol{x} \in \mathbb{R}^2 \backslash \{\Sigma_1 \cup \Sigma_2\}$ and $\dot{v} = [-2, -1]$ for $\boldsymbol{x} \in \{\Sigma_1 \cup \Sigma_2\} \backslash \boldsymbol{0}$. Solutions stay only at one moment of time on $\Sigma_{1,2} \backslash \boldsymbol{0}$, because the switching boundaries are crossed transversally outside the origin. Consequently, $\dot{v}(t) = -2$ for almost all t as long as $\boldsymbol{x}(t) \in \mathbb{R}^2 \backslash \boldsymbol{0}$. It therefore must hold that solutions reach the equilibrium in finite time Δt,

$$\Delta t = \frac{1}{2}(|x_1(t_0)| + |x_2(t_0)|). \tag{3.31}$$

The equilibrium can only be reached in an infinite number of mode switches (see Figure 3.4). An infinite number of mode switches occurs in a finite time. The equilibrium is therefore called a *(right) accumulation point.*

3.4 Summary

This chapter briefly discusses the behaviour of solutions to differential inclusions. It has been shown in Section 3.2 how Filippov's convex method extends a discontinuous differential equation to a differential inclusion. Differential inclusions of Filippov-type (i.e. obtained by a convexification of the discontinuous differential equation) obey existence of solutions. The state space of differential inclusions is divided in smooth subspaces by switching boundaries. Section 3.3 discusses the attractive and repulsive sliding modes of trajectories along a switching boundary. Repulsive sliding modes lead to non-uniqueness of solutions. Filippov's theory for differential inclusions will turn out to be very important to understand periodic solutions for which part of the orbit is a sliding mode. The mathematical modelling of mechanical systems with dry friction leads often to differential inclusions of Filippov-type and will be discussed in the next chapters.

4

Modelling of Dry Friction

Dry friction plays an important role in the dynamic behaviour of mechanical systems. Mechanical systems can come to rest in a finite time due to the presence of dry friction. Moreover, dry friction is also responsible for stick-slip vibrations, which occur for instance in brakes, wheel-rail contact and bowed instruments such as violins. We briefly discuss the basic tribological phenomena related to dry friction in Section 4.1. Modelling of dry friction is far from trivial due to the intrinsically different nature of the sticking phenomenon and the sliding process of bodies in contact. Commonly used friction models are discussed in Section 4.2. Some friction models can be described by inclusions and form together with the equation of motion a differential inclusion, for which a well-defined solution concept exists (see Chapter 3). Dry friction models of this type, which lead to differential inclusions, will be used in Chapter 5 to set up a mathematical formulation for mechanical multibody systems with dry friction.

4.1 Dry Friction Phenomena

In tribology, dry friction is defined as a force that resists relative motion between contacting surfaces of solid bodies. Although surfaces of solids look smooth on a macroscopic scale, they are microscopically rough. The microscopic protuberant features of the surfaces are called asperities and have a typical width of 10 μm. When we consider two flat surfaces in contact (a block on a table for instance), then we speak of an apparent area of contact being the macroscopic area where the objects touch each other (e.g. the area of the bottom plane of the block). True contact between two surfaces only occurs where the asperities touch each other. The true contact area is therefore much smaller than the apparent area of contact. The true contact area is proportional to the total normal load and therefore independent of the apparent contact area. While loading, the asperities deform to generate the contact area necessary to take up the total load. The asperities are sheared when the surfaces are moved tangentially over each other, giving rise to the resisting friction force. Friction is a partly abrasive process where the asperities are torn through each other, but is highly dependent on oxide films on the surfaces. Lubricant between the surfaces greatly changes the friction process and diminishes the friction. The friction force F_T, during relative sliding of the contacting surfaces, is approximately proportional to the true contact area and is therefore proportional to the normal load F_N and independent of the apparent contact area

$$|F_T| = \mu F_N, \qquad v_{\mathrm{rel}} \neq 0, \tag{4.1}$$

where μ denotes the coefficient of friction and v_{rel} is the relative velocity between the contacting surfaces. This relation was already stated in words by Leonardo da Vinci (Codex Madrid I), but he considered the friction coefficient to be $\mu = \frac{1}{3}$ by definition (which is, in fact, a good approximation of the experimentally found $\mu = 0.1 - 0.3$ for surfaces without lubrication). The relation between normal contact force and tangential friction force was later stated and experimentally studied by Amontons. Coulomb considered the influence of the sliding velocity and direction and stated that the friction coefficient is almost independent of the sliding velocity and that the friction force opposes the sliding direction

$$F_T = -\mu F_N \operatorname{sign}(v_{\mathrm{rel}}), \qquad v_{\mathrm{rel}} \neq 0, \tag{4.2}$$

which is now known as Amontons-Coulomb's law, or more briefly denoted as Coulomb's law. The tangential contact law of Amontons-Coulomb makes a clear distinction between 'stick' and 'slip/sliding' of the contacting bodies. The bodies 'stick' to each other when the relative velocity between the contacting surfaces is zero. If the bodies slide over each other with a non-zero velocity, we speak of 'slip' or 'sliding'. In rolling contact of two bodies, 'stick' should be regarded as pure rolling without slip. Amontons-Coulomb's law is characterized by a different behaviour in the slip and stick phase. The friction force in the stick phase adjusts itself to make equilibrium with external forces on the bodies as well as with possible d'Alembert forces. The bodies remain sticking as long as equilibrium is ensured. If the friction force in the stick phase exceeds a threshold, called break-away friction force or maximum static friction force, then the bodies will begin to slip over each other. The maximum static friction force will be denoted by $F_{\mathrm{s}} = \mu_{\mathrm{s}} F_N$ and the friction force F_T must be in the interval $-F_{\mathrm{s}} \leq F_T \leq F_{\mathrm{s}}$ when the bodies stick. In the slip phase, the friction force is idealized to be a single-valued function of the relative velocity v_{rel}, i.e. $F_T = -\mu_{\mathrm{d}} F_N \operatorname{sign}(v_{\mathrm{rel}})$ where μ_{d} is the friction coefficient in the slip phase. The friction force during slip is often called the 'dynamic' friction force. Pure viscous friction does not contain a stick phase and can be regarded as dynamic friction. The classical Amontons-Coulomb's law defines $\mu = \mu_{\mathrm{d}} = \mu_{\mathrm{s}}$.

Amontons-Coulomb's law is an idealization of the complex physical contact behaviour. A microscopic tangential relative displacement will occur when the

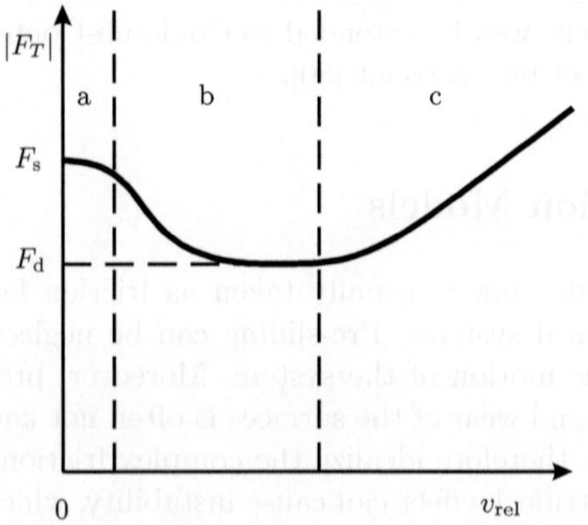

Fig. 4.1. Stribeck friction curve; a) boundary lubrication, b) partial lubrication (Stribeck effect), c) full lubrication, not on scale.

bodies are macroscopically sticking to each other. This pre-sliding displacement is caused by the elastic deformation of the asperities due to shear loading of the external forces on the bodies. The asperities will start to deform plastically when the external forces are increased above the break-away friction force, which gives the onset to sliding. The surfaces of solids are to some extent polluted with dirt or grease or are deliberately lubricated with oil. For a small relative velocity, the surfaces will slide with touching asperities, even if the surfaces are lubricated. This low-velocity regime is called boundary lubrication. Increasing the relative velocity will increase the separation of the surfaces and allow partial lubrication which diminishes the friction. The decrease of friction due to increasing sliding velocity is called the Stribeck effect (see Figure 4.1). Full lubrication will be established for high sliding velocities. The friction in the full fluid lubricated regime is dominated by the viscosity of the lubricant and is a form of viscous friction which rises with increasing velocity. The friction coefficient is therefore in general dependent on the relative velocity. Tribological experiments have also shown that the friction force during tangential acceleration is higher than during deceleration, yielding a kind of hysteresis phenomenon. The hysteresis phenomenon is not taken into account by the classical Amontons-Coulomb law.

Contacting bodies do not only slide over each other but generally also spin relative to each other. The relative spin is accompanied by a friction torque, sometimes denoted by ‘drilling friction’. The coupling between sliding friction forces and friction torque is dependent on the apparent contact area, which is for hard spherical bodies very small. However, the influence of spin and friction torque on the sliding friction can not be neglected if the spinning velocity is very large (such as in gyros and bearing devices). The coupling between sliding friction and spinning friction is known as the Contensou effect [16]. We will restrict ourselves to Coulomb friction in the sequel, but the methods presented in this book can very well be extended to Coulomb-Contensou friction taking the Contensou effect into account [80].

4.2 Dry Friction Models

Amontons-Coulomb's law is usually taken as friction law to study the dynamics of mechanical systems. Pre-sliding can be neglected when compared to the macroscopic motion of the system. Moreover, precise information on surface roughness and wear of the surfaces is often not known in practical situations. Engineers therefore idealize the complex friction behaviour as much as possible. The Stribeck effect can cause instability, which gives rise to stick-slip vibrations, and is therefore often also included in the friction model by defining a velocity dependent friction coefficient $\mu(v_{\mathrm{rel}})$.

Many different models are proposed for the mathematical description of dry friction which mostly differ in the way the stick phase is modelled. Three different friction models for unidirectional motion are depicted in Figure 4.2. The friction models (a), (b) and (c) are ordered to their degree of smoothness. All these models are antisymmetric, i.e. $F_T(v_{\mathrm{rel}}) = -F_T(-v_{\mathrm{rel}})$ although friction models can in general be non-symmetric. We will study the differences between these friction models when they are applied to the simple mechanical system depicted in Figure 4.3. The system consists of a rigid block with mass m that can slide over a floor with velocity v relative to the floor under the action of an external force F_{ex}. An important remark has to be made concerning the meaning of friction models (b) and (c) from a mechanical point of view. Friction models (b) and (c) should be understood such that only a transition from stick to slip can take place if $|F_{\mathrm{ex}}|$ exceeds F_{s}. This mechanical understanding of the friction models can conflict with the mathematical interpretation where the systems with friction models are interpreted as differential inclusions. This problem will be further explained in the following paragraphs.

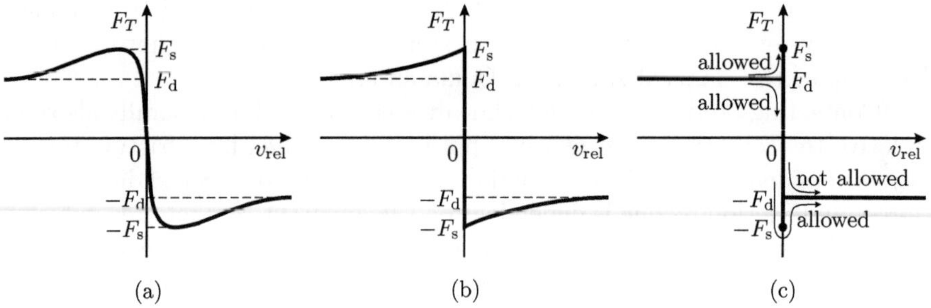

Fig. 4.2. Friction models [51].

Friction model (a) describes the friction force F_T as a smooth single-valued function of v_{rel} on the whole domain. We can also regard a system with friction model (a) as an approximation of the differential inclusion (b) with the smoothing method (see Section 6.1). The equation of motion of the block with friction model (a) reads as

$$m\dot{v}(t) = F_{\mathrm{ex}} + F_T(v(t)). \tag{4.3}$$

A suitable choice for the function F_T, which has the characteristics of friction model (a), is

$$F_T(v(t)) = -\frac{2}{\pi}\arctan(\varepsilon v(t))\Big[\frac{F_{\mathrm{s}} - F_{\mathrm{d}}}{1 + \delta|v(t)|} + F_{\mathrm{d}}\Big], \tag{4.4}$$

where $\varepsilon \gg 1$ and $\delta > 0$. The slope of the function is very steep for small $|v(t)|$. The friction force (4.4) has a maximum which is approximately F_{s} for large values of ε. This maximum was already introduced as the static friction force. The upward ramp is followed by a downward ramp which brings the friction force to an almost constant value F_{d}. Friction model (a) has all advantages and disadvantages of the smoothing method (see Section 6.1). The friction model results in a smooth ordinary differential equation for the equation of motion of the system. All standard integration routines can therefore be directly applied and the solution of the IVP always exists and is unique. The disadvantages are twofold. Firstly, the friction model can not describe the stick phase properly. The friction force is zero at zero velocity. The block will therefore always slip over the floor if an external force is present. We could define 'stick' for this model as the phase where the friction is on the upward ramp but this notion of 'stick' becomes unclear if the friction force does not contain a maximum (for instance if $F_{\mathrm{s}} = F_{\mathrm{d}}$). The second disadvantage is the steep slope of the upward and downward ramps. The equation of motion with friction model (a) will be a very stiff differential equation which is inconvenient from a numerical point of view.

In friction model (b), the upward ramp of friction model (a) is replaced by a set-valued function at $v = 0$. The equation of motion together with friction model (b) can be regarded as a differential inclusion, i.e.

$$m\dot{v}(t) = F_{\mathrm{ex}} + F_T(v(t)), \qquad F_T(v) \in \begin{cases} -\frac{F_{\mathrm{s}}-F_{\mathrm{d}}}{1+\delta|v|} - F_{\mathrm{d}}, & v > 0, \\ [-F_{\mathrm{s}}, F_{\mathrm{s}}], & v = 0, \\ \frac{F_{\mathrm{s}}-F_{\mathrm{d}}}{1+\delta|v|} + F_{\mathrm{d}}, & v < 0, \end{cases} \tag{4.5}$$

where the choice for the friction force in the slip phase is similar to the one in friction model (a). Friction model (b) describes the stick phase correctly as it allows the body to stick to the floor with zero velocity under the action of an external force. Friction model (b) is therefore often used in rigid multi-body dynamics and can well be handled by integration techniques developed in the Non-smooth Mechanics community (see Chapter 6). The solution of the differential inclusion (4.5) exists as $F_T(v)$ of (4.5) is upper semi-continuous, non-empty and bounded and has a closed, convex image (Theorem 3.2). Moreover, the set $[-F_s, F_s]$ at $v = 0$ is the *smallest* closed convex set that contains the left and right limits, $\lim v \downarrow 0$ and $\lim v \uparrow 0$ of $F_T(v)$. If $-F_s \leq F_{ex} \leq F_s$, then $\dot{v} < 0$ for $v > 0$ and $\dot{v} > 0$ for $v < 0$. Consequently, there exists an attractive sliding mode at $v = 0$ if $-F_s \leq F_{ex} \leq F_s$. In the same way we can infer that a transversal intersection exists at $v = 0$ if $|F_{ex}| > F_s$. A repulsive sliding mode is not possible with friction model (b). Uniqueness of the solution of the differential inclusion (4.5) is therefore always ensured (under the assumption of a bounded normal contact force). This is due to the fact that the friction force is dissipating energy and because the set-valued part of F_T is the smallest closed convex set that obeys upper semi-continuity. Provided that an efficient integration algorithm is available to solve the IVP for a differential inclusion, then friction model (b) can be more efficient than friction model (a). The slope of the downward ramp, however, can still be steep for large values of δ.

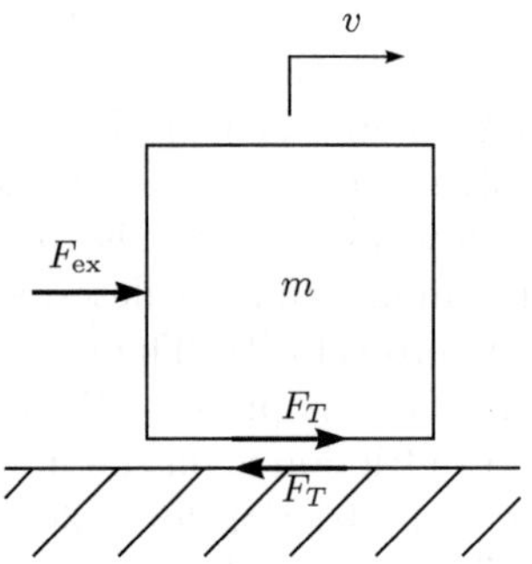

Fig. 4.3. Block on a floor with friction force F_T and external force F_{ex}.

Friction model (c) also folds the downward ramp into the set-valued part at $v = 0$. This friction model is known as the *stiction friction model*. The mechanical understanding of the stiction friction model is that the transition from stick to slip has to occur via the maximum static friction. In the transition from slip to stick, any value of the friction force is allowed within the stick force interval $[-F_s, F_s]$. This restriction is depicted with the arrows 'allowed' and 'not allowed' in Figure 4.2c.

If we would regard the system with friction model (c) as a differential inclusion, we obtain

$$m\dot{v}(t) = F_{ex} + F(v(t)), \quad F_T(v) \in \begin{cases} -F_d, & v > 0, \\ [-F_s, F_s], & v = 0, \\ F_d, & v < 0, \end{cases} \quad \text{mechanically not consistent!} \tag{4.6}$$

where $F_\mathrm{d} < F_\mathrm{s}$. The solution of the differential inclusion (4.6) exists as $F_T(v)$ is upper semi-continuous, non-empty and bounded and has a closed convex image. Remark that the set $[-F_\mathrm{s}, F_\mathrm{s}]$ at $v = 0$ is *not* the smallest closed convex set that contains the left and right limits, $\lim v \downarrow 0$ and $\lim v \uparrow 0$ of $F_T(v)$. The smallest closed convex set would be $[-F_\mathrm{d}, F_\mathrm{d}]$. This has important consequences for the uniqueness of the solution. There exists an attractive sliding mode at $v = 0$ if $-F_\mathrm{d} \leq F_\mathrm{ex} \leq F_\mathrm{d}$. A transversal intersection exists at $v = 0$ if $|F_\mathrm{ex}| > F_\mathrm{s}$.

If $F_\mathrm{d} < F_\mathrm{ex} < F_\mathrm{s}$, then $\dot{v}(t) > 0$ for $v(t) \neq 0$. It holds that $\dot{v}(t) \in F_\mathrm{ex} + [-F_\mathrm{s}, F_\mathrm{s}]$ for $v(t) = 0$ and $0 \in F_\mathrm{ex} + [-F_\mathrm{s}, F_\mathrm{s}]$. Consequently, the IVP with $v(0) = 0$ as initial condition has two distinct solutions $v(t) = 0$ and $v(t) = (F_\mathrm{ex} - F_\mathrm{d})t$ for $t \geq 0$ if $F_\mathrm{d} < F_\mathrm{ex} < F_\mathrm{s}$. A similar reasoning can be held for $-F_\mathrm{d} > F_\mathrm{ex} > -F_\mathrm{s}$. The solution of the differential inclusion (4.6) is therefore not unique if $F_\mathrm{d} < |F_\mathrm{ex}| < F_\mathrm{s}$.

In friction model (b) and (c), the external force $|F_\mathrm{ex}|$ has to exceed F_s to force the block from stick to slip. Regarding a system with friction model (b) as a differential inclusion is consistent with the mechanical concept of friction. If, however, a system with friction model (c) is regarded as a differential inclusion, then a transition from stick to slip is possible for $|F_\mathrm{ex}| < F_\mathrm{s}$ which results in non-uniqueness of solutions. A system with friction model (c) can therefore *not* be regarded as a differential inclusion as the solution concept is clearly inconsistent with the mechanical concept of friction[1]. As a consequence, it is generally not possible to integrate a system with friction model (c) using the event-driven or time-stepping integration methods which will be presented in Chapter 6. The Switch Model (see Section 6.2), however, forms an exception and is capable to simulate the dynamics of a system with friction model (c).

Friction models of type (c) are frequently used in literature [3,8,34,58,59, 62,70,108,124,146], because it yields a piecewise linear system with Stribeck effect and can therefore be used to describe stick-slip motion. Care should be taken how friction models of type (c) are to be understood.

4.3 Summary

In this chapter, we discussed the basic physical phenomena related to dry friction. The ability of dry friction to resist relative motion of contacting bodies (sticking) as well as the Stribeck effect greatly influence the dynamics of multibody systems. Dry friction models, describing the sticking phenomenon and the Stribeck effect during sliding, were discussed in Section 4.2. A proper modelling of the sticking phenomenon requires the dry friction model to be set-valued for zero relative velocity. The Stribeck effect is either modelled by

[1] A remark related to this problem has been made by [46]. An equivalent description of friction model (b) can be given by a non-smooth potential (see Chapter 5). A non-smooth potential does not exist for friction models of type (c).

a difference between the static friction coefficient and the initial dynamic friction coefficient (the stiction friction model), or by a continuous decrease of the friction coefficient during sliding. The later friction model is described by a set-valued mapping with a *minimal* closed convex upper semi-continuous image and therefore leads to a differential inclusion of Filippov-type. This type of set-valued friction model will be used in Chapter 5 to set up a unified mathematical formulation for mechanical multibody systems with dry friction. The stiction friction model will not be considered in Chapter 5, but is frequently used by engineers and scientists. The complications which arise when using the stiction friction model will be demonstrated in Sections 6.5.2, 7.3, 9.9.1 and 9.9.2.

5

Mechanical Systems with Set-valued Force-laws

Unilateral contact and friction are often modelled with set-valued force laws. Mechanical systems with set-valued force laws lead to differential inclusions and form an important class of non-smooth systems. In this chapter we will discuss the mathematical formulation of mechanical multibody systems with unilateral contact and friction modelled with set-valued force laws. The contact laws for unilateral contact and dry friction will be expressed by normal cone formulations which allows for a unified approach for the treatment of normal and tangential contact forces. The normal cone formulations put the set-valued force laws in a framework of Convex Analysis which opens the way to a description by non-smooth potentials. The mathematical framework developed in this chapter will give the basis for the numerical simulation methods presented in Chapter 6. The notation in this chapter is kept as close as possible to the notation of Glocker [51].

5.1 Equation of Motion with Contact Forces

The dynamics of a mechanical multibody system with only frictionless bilateral constraints can be described by a minimal set of generalized coordinates $\boldsymbol{q} \in \mathbb{R}^n$ and the equation of motion

$$\boldsymbol{M}(\boldsymbol{q}, t)\ddot{\boldsymbol{q}} - \boldsymbol{h}(\boldsymbol{q}, \dot{\boldsymbol{q}}, t) = \boldsymbol{0}, \tag{5.1}$$

where $\boldsymbol{M} \in \mathbb{R}^{n \times n}$ is the symmetric and positive definite mass matrix and $\boldsymbol{h} \in \mathbb{R}^n$ is the vector containing all smooth forces and gyroscopic accelerations. A linear system would be described by $\boldsymbol{h}(\boldsymbol{q}, \dot{\boldsymbol{q}}, t) = -\boldsymbol{C}\dot{\boldsymbol{q}} - \boldsymbol{K}\boldsymbol{q} + \boldsymbol{f}(t)$ in which $\boldsymbol{C}$ is the viscous damping matrix, $\boldsymbol{K}$ is the stiffness matrix and $\boldsymbol{f}(t)$ is the column vector with externally applied forces. Contact forces of the bilateral constraints do not appear in the equation of motion (5.1) because the coordinates $\boldsymbol{q}$ are minimal with respect to the bilateral constraints. If the system is additionally subjected to unilateral constraints or friction forces, then the equation of motion can be extended with normal and tangential contact forces

$$\boldsymbol{M}(\boldsymbol{q},t)\dot{\boldsymbol{u}} - \boldsymbol{h}(\boldsymbol{q},\boldsymbol{u},t) - \sum_{i\in I_x} (\boldsymbol{w}_{Ni}(\boldsymbol{q},t)\lambda_{Ni} - \boldsymbol{w}_{Ti}(\boldsymbol{q},t)\lambda_{Ti}) = \boldsymbol{0}, \tag{5.2}$$

where λ_{Ni} is the normal contact force of contact point i and λ_{Ti} is the tangential contact force due to (planar) friction. The vectors $\boldsymbol{w}_{Ni} \in \mathbb{R}^n$ and $\boldsymbol{w}_{Ti} \in \mathbb{R}^n$ define the force direction of the contact forces. The set I_x denotes the set of n_x contact points that are considered in (5.2). We will sometimes consider all closed and open contact points, denoted by the set I_C with n_C elements, or the only all closed contacts, denoted by the set I_N with n_N elements. The normal and tangential contact forces can be assembled in vectors $\boldsymbol{\lambda}_N = \mathrm{col}\{\lambda_{Ni}\} \in \mathbb{R}^{n_x}$ and $\boldsymbol{\lambda}_T = \mathrm{col}\{\lambda_{Ti}\} \in \mathbb{R}^{n_x}$. Similarly, the force vectors are assembled in the matrices $\boldsymbol{W}_N = \mathrm{mat}\{\boldsymbol{w}_{Ni}\} \in \mathbb{R}^{n\times n_x}$ and $\boldsymbol{W}_T = \mathrm{mat}\{\boldsymbol{w}_{Ti}\} \in \mathbb{R}^{n\times n_x}$. The equation of motion for mechanical systems with frictional unilateral contact without impulses can be written as

$$\boldsymbol{M}(\boldsymbol{q},t)\dot{\boldsymbol{u}} - \boldsymbol{h}(\boldsymbol{q},\boldsymbol{u},t) - \boldsymbol{W}_N(\boldsymbol{q},t)\boldsymbol{\lambda}_N - \boldsymbol{W}_T(\boldsymbol{q},t)\boldsymbol{\lambda}_T = \boldsymbol{0}, \tag{5.3}$$

for almost all t, with

$$\boldsymbol{u} = \dot{\boldsymbol{q}}, \ \forall t. \tag{5.4}$$

Motion without impulses implies that $\boldsymbol{\lambda}_N(t)$ is (locally) bounded and time-continuous. The friction force $\boldsymbol{\lambda}_T(t)$ is discontinuous when a slip-to-stick transition takes place or when the relative sliding velocity of a frictional contact reverses its sign. The acceleration $\dot{\boldsymbol{u}}$ is not defined when $\boldsymbol{\lambda}_T(t)$ is discontinuous. The set of time instances for which $\boldsymbol{\lambda}_T(t)$ is discontinuous is of measure zero and (5.3) therefore holds for almost all t. The choice of a different symbol for positions $\boldsymbol{q}$ and velocities $\boldsymbol{u}$ will be useful for studying non-smooth systems. Note that the coordinates $\boldsymbol{q}$ in (5.3) are still minimal with respect to the bilateral constraints. The unilateral contact and friction forces $\boldsymbol{\lambda}_N$ and $\boldsymbol{\lambda}_T$ can be envisaged as Lagrangian multipliers to force the system to fulfil the frictional unilateral constraints (see Section 5.5.2).

5.2 Contact Laws

The equation of motion (5.3) has to be complemented with constitutive laws for the normal contact forces $\boldsymbol{\lambda}_N$ and the friction forces $\boldsymbol{\lambda}_T$. Mechanical constitutive laws are commonly expressed by derivatives of potentials. Both conservative and nonconservative forces can be derived from potentials. However, nonconservative potentials, also called dissipation functions, are non-integrable, i.e. the value of the integral is path dependent.

The potential of a linear mechanical spring is given by $V(q) = \frac{1}{2}kq^2$, where k is the spring stiffness and q is the elongation of the spring. The force law of a linear spring can be expressed as

$$-\lambda = \frac{\partial V(q)}{\partial q} = kq, \tag{5.5}$$

where λ is the force exerted by the spring. The potential $V(q)$ of a linear spring is a smooth function of the displacement q. However, many force laws are expressed by non-smooth potentials [50, 51].

5.2.1 Set-valued Normal Contact Law

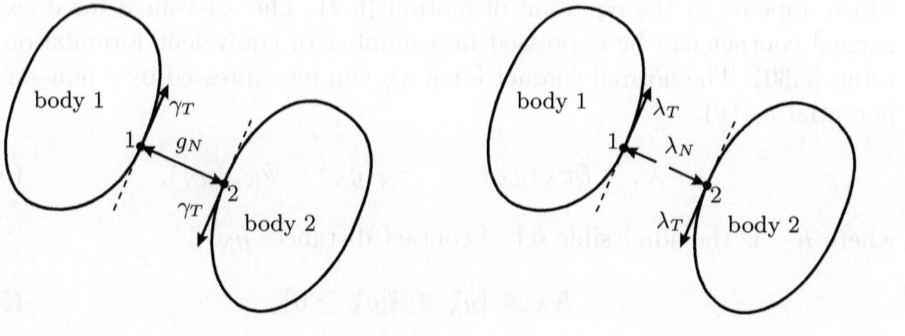

Fig. 5.1. Contact distance g_N and tangential velocity γ_T between two rigid bodies.

Normal contact between rigid bodies is described by a set-valued force law called Signorini's law. Consider two convex rigid bodies at a relative distance g_N from each other (Figure 5.1). The normal contact distance g_N is for convex bodies uniquely defined and is such, that the points 1 and 2 have parallel tangent planes (shown as dashed lines in Figure 5.1). The normal contact distance g_N is nonnegative because the bodies do not penetrate into each other. The bodies touch when $g_N = 0$. The normal contact force λ_N between the bodies is nonnegative because the bodies can only push on each other, i.e. the constraint is unilateral. The normal contact force vanishes when there is no contact $g_N > 0$ and can only be positive when contact is present $g_N = 0$. Under the assumption of impenetrability $g_N \geq 0$, only two situations may occur:

$$\begin{aligned} g_N = 0 \wedge \lambda_N \geq 0 \text{ contact} \\ g_N > 0 \wedge \lambda_N = 0 \text{ no contact} \end{aligned} \tag{5.6}$$

From (5.6) we see that the normal contact law shows a complementarity behaviour: the product of the contact force and normal contact distance is always zero:

$$g_N \lambda_N = 0. \tag{5.7}$$

The relation between the normal contact force and the normal contact distance is therefore

$$g_N \geq 0, \quad \lambda_N \geq 0, \quad g_N \lambda_N = 0, \tag{5.8}$$

which is the complementarity condition between g_N and λ_N. The complementarity behaviour of the normal contact law is depicted in Figure 5.2 and shows a corner of admissible combinations of g_N and λ_N. The magnitude of the contact force is denoted by λ_N and the direction of the contact force is normal to the bodies, i.e. along the line 1–2. The generalized force direction of λ_N with respect to the minimal coordinates $\boldsymbol{q}$ is given by a vector $\boldsymbol{w}_N \in \mathbb{R}^n$,

$$\boldsymbol{w}_N = \left(\frac{\partial g_N}{\partial \boldsymbol{q}}\right)^{\mathrm{T}}, \tag{5.9}$$

which appears in the equation of motion (5.2). The set-valued force law for normal contact can be expressed in a number of equivalent formulations by using (2.30). The normal contact force λ_N can be expressed by a non-smooth potential $\pi_N(y)$:

$$-\lambda_N \in \partial \pi_N(g_N), \qquad \pi_N(g_N) = \Psi_{K_N}(g_N), \tag{5.10}$$

where K_N is the admissible set of contact distances g_N

$$K_N = \{g_N^* \in \mathbb{R} | g_N^* \geq 0\}, \tag{5.11}$$

and Ψ_{K_N} is the indicator function of K_N. Alternatively we can formulate the contact law in a compact form by means of the normal cone of K_N (see Definition 2.10):

$$-\lambda_N \in N_{K_N}(g_N). \tag{5.12}$$

Similarly, we can formulate the contact law as

$$-g_N \in N_{C_N}(\lambda_N), \qquad C_N = \{\lambda^* \in \mathbb{R} | \lambda^* \geq 0\}, \tag{5.13}$$

where C_N is the admissible set of contact forces.

The normal contact law, also called Signorini's law, expresses impenetrability of the contact and can formally be stated for a number of contact points $i = 1, \ldots, n_C$ as

$$-\boldsymbol{g}_N \in N_{C_N}(\boldsymbol{\lambda}_N), \qquad C_N = \{\boldsymbol{\lambda}_N^* \in \mathbb{R}^n | \boldsymbol{\lambda}_N^* \geq \boldsymbol{0}\}, \tag{5.14}$$

where $\boldsymbol{\lambda}_N$ is the vector containing the normal contact forces λ_{Ni} and $\boldsymbol{g}_N$ is the vector of normal contact distances g_{Ni}.

5.2.2 Set-valued Coulomb Friction Law

Coulomb's friction law is another classical example of a force law that can be described by a non-smooth potential. Consider again the two bodies of Figure 5.1 but with Coulomb friction at the contact point. We denote the relative velocity of point 1 with respect to point 2 along their tangent plane by γ_T. If contact is present between the bodies, i.e. $g_N = 0$, then the friction between the bodies imposes a force λ_T along the tangent plane of the contact point. The friction force λ_T is, when the bodies are sliding over the each other, equal to $\mu\lambda_N$ and in the direction of $-\gamma_T$

$$-\lambda_T = \mu\lambda_N \operatorname{sign}(\gamma_T), \quad \gamma_T \neq 0, \tag{5.15}$$

where μ is the friction coefficient and λ_N is the normal contact force. If the relative tangential velocity vanishes, i.e. $\gamma_T = 0$, then the bodies purely roll over each other without slip. Pure rolling, or no slip for locally flat objects, is denoted by *stick*. When the bodies stick, then the friction force must lie in the interval $-\mu\lambda_N \leq \lambda_T \leq \mu\lambda_N$. For unidirectional friction, i.e. for planar contact problems, the following three cases are possible:

$$\begin{array}{ll} \gamma_T = 0 \Rightarrow |\lambda_T| \leq \mu\lambda_N & \text{sticking} \\ \gamma_T < 0 \Rightarrow \lambda_T = +\mu\lambda_N & \text{negative sliding} \\ \gamma_T > 0 \Rightarrow \lambda_T = -\mu\lambda_N & \text{positive sliding} \end{array} \tag{5.16}$$

We can express the friction force by a potential $\pi_T(\gamma_T)$

$$-\lambda_T \in \partial_{\gamma_T}\pi_T(\gamma_T), \quad \pi_T(\gamma_T) = \mu\lambda_N|\gamma_T|, \tag{5.17}$$

from which follows the set-valued force law

$$\lambda_T \in \begin{cases} -\mu\lambda_N, & \gamma_T > 0, \\ [-1,1]\mu\lambda_N, & \gamma_T = 0, \\ \mu\lambda_N, & \gamma_T < 0. \end{cases} \tag{5.18}$$

A non-smooth convex potential therefore leads to a closed convex set-valued force law. The graph of Coulomb's friction law (5.18) is depicted in Figure 5.2. The admissible values of the tangential force λ_T form a convex set C_T which is bounded by the values of the normal force [118]:

$$C_T = \{\lambda_T \mid -\mu\lambda_N \leq \lambda_T \leq +\mu\lambda_N\}. \tag{5.19}$$

Coulomb's law can be expressed with the aid of the indicator function of C_T as

$$-\gamma_T \in \partial\Psi_{C_T}(\lambda_T) \Leftrightarrow -\gamma_T \in N_{C_T}(\lambda_T), \tag{5.20}$$

where the indicator function Ψ_{C_T} is the conjugate potential of π_T [51], i.e. $\pi_T(\gamma_T) = \Psi^*_{C_T}(\gamma_T)$. The generalized force direction of λ_T with respect to the minimal coordinates $\boldsymbol{q}$ is given by a vector $\boldsymbol{w}_T$,

$$\boldsymbol{w}_T = \left(\frac{\partial\gamma_T}{\partial\boldsymbol{u}}\right)^{\mathrm{T}}, \tag{5.21}$$

where $\boldsymbol{u}$ are the generalized velocities.

Coulomb's law can be formally stated for a number of contact points $i = 1, \ldots, n_C$ as

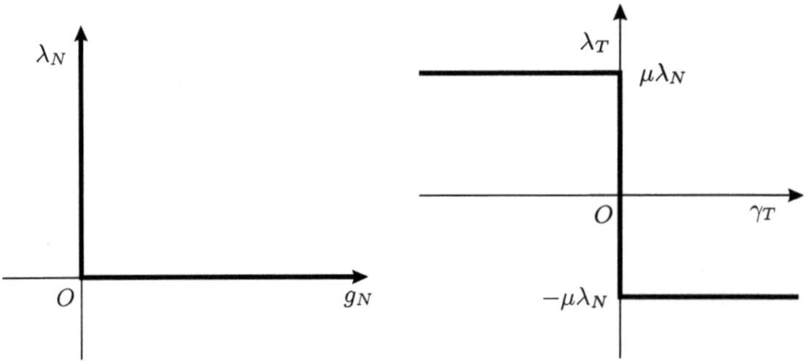

Fig. 5.2. Signorini's normal contact law and Coulomb's friction law.

$$-\gamma_T \in N_{C_T}(\boldsymbol{\lambda}_T), \tag{5.22}$$

with

$$C_T = \{\boldsymbol{\lambda}_T \mid -\mu\boldsymbol{\lambda}_N \leq \boldsymbol{\lambda}_T \leq +\mu\boldsymbol{\lambda}_N\}, \tag{5.23}$$

where $\boldsymbol{\gamma}_T = \text{col}\{\gamma_{Ti}\}$ is the vector of tangential contact velocities. The set-valued force law for unidirectional friction can also be formulated by complementarity conditions [46, 118] as will be shown in Section 6.3.

The friction law of Coulomb, as defined above, assumes the friction forces to be a function of the unilateral normal forces. Both the normal contact forces and the friction forces have to be determined. However, in many applications the situation is less complicated as the normal force is constant or at least a given function of time. A known normal contact force allows for a simplified contact law. The tangential friction forces are assumed to obey either one of the following friction laws:

1. Non-associated Coulomb's law for which the normal force is dependent on the solution $(\boldsymbol{q}, \boldsymbol{u})$ and therefore not known in advance. The set of admissible contact forces is given by
$$C_T(\boldsymbol{\lambda}_N) = \{\boldsymbol{\lambda}_T \mid \|\lambda_{Ti}\| \leq \mu_i\lambda_{Ni}; \forall i\}, \tag{5.24}$$
which is dependent on the normal contact forces $\boldsymbol{\lambda}_N$ and friction coefficients $\boldsymbol{\mu} = \text{col}\{\mu_i\}$.
2. Associated Coulomb's law for which the normal force is known in advance. The set of admissible contact forces is given by
$$C_D = \{\boldsymbol{\lambda}_T \mid \|\lambda_{Ti}\| \leq \mu_i F_{Ni}; \forall i\}, \tag{5.25}$$
which is dependent on the known normal forces $\boldsymbol{F}_N$ and friction coefficients $\boldsymbol{\mu} = \text{col}\{\mu_i\}$.

5.3 Kinematic Levels of Set-valued Force Laws

The set of all contact point $I_C = \{1, 2, \ldots, n_C\}$ contains a subset $I_N \subset I_C$ of all closed contact points, i.e. $I_N = \{i \in I_C \mid g_{Ni} = 0, \ \gamma_{Ni} = 0\}$. Each closed contact $i \in I_N$ is characterized by a vanishing contact distance g_{Ni} and normal relative velocity γ_{Ni}. Under the assumption of impenetrability $g_{Ni} \geq 0$, only two situations may occur:

$$\begin{array}{ll} \dot{\gamma}_{Ni} = 0 \wedge \lambda_{Ni} \geq 0 & \text{contact is maintained} \\ \dot{\gamma}_{Ni} > 0 \wedge \lambda_{Ni} = 0 & \text{detachment} \end{array}, \quad i \in I_N. \tag{5.26}$$

The product of the contact force and acceleration is always zero:

$$\dot{\gamma}_{Ni}\lambda_{Ni} = 0, \qquad i \in I_N. \tag{5.27}$$

Consequently, a closed contact point $i \in I_N$ exhibits a complementarity behaviour on acceleration level.

The set-valued force laws introduced in the previous section can be expressed on different kinematic levels. Stating the force law on a lower kinematic level can be regarded as an index reduction of the associated differential algebraic equation. The force law for the normal contact forces can be expressed on three different kinematic levels

$$\begin{array}{ll} \text{displacement level:} & -\boldsymbol{g}_N \in N_{C_N}(\boldsymbol{\lambda}_N), \\ \text{velocity level:} & -\boldsymbol{\gamma}_N \in N_{C_N}(\boldsymbol{\lambda}_N),\ \boldsymbol{g}_N = 0, \\ \text{acceleration level:} & -\dot{\boldsymbol{\gamma}}_N \in N_{C_N}(\boldsymbol{\lambda}_N),\ \boldsymbol{g}_N = 0,\ \boldsymbol{\gamma}_N = 0, \end{array} \tag{5.28}$$

where $\boldsymbol{g}_N(\boldsymbol{q}, t)$ and $\boldsymbol{\gamma}_N(\boldsymbol{q}, \boldsymbol{u}, t) = \dot{\boldsymbol{g}}_N(\boldsymbol{q}, t)$ are the vectors with relative contact distances and velocities, respectively.

Similarly, Coulomb's law can be formulated on acceleration level. For a closed sticking contact $i \in I_H$, with $I_H = \{i \in I_N \mid \gamma_{Ti} = 0\}$, the following three cases are possible

$$\begin{array}{ll} \dot{\gamma}_{Ti} = 0 \Rightarrow |\lambda_{Ti}| \leq \mu_i\lambda_{Ni} & \text{remains sticking} \\ \dot{\gamma}_{Ti} < 0 \Rightarrow \lambda_{Ti} = +\mu_i\lambda_{Ni} & \text{commences negative sliding} \\ \dot{\gamma}_{Ti} > 0 \Rightarrow \lambda_{Ti} = -\mu_i\lambda_{Ni} & \text{commences positive sliding} \end{array} \quad i \in I_H. \tag{5.29}$$

The force law for the non-associated and associated Coulomb friction can therefore be expressed on two different kinematic levels

$$\begin{array}{ll} \text{velocity level:} & -\boldsymbol{\gamma}_T \in N_{C_T(\boldsymbol{\lambda}_N)}(\boldsymbol{\lambda}_T) \text{ or } -\boldsymbol{\gamma}_T \in N_{C_D}(\boldsymbol{\lambda}_T), \\ \text{acceleration level:} & -\dot{\boldsymbol{\gamma}}_T \in N_{C_T(\boldsymbol{\lambda}_N)}(\boldsymbol{\lambda}_T) \text{ or } -\dot{\boldsymbol{\gamma}}_T \in N_{C_D}(\boldsymbol{\lambda}_T),\ \boldsymbol{\gamma}_T = 0. \end{array} \tag{5.30}$$

Note that the sets C_N, $C_T(\boldsymbol{\lambda}_N)$ and C_D are convex sets and that the above contact laws are set-valued relations between the contact forces $\boldsymbol{\lambda}$ and primal variables $\boldsymbol{s}$, e.g. $\boldsymbol{s} = -\boldsymbol{g}_N$ or $\boldsymbol{s} = -\boldsymbol{\gamma}_N$ etc.

The vectors of normal and tangential contact velocities and accelerations are a function of the generalized positions $\boldsymbol{q}$, velocities $\boldsymbol{u}$, accelerations $\dot{\boldsymbol{u}}$ and time t. The following notation will be used throughout this book:

$$\begin{aligned} \boldsymbol{\gamma}_N(t, \boldsymbol{q}, \boldsymbol{u}) &= \boldsymbol{W}_N^{\mathrm{T}}(t, \boldsymbol{q})\boldsymbol{u} + \tilde{\boldsymbol{w}}_N(t, \boldsymbol{q}), \\ \dot{\boldsymbol{\gamma}}_N(t, \boldsymbol{q}, \boldsymbol{u}) &= \boldsymbol{W}_N^{\mathrm{T}}(t, \boldsymbol{q})\dot{\boldsymbol{u}} + \bar{\boldsymbol{w}}_N(t, \boldsymbol{q}, \boldsymbol{u}), \end{aligned} \tag{5.31}$$

and

$$\begin{aligned} \gamma_T(t,\boldsymbol{q},\boldsymbol{u}) &= \boldsymbol{W}_T^{\mathrm{T}}(t,\boldsymbol{q})\boldsymbol{u} + \tilde{\boldsymbol{w}}_T(t,\boldsymbol{q}), \\ \dot{\gamma}_T(t,\boldsymbol{q},\boldsymbol{u}) &= \boldsymbol{W}_T^{\mathrm{T}}(t,\boldsymbol{q})\dot{\boldsymbol{u}} + \bar{\boldsymbol{w}}_T(t,\boldsymbol{q},\boldsymbol{u}), \end{aligned} \tag{5.32}$$

with

$$\boldsymbol{W}_N(t,\boldsymbol{q}) = \left(\frac{\partial \boldsymbol{g}_N(t,\boldsymbol{q})}{\partial \boldsymbol{q}}\right)^{\mathrm{T}}, \qquad \boldsymbol{W}_T(t,\boldsymbol{q}) = \left(\frac{\partial \gamma_T(t,\boldsymbol{q},\boldsymbol{u})}{\partial \boldsymbol{u}}\right)^{\mathrm{T}}. \tag{5.33}$$

The normal contact velocities γ_N are split in a part $\boldsymbol{W}_N^{\mathrm{T}}(t,\boldsymbol{q})\boldsymbol{u}$ dependent on $\boldsymbol{u}$ and a rest term $\tilde{\boldsymbol{w}}_N(t,\boldsymbol{q})$. The tangential contact velocities are split in a similar fashion. The subdivision of the contact accelerations yield the rest terms $\bar{\boldsymbol{w}}_N(t,\boldsymbol{q},\boldsymbol{u})$ and $\bar{\boldsymbol{w}}_T(t,\boldsymbol{q},\boldsymbol{u})$ which are not dependent on $\dot{\boldsymbol{u}}$. Alternatively, we can derive the rest terms from

$$\tilde{\boldsymbol{w}}_N(t,\boldsymbol{q}) = \frac{\partial \boldsymbol{g}_N(t,\boldsymbol{q})}{\partial t}, \qquad \bar{\boldsymbol{w}}_N(t,\boldsymbol{q},\boldsymbol{u}) = \dot{\boldsymbol{W}}_N^{\mathrm{T}}(t,\boldsymbol{q})\boldsymbol{u} + \dot{\tilde{\boldsymbol{w}}}_N(t,\boldsymbol{q}), \tag{5.34}$$

$$\bar{\boldsymbol{w}}_T(t,\boldsymbol{q},\boldsymbol{u}) = \dot{\boldsymbol{W}}_T^{\mathrm{T}}(t,\boldsymbol{q})\boldsymbol{u} + \dot{\tilde{\boldsymbol{w}}}_T(t,\boldsymbol{q}). \tag{5.35}$$

The arguments $(t,\boldsymbol{q},\boldsymbol{u})$ will in the sequel often be dropped to abbreviate the notation.

5.4 Non-smooth Potentials

In Section 5.5 it is convenient to express the contact laws in the framework of non-smooth potential theory. Independent of the kinematic level we choose, we can express a set-valued contact law, such as (5.28) and (5.30), by subdifferentials of (quasi) potentials (see [51])

$$\boldsymbol{s} \in \partial\Psi_C(\boldsymbol{\lambda}) \iff \boldsymbol{\lambda} \in \partial\Psi_C^*(\boldsymbol{s}), \tag{5.36}$$

where Ψ_C is the indicator function of the convex set C and $\partial\Psi_C(\boldsymbol{\lambda}) = N_C(\boldsymbol{\lambda})$ is the subdifferential of the indicator function which agrees with the normal cone to C at $\boldsymbol{\lambda}$. The support function $\Psi_C^*(\boldsymbol{s})$ (Definition 2.14) is the conjugate potential of the indicator function of C. Note that for normal contact on displacement level we have the convention $\boldsymbol{s} = -\boldsymbol{g}_N$, $\boldsymbol{\lambda} = \boldsymbol{\lambda}_N$, $C = C_N$ (see (5.14)). Non-associated friction leads to a quasi potential $\Psi_{C_T(\boldsymbol{\lambda}_N)}(\boldsymbol{\lambda})$, due to the dependence on the normal contact force. The potentials, conjugate potentials and subdifferentials are shown in Figures 5.3 and 5.4 of the normal contact problem and the tangential contact problem.

The force law (5.36) is expressed in terms of an inclusion. However, by making use of the definition of the proximal point to the set C, denoted by $\mathrm{prox}_C(\boldsymbol{x})$ (see Definition 2.15), we can write the force law as an equality

$$\boldsymbol{\lambda} = \mathrm{prox}_C(\boldsymbol{\lambda} + r\boldsymbol{s}),\ r > 0 \iff \boldsymbol{\lambda} \in \partial\Psi_C^*(\boldsymbol{s}) \iff \boldsymbol{s} \in \partial\Psi_C(\boldsymbol{\lambda}). \tag{5.37}$$

The arbitrary auxiliary parameter $r > 0$ is basically used to set up a cone normal to C. Hence, there are at least three different ways to express a contact law: by means of the indicator function, by means of the support function or by means of the proximal point. Solving the equation of motion (stated as a differential equation or as an equality of measures), together with the set-valued algebraic relations between the contact forces $\boldsymbol{\lambda}$ and the primal variables $\boldsymbol{s}$, leads for each time instance to a set of algebraic inclusions in the accelerations $\dot{\boldsymbol{u}}$. For instance, for the acceleration problem with non-associated friction we obtain

$$\boldsymbol{M}\dot{\boldsymbol{u}} - \boldsymbol{h} - \boldsymbol{W}_N \partial \Psi^*_{C_N}(\boldsymbol{s}_N(\dot{\boldsymbol{u}})) - \boldsymbol{W}_T \partial \Psi^*_{C_T(\boldsymbol{\lambda}_N(\dot{\boldsymbol{u}}))}(\boldsymbol{s}_T(\dot{\boldsymbol{u}})) \ni \boldsymbol{0}, \tag{5.38}$$

where C_T depends through $\boldsymbol{\lambda}_N$ on the solution $\dot{\boldsymbol{u}}$, or for associated friction

$$\boldsymbol{M}\dot{\boldsymbol{u}} - \boldsymbol{h} - \boldsymbol{W}_N \partial \Psi^*_{C_N}(\boldsymbol{s}_N(\dot{\boldsymbol{u}})) - \boldsymbol{W}_T \partial \Psi^*_{C_D}(\boldsymbol{s}_T(\dot{\boldsymbol{u}})) \ni \boldsymbol{0}, \tag{5.39}$$

with $\boldsymbol{s}_N = -\dot{\boldsymbol{\gamma}}_N$ and $\boldsymbol{s}_T = -\dot{\boldsymbol{\gamma}}_T$. Equations (5.38) or (5.39) are algebraic inclusions in the accelerations $\dot{\boldsymbol{u}}$ for each time instance t and therefore a set of differential inclusions for $\boldsymbol{u}$. Numerical integration of such a differential inclusion requires that $\dot{\boldsymbol{u}}$ is calculated in each time-step for fixed t, $\boldsymbol{q}$ and $\boldsymbol{u}$. The time-incremental problem is therefore an algebraic inclusion. Only in very special cases, we have the simplification that the inclusion can be transformed to a linear complementarity problem, for which numerical solvers are readily available. For instance, the acceleration problem of mechanical system with frictionless unilateral constraints can be written as

$$\begin{gathered} \boldsymbol{M}\dot{\boldsymbol{u}} - \boldsymbol{h} - \boldsymbol{W}_N \boldsymbol{\lambda}_N = \boldsymbol{0}, \\ \boldsymbol{\lambda}_N \geq \boldsymbol{0}, \quad \dot{\boldsymbol{\gamma}}_N(\dot{\boldsymbol{u}}) \geq \boldsymbol{0}, \quad \boldsymbol{\lambda}_N^{\mathrm{T}} \dot{\boldsymbol{\gamma}}_N(\dot{\boldsymbol{u}}) = 0. \end{gathered} \tag{5.40}$$

with $\dot{\boldsymbol{\gamma}}_N(\dot{\boldsymbol{u}})$ being a linear affine function. Spatial frictional contact problems can be transformed to nonlinear complementarity problems (NCPs) but the numerical solution of the resulting type of NCPs is very cumbersome.

5.5 Augmented Lagrangian Method

A solution concept for differential inclusions has been presented in Chapter 3. Filippov's solution concept allows switches between modes of the system and also sliding modes along switching boundaries. Chapter 6 will discuss numerical integration techniques for differential inclusions with special attention to differential inclusions arising from mechanical systems with set-valued friction laws. An integration method for differential inclusions has to decide in which mode the system will pursue its motion when the state is on a switching boundary. Some integration methods determine the next mode only on the switching boundaries while other integration methods take this decision in every integration step. The search for the mode in the next time instant

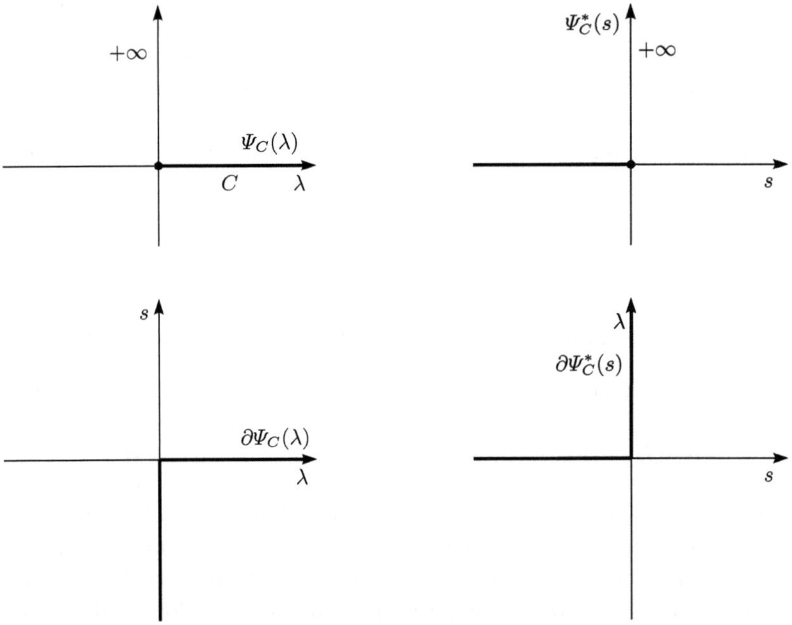

Fig. 5.3. Potential, conjugate potential and subdifferential of the normal contact problem $C = C_N = \mathbb{R}^+$.

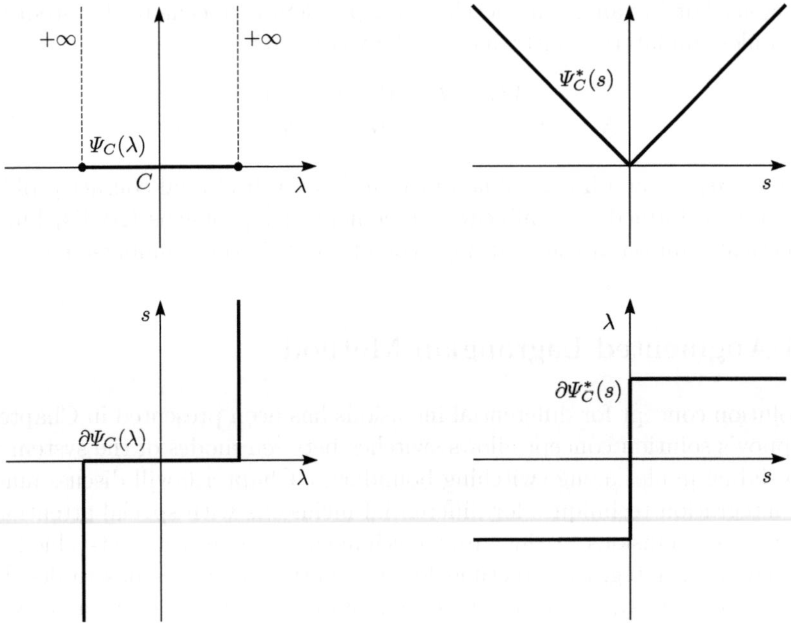

Fig. 5.4. Potential, conjugate potential and subdifferential of the tangential contact problem $C = C_D$.

leads to a mathematical problem in the form of an algebraic inclusion, which we will denote by the *contact problem*.

The formulation of frictional contact problems of mechanical rigid multibody systems with set-valued contact laws leads to a nonlinear algebraic inclusion, which is complicated to solve. The first (and often worst) method is to regularize the set-valued force laws and obtain a stiff problem with an inexact solution. Instead, one simply has to accept the set-valued nature of the problem and deal with the nonlinear algebraic equation. For planar frictional contact problems (unidirectional friction) and for frictionless spatial problems, the contact problem can conveniently be formulated as a linear complementarity problem [46, 118]. A linear complementarity formulation of planar frictional contact problems will be given in Chapter 6. Spatial frictional contact, for which the friction force is two-dimensional, can be formulated as a nonlinear complementarity problem [49, 51, 139], but this has a number of drawbacks. First of all, the numerical solution of the resulting type of NCPs is cumbersome. Secondly, the formulation of spatial anisotropic Coulomb friction or more elaborate friction models is complicated and not suitable from a practical point of view.

Alart and Curnier [2] have presented an augmented Lagrangian approach to the solution of static frictional contact problems. The augmented Lagrangian method has originally been proposed by Hestenes [57] and Powell [125] to solve nonlinear programming problems with equality constraints. Rockafellar [129] has extended the method to inequality constraints (such as frictionless unilateral contact problems). Fortin [40] has applied the method to a non-differentiable problem (such as the pure friction problem with known normal force). Alart and Curnier [2] have given a complete mixed formulation of the combined normal-tangential frictional contact problem for static mechanical systems.

In this section we will discuss the augmented Lagrangian method to solve the contact problem of mechanical frictional contact problems. The augmented Lagrangian method has the advantage over (non)-linear complementarity methods to be applicable to more general set-valued force laws. This section is for a large part based on the work of Alart and Curnier [2], but gives an extension to dynamical systems and gives a novel interpretation of the method. The results of [2] will be extended to dynamical mechanical systems through the principle of Gauß and a clear interpretation will be given of the augmented Lagrangian method in terms of non-smooth potential theory [51].

5.5.1 Exact Regularization

The idea of a regularization is to replace the set-valued contact laws by single-valued contact laws for which the contact forces $\boldsymbol{\lambda}$ can be simply a single-valued function of the primal variables $\boldsymbol{s}$. Examples of regularizations are unilateral springs for normal contact, which replace the rigid contact by a compliant contact. Coulomb's friction law is usually regularized by a finite

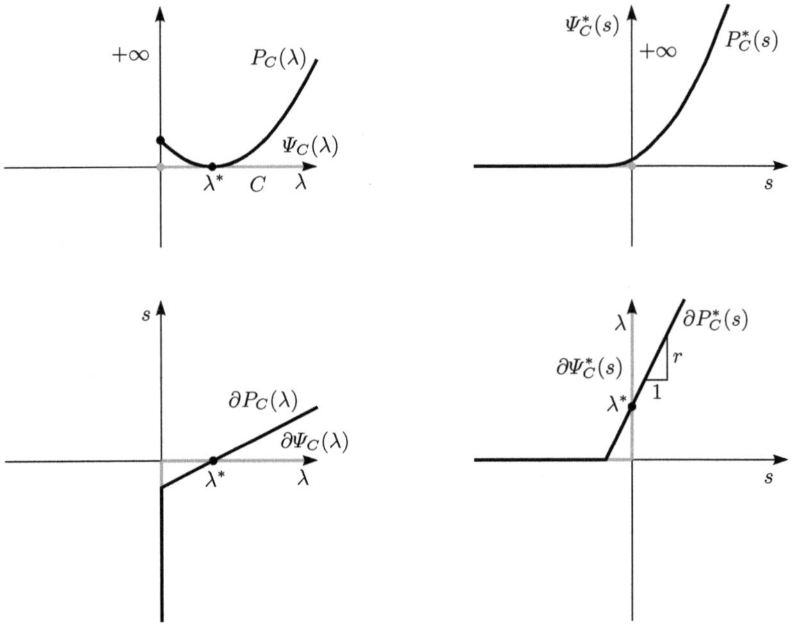

Fig. 5.5. Potential, conjugate potential and subdifferential of the normal contact problem $C = C_N = \mathbb{R}^+$; (set-valued force law = gray; regularized force law = black).

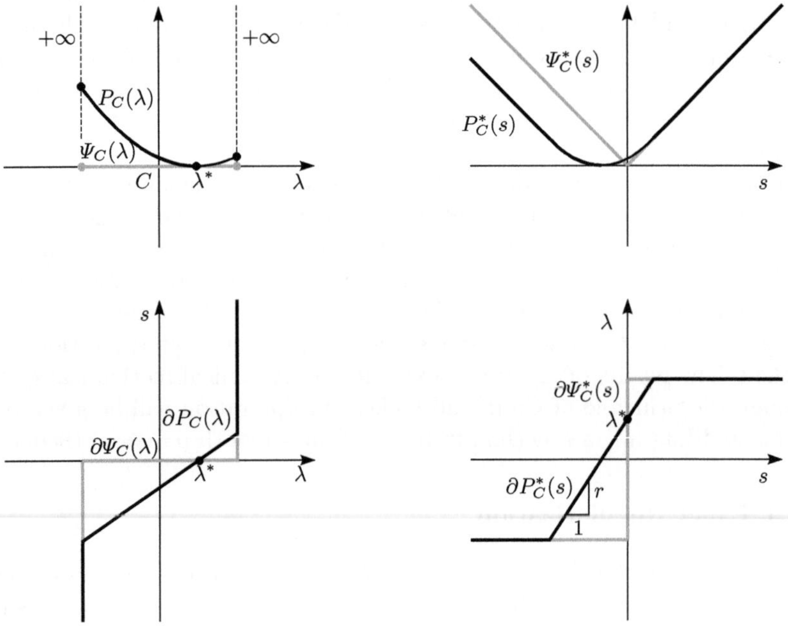

Fig. 5.6. Potential, conjugate potential and subdifferential of the tangential contact problem $C = C_D$; (set-valued force law = gray; regularized force law = black).

slope of the force law within the stick phase. Such regularizations are of course generally not exact, i.e. the compliant contact will cause an indentation of the contacting bodies and 'stick' will be accompanied by a non-zero sliding velocity.

The idea of an exact regularization is to shift the regularized contact law such that the exact solution is obtained. This seems a paradox, because the solution has to be known in advance. Indeed, we have to solve for the shift together with the equation of motion and the regularized contact law, such that we obtain the exact solution. So, instead of determining the contact forces from a set-valued contact law, we calculate the shift from a single-valued contact law. The latter is much easier because we can solve for it with a root-finding algorithm.

As regularization we will use the following single-valued relation between $\boldsymbol{\lambda}$ and $\boldsymbol{s}$, $\boldsymbol{s} \rightarrow \boldsymbol{\lambda}$,

$$\boldsymbol{\lambda}(\boldsymbol{s}) = \operatorname{prox}_C(\boldsymbol{\lambda}^* + r\boldsymbol{s}), \quad r > 0, \tag{5.41}$$

where $\boldsymbol{\lambda}^*$ is the shift of the regularization and $r > 0$ is now a steepness parameter of the regularization. If the regularization is exact then we must have $\boldsymbol{\lambda} = \boldsymbol{\lambda}^*$, see (5.37) and Figures 5.5 and 5.6 for contact law characteristics.

Consider for instance a mechanical system with the force laws for the normal and tangential contact forces on acceleration level. We put (5.41) in (5.3) and obtain the first line of (5.42). Subsequently, we replace $\boldsymbol{\lambda}$ by $\boldsymbol{\lambda}^*$ in (5.41) to obtain the shift, which gives the last two lines of (5.42)

$$\begin{aligned}
&\boldsymbol{M}\dot{\boldsymbol{u}} - \boldsymbol{h} - \boldsymbol{W}_N \operatorname{prox}_{C_N}(\boldsymbol{\lambda}_N^* - r\dot{\boldsymbol{\gamma}}_N) - \boldsymbol{W}_T \operatorname{prox}_{C_T}(\boldsymbol{\lambda}_T^* - r\dot{\boldsymbol{\gamma}}_T) = \boldsymbol{0},\\
&\boldsymbol{\lambda}_N^* - \operatorname{prox}_{C_N}(\boldsymbol{\lambda}_N^* - r\dot{\boldsymbol{\gamma}}_N) = \boldsymbol{0},\\
&\boldsymbol{\lambda}_T^* - \operatorname{prox}_{C_T(\boldsymbol{\lambda}_N^*)}(\boldsymbol{\lambda}_T^* - r\dot{\boldsymbol{\gamma}}_T) = \boldsymbol{0}.
\end{aligned} \tag{5.42}$$

The last line of (5.42) contains the tangential contact law for non-associated Coulomb friction, but can be replaced by the associated law using C_D instead of $C_T(\boldsymbol{\lambda}_N^*)$. We now have three non-smooth continuous algebraic equations for the three unknowns $(\dot{\boldsymbol{u}}, \boldsymbol{\lambda}_N^*, \boldsymbol{\lambda}_T^*)$. Or, if we regard the first line of (5.42) as a differential equation, then we can look upon (5.42) as a set of differential algebraic equations [27,55]. These equations can be solved with a root-finding algorithm like a generalized Newton method. We therefore solve for the shift $\boldsymbol{\lambda}^*$, but the difference between the shift and $\boldsymbol{\lambda}$ is formal as they agree at the solution, i.e. $\boldsymbol{\lambda} = \boldsymbol{\lambda}^*$. A very elegant method is therefore obtained to solve an inclusion of type (5.38), which describes a frictional contact problem. Note that this approach works for arbitrary convex sets C of admissible contact forces. The convex sets C are also allowed to have a non-smooth boundary. At this point we could stop our discussion. But still, the method seems almost like a trick and is lacking a fundamental background.

5.5.2 Stationarity Conditions of the Augmented Lagrangian

In this section we will show that the exact regularization method is identical to the augmented Lagrangian method in optimization theory. Non-smooth potentials will be derived for the regularized force law in order to state the mechanical problem as an optimization problem. It will turn out that the augmented Lagrangian method yields an unconstrained min-max problem and that the exact regularization method corresponds to the stationarity conditions for the saddle-point of the augmented Lagrangian function.

The regularized force law (5.41) for an arbitrary convex set C can also be expressed in terms of potential functions and conjugate potential functions. Using (2.18) it can be shown that (for fixed $\boldsymbol{\lambda}^*$)

$$\begin{aligned} \boldsymbol{\lambda} = \operatorname{prox}_C(\boldsymbol{\lambda}^* + r\boldsymbol{s}), \quad r > 0 \quad &\Longleftrightarrow \quad \boldsymbol{\lambda}^* + r\boldsymbol{s} - \boldsymbol{\lambda} \in N_C(\boldsymbol{\lambda}) \\ &\Longleftrightarrow \quad \boldsymbol{s} \in N_C(\boldsymbol{\lambda}) + \frac{1}{r}(\boldsymbol{\lambda} - \boldsymbol{\lambda}^*). \end{aligned} \tag{5.43}$$

The regularized force law can therefore be stated as a subdifferential of a non-smooth potential $P_C(\boldsymbol{\lambda})$

$$\boldsymbol{s} \in \partial P_C(\boldsymbol{\lambda}), \tag{5.44}$$

from which we obtain by integration

$$P_C(\boldsymbol{\lambda}) = \Psi_C(\boldsymbol{\lambda}) + \frac{1}{2r}\|\boldsymbol{\lambda} - \boldsymbol{\lambda}^*\|^2. \tag{5.45}$$

The potential function $P_C(\boldsymbol{\lambda})$ is convex and has a minimum at $\boldsymbol{\lambda} = \boldsymbol{\lambda}^*$ if $\boldsymbol{\lambda}^* \in C$. The integration constant has be chosen such that $P_C(\boldsymbol{\lambda}^*) = 0$. Subsequently, the conjugate potential $P_C^*(\boldsymbol{s})$ is derived from the potential function $P_C(\boldsymbol{\lambda})$ (see Definition 2.13)

$$\begin{aligned} P_C^*(\boldsymbol{s}) &= \sup_{\boldsymbol{\lambda}}(\boldsymbol{\lambda}^{\mathrm{T}}\boldsymbol{s} - P_C(\boldsymbol{\lambda})) \\ &= \sup_{\boldsymbol{\lambda}}(\boldsymbol{\lambda}^{\mathrm{T}}\boldsymbol{s} - \Psi_C(\boldsymbol{\lambda}) - \frac{1}{2r}\|\boldsymbol{\lambda} - \boldsymbol{\lambda}^*\|^2). \end{aligned} \tag{5.46}$$

The argument of the supremum is maximal when

$$\boldsymbol{s} - \partial\Psi_C(\boldsymbol{\lambda}) - \frac{1}{r}(\boldsymbol{\lambda} - \boldsymbol{\lambda}^*) \ni \boldsymbol{0} \quad \Longleftrightarrow \quad \boldsymbol{\lambda} = \operatorname{prox}_C(\boldsymbol{\lambda}^* + r\boldsymbol{s}), \tag{5.47}$$

where the relation (5.43) has been used. The conjugate potential therefore becomes

$$\begin{aligned} P_C^*(\boldsymbol{s}) = \operatorname{prox}_C^{\mathrm{T}}(\boldsymbol{\lambda}^* + r\boldsymbol{s})\boldsymbol{s} - \frac{1}{2r}\|\operatorname{prox}_C(\boldsymbol{\lambda}^* + r\boldsymbol{s})\|^2 + \\ \frac{1}{r}\operatorname{prox}_C^{\mathrm{T}}(\boldsymbol{\lambda}^* + r\boldsymbol{s})\boldsymbol{\lambda}^* - \frac{1}{2r}\|\boldsymbol{\lambda}^*\|^2 \\ = \frac{1}{r}\operatorname{prox}_C^{\mathrm{T}}(\boldsymbol{\lambda}^* + r\boldsymbol{s})(\boldsymbol{\lambda}^* + r\boldsymbol{s}) - \frac{1}{2r}\|\operatorname{prox}_C(\boldsymbol{\lambda}^* + r\boldsymbol{s})\|^2 - \frac{1}{2r}\|\boldsymbol{\lambda}^*\|^2 \\ = -\frac{1}{2r}\|\boldsymbol{\lambda}^* + r\boldsymbol{s} - \operatorname{prox}_C(\boldsymbol{\lambda}^* + r\boldsymbol{s})\|^2 + \frac{1}{2r}\|\boldsymbol{\lambda}^* + r\boldsymbol{s}\|^2 - \frac{1}{2r}\|\boldsymbol{\lambda}^*\|^2 \\ = \boldsymbol{\lambda}^{*\mathrm{T}}\boldsymbol{s} + \frac{r}{2}\|\boldsymbol{s}\|^2 - \frac{1}{2r}\operatorname{dist}_C^2(\boldsymbol{\lambda}^* + r\boldsymbol{s}), \end{aligned} \tag{5.48}$$

where we have used $\Psi_C(\text{prox}_C(\boldsymbol{\lambda}^* + r\boldsymbol{s})) = 0$. The conjugate potential $P_C^*(\boldsymbol{s})$ could also have been found more directly by making use of (2.26).

At this point, the potential $P_C(\boldsymbol{\lambda})$ and conjugate potential $P_C^*(\boldsymbol{s})$ are derived and we have all the ingredients to transform an inclusion of the form (5.38) into an optimization problem. We use the principle of least constraints, which is basically equivalent to the principle of Gauß. The 'Zwang'[1] of the system is defined as

$$Z(\dot{\boldsymbol{u}}) = \frac{1}{2}\|\dot{\boldsymbol{u}}_R - \dot{\boldsymbol{u}}\|^2_{\boldsymbol{M}}, \qquad \text{with } \boldsymbol{M}\dot{\boldsymbol{u}}_R = \boldsymbol{h}. \tag{5.49}$$

We now define the *extended Zwang* as the sum of the Zwang and the support function $\Psi_C^*(\boldsymbol{s})$ (see [51])

$$Z_C(\dot{\boldsymbol{u}}) = Z(\dot{\boldsymbol{u}}) + \Psi_C^*(\boldsymbol{s}), \quad \boldsymbol{s} = \boldsymbol{s}(\dot{\boldsymbol{u}}), \tag{5.50}$$

where $\boldsymbol{s}$ is a function of the minimization variable $\dot{\boldsymbol{u}}$.

The strictly convex function $Z_C(\dot{\boldsymbol{u}})$ allows to express the equilibrium (in the sense of d'Alembert) of the multibody system with associated set-valued force laws by the non-smooth *unconstrained* optimization problem

$$\dot{\boldsymbol{u}} = \text{argmin}\{Z_C(\dot{\boldsymbol{u}}^*)\} \iff Z_C(\dot{\boldsymbol{u}}) \leq Z_C(\dot{\boldsymbol{u}}^*) \quad \forall \dot{\boldsymbol{u}}^*, \tag{5.51}$$

which is the extended principle of least constraints. A necessary and sufficient condition for the minimum is

$$\partial Z_C(\dot{\boldsymbol{u}}) \ni \boldsymbol{0} \Longrightarrow \boldsymbol{M}\dot{\boldsymbol{u}} - \boldsymbol{h} - \boldsymbol{W}\boldsymbol{\lambda} = \boldsymbol{0}, \tag{5.52}$$

with

$$-\boldsymbol{\lambda} \in \partial\Psi_C^*(\boldsymbol{s}), \tag{5.53}$$

and

$$\boldsymbol{W} = -\left(\frac{\partial \boldsymbol{s}(\dot{\boldsymbol{u}})}{\partial \dot{\boldsymbol{u}}}\right)^{\mathrm{T}}. \tag{5.54}$$

The minimization problem (5.51) is able to describe the dynamics of systems with associated set-valued force laws, e.g.

1. Associated Coulomb friction. By this we mean Coulomb's friction law with a *known normal force*. The friction law is therefore associated to a potential which serves as dissipation function. The contact can be spatial (having a 2-dimensional tangent plane) or be planar (with a 1-dimensional tangent plane).
2. Unilateral perfect constraints. Among these are unilateral normal contact laws for rigid bodies.
3. Bilateral perfect constraints. With $C = \mathbb{R}^m$ we obtain $\boldsymbol{s} = \boldsymbol{0}$ and $\boldsymbol{\lambda} \in \mathbb{R}^m$, where m is the number of constraints.

[1] The *Zwang* is the German word for 'constraint'. The English term is however too confusing.

4. No constraints or no set-valued forces. This is the trivial case which is obtained from $C = \{\mathbf{0}\}$ giving rise to $\boldsymbol{s} \in \mathbb{R}^m$ and $\boldsymbol{\lambda} = \mathbf{0}$.

All four force laws can be directly associated with a potential. For associated Coulomb friction, the convex set C is the set C_D, which leads to a standard minimization problem (5.51). For non-associated Coulomb friction, the set C_T depends on the normal contact forces $\boldsymbol{\lambda}_N$, which in turn depend on the solution $\dot{\boldsymbol{u}}$, i.e. $C_T(\boldsymbol{\lambda}_N(\dot{\boldsymbol{u}}))$. The dependency of C_T on $\boldsymbol{\lambda}_N(\dot{\boldsymbol{u}})$ causes a 'quasi'-potential $\Psi^*_{C_T(\boldsymbol{\lambda}_N(\dot{\boldsymbol{u}}))}(\boldsymbol{s})$ and a 'quasi'-optimization problem [2]

$$\dot{\boldsymbol{u}} = \operatorname{argmin}\{Z_{C_T(\boldsymbol{\lambda}_N(\dot{\boldsymbol{u}}^*))}(\dot{\boldsymbol{u}}^*)\}. \tag{5.55}$$

At this point it is essential to note that the extended Zwang $Z_C(\dot{\boldsymbol{u}})$ can be non-smooth and also not be locally bounded due to the fact that Ψ^*_C can contain indicator functions (usually of the form $\Psi_K(\boldsymbol{s})$ with $K = C^\perp$). This limits the practical use of the minimization problem (5.51) or quasi-minimization problem (5.55) to find $\dot{\boldsymbol{u}}$.

In the special case that the convex set C is a cone, we can express the support function by an indicator function $\Psi^*_C(\boldsymbol{s}) = \Psi_K(\boldsymbol{s})$, where K is the set of kinematically admissible values of $\boldsymbol{s}$. This is for instance the case for purely normal contact ($C_N = \mathbb{R}^{m+}$ is a cone, $K_N = \mathbb{R}^{m-}$). It can be proven that K is the polar cone to C, i.e. $K = C^\perp$. For a cone C we could also write the minimization problem as a constrained minimization problem without making use of the support function

$$\dot{\boldsymbol{u}} = \operatorname*{argmin}_{\boldsymbol{s}(\dot{\boldsymbol{u}}^*)\in K}\{Z(\dot{\boldsymbol{u}}^*)\}, \qquad K = C^\perp, \qquad C \text{ is a cone.} \tag{5.56}$$

The indicator function can therefore be envisaged as a tool to write an intrinsically constrained minimization problem (5.56) as an unconstrained minimization problem (5.51). Bilateral constraints can be regarded as a special case where the cone C is a linear space, i.e. $C = \mathbb{R}^m$ and $K = C^\perp = \{\mathbf{0}\}$. Similarly, if no constraints or set-valued forces are present, then the cone C is the origin and K is a linear space, i.e. $C = \{\mathbf{0}\}$ and $K = C^\perp = \mathbb{R}^m$.

In the special case that $Z_C(\dot{\boldsymbol{u}})$ does not contain indicator functions, it holds that $Z_C(\dot{\boldsymbol{u}})$ is a (non-smooth) continuous function in $\dot{\boldsymbol{u}}$. Associated Coulomb's law, for instance, yields a standard minimization problem (5.51) for which $Z_C(\dot{\boldsymbol{u}})$ remains locally bounded.

A usual way to handle bilateral constraints is to introduce the Lagrangian function

$$L(\dot{\boldsymbol{u}}, \boldsymbol{\lambda}) = Z(\dot{\boldsymbol{u}}) + \boldsymbol{\lambda}^{\mathrm{T}}\boldsymbol{s}(\dot{\boldsymbol{u}}), \tag{5.57}$$

with Lagrangian multiplier $\boldsymbol{\lambda}$ which we immediately associate with the contact forces. The Lagrangian problem allows to transform constrained minimization problems with *bilateral* constraints, such as (5.56) with $C = \mathbb{R}^m$, to an *unconstrained* min-max problem. For *unilateral* constraints we can pose the minimization problem (5.51) as a *constrained* min-max problem by making use of the Lagrangian $L(\dot{\boldsymbol{u}}, \boldsymbol{\lambda})$

$$\min_{\dot{\boldsymbol{u}}} \max_{\boldsymbol{\lambda} \in C} L(\dot{\boldsymbol{u}}, \boldsymbol{\lambda}). \tag{5.58}$$

Note that (5.58) is a *constrained* min-max problem, regardless of $\Psi_C^*(\boldsymbol{s})$ having indicator parts or not. The Lagrangian min-max problem (5.58) is therefore also constrained for associated and non-associated friction laws. The Lagrangian min-max problem (5.58) is only unconstrained if $C = \mathbb{R}^m$, which means a bilateral constraint.

The Lagrangian function $L(\dot{\boldsymbol{u}}, \boldsymbol{\lambda})$ is convex-concave and admits a unique saddle-point. The saddle-point does in general *not* agree with the solution of the constrained min-max problem. The saddle-point only agrees with the min-max solution if the saddle-point is located in the admissible region of $\boldsymbol{\lambda}$. The stationarity conditions for the saddle-point of $L(\dot{\boldsymbol{u}}, \boldsymbol{\lambda})$ are

$$\begin{aligned} L_{\dot{\boldsymbol{u}}} &= \boldsymbol{M}\dot{\boldsymbol{u}} - \boldsymbol{h} - \boldsymbol{W}\boldsymbol{\lambda} = \boldsymbol{0}, \\ L_{\boldsymbol{\lambda}} &= \boldsymbol{s} = \boldsymbol{0}, \end{aligned} \tag{5.59}$$

and correspond with a solution $(\dot{\boldsymbol{u}}, \boldsymbol{\lambda})$ which is constrained $\boldsymbol{s}(\dot{\boldsymbol{u}}) = 0$, but not always admissible, i.e. $\boldsymbol{\lambda} \notin C$.

Non-associated Coulomb friction yields a quasi min-max problem

$$\min_{\dot{\boldsymbol{u}}} \max_{\substack{\boldsymbol{\lambda}_N \in C_N \\ \boldsymbol{\lambda}_T \in C_T(\boldsymbol{\lambda}_N)}} L(\dot{\boldsymbol{u}}, \boldsymbol{\lambda}_N, \boldsymbol{\lambda}_T). \tag{5.60}$$

The solution of an inclusion describing non-associated frictional contact problems (or their related (N)LCP-form) is generally not unique, which is expressed by the fact that the quasi min-max solution of (5.60) is generally not unique. The non-uniqueness or even non-existence of solutions of non-associated frictional contact problems is due to the Painlevé paradox (see Section 5.6). However, the saddle-point of $L(\dot{\boldsymbol{u}}, \boldsymbol{\lambda})$ is unique, which agrees with the fact that a frictional contact problem has only one solution for which for all contacts are sticking ($\gamma_T = 0$).

The fact that the min-max problem of the Lagrangian is constrained makes it not directly suitable for numerical optimization techniques. The idea of the augmented Lagrangian method is to formulate an augmented Lagrangian function L_A such that the unconstrained min-max problem on L_A agrees with the constrained min-max problem on L. Instead of adding the support function $\Psi_C^*(\boldsymbol{s})$ to the Zwang, we add the conjugate potential $P_C^*(\boldsymbol{s})$ of the regularized force law

$$\begin{aligned} L_A(\dot{\boldsymbol{u}}, \boldsymbol{\lambda}^*) &= Z(\dot{\boldsymbol{u}}) + P_C^*(\boldsymbol{s}) \\ &= Z(\dot{\boldsymbol{u}}) + \boldsymbol{\lambda}^{*\mathrm{T}}\boldsymbol{s} + \frac{r}{2}\|\boldsymbol{s}\|^2 - \frac{1}{2r}\operatorname{dist}_C^2(\boldsymbol{\lambda}^* + r\boldsymbol{s}). \end{aligned} \tag{5.61}$$

The function $L_A(\dot{\boldsymbol{u}}, \boldsymbol{\lambda}^*)$ can be recognized to be the augmented Lagrangian for set-valued force laws (see [2]). The min-max problem of the augmented Lagrangian is *unconstrained*

$$\min_{\dot{\boldsymbol{u}}} \max_{\boldsymbol{\lambda}^*} L_A(\dot{\boldsymbol{u}}, \boldsymbol{\lambda}^*), \tag{5.62}$$

of which the solution is a saddle-point. The stationarity conditions of a saddle-point belonging to the augmented Lagrangian are

$$\begin{aligned} L_{A,\dot{\boldsymbol{u}}} &= \boldsymbol{M}\dot{\boldsymbol{u}} - \boldsymbol{h} - \boldsymbol{W}\,\mathrm{prox}_C(\boldsymbol{\lambda}^* + r\boldsymbol{s}) = \boldsymbol{0}, \\ L_{A,\boldsymbol{\lambda}^*} &= -\tfrac{1}{r}\big(\boldsymbol{\lambda}^* - \mathrm{prox}_C(\boldsymbol{\lambda}^* + r\boldsymbol{s})\big) = \boldsymbol{0}, \end{aligned} \tag{5.63}$$

which we recognize to be the set of algebraic equations of the exact regularization (5.42). The exact regularization method can therefore be interpreted as finding the saddle point of the augmented Lagrangian. The fact that the augmented Lagrangian is a smooth differentiable function, leading to an unconstrained min-max problem, explains why the exact regularization leads to a set of non-smooth continuous equations which can simply be solved with a root-finding algorithm. The augmented Lagrangian gives therefore the theoretical framework for the exact regularization method.

5.5.3 Example: Block with Dry Friction

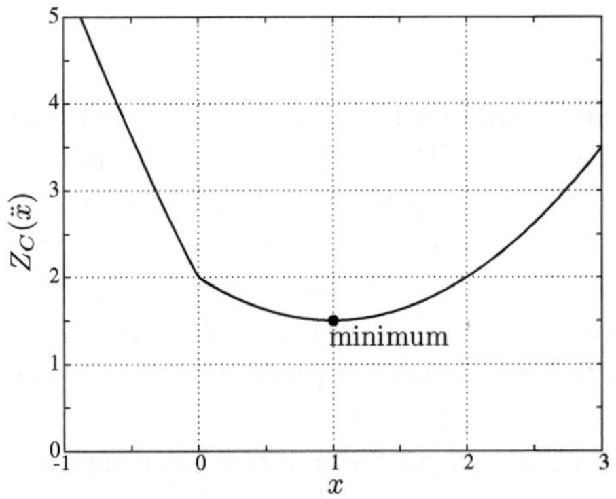

Fig. 5.7. The cost function $Z_C(\ddot{x})$ for the block-on-floor example.

We will treat an almost trivial example which demonstrates well the above theory. Consider a block on a floor with mass $m = 1$ kg which is pulled by a force $f = 2$ N

$$m\ddot{x} = f + \lambda_T, \tag{5.64}$$

where λ_T is the dry friction force between the floor and the mass

$$\lambda_T \in -\mu m g\,\mathrm{Sign}(\gamma_T), \quad \gamma_T = \dot{x}, \tag{5.65}$$

with friction coefficient $\mu = 0.1$ and $g = 10$ m/s^2. The friction law (5.65) is an associated Coulomb's law because the normal force is known in advance, i.e. $F_N = mg$. Assume that the block is momentarily sticking, i.e. $\gamma_T = \dot{x} = 0$ m/s. We want to know whether the block will start to slip. The force law on acceleration level of the sticking block is

$$\lambda_T \in -\mu mg\,\mathrm{Sign}(\dot{\gamma}_T), \quad \gamma_T = 0, \tag{5.66}$$

which is basically an index reduction of (5.65). The force law can be expressed more conveniently by the support function

$$\lambda_T \in \partial \Psi^*_{C_D}(s), \quad \Psi^*_{C_D}(s) = \mu mg|s|, \tag{5.67}$$

where $s = -\dot{\gamma}_T = -\ddot{x}$ and $C_D = \{\lambda_T \mid |\lambda_T| \leq \mu mg\}$ is the set of admissible friction forces. Adding the support function to the Zwang gives the convex function Z_C and the unconstrained minimization problem

$$\ddot{x} = \operatorname{argmin} Z_C(\ddot{x}^*), \qquad Z_C(\ddot{x}) = \frac{1}{2}m\ddot{x}^2 - f\ddot{x} + \frac{1}{2}mf^2 + \mu mg|\ddot{x}|, \tag{5.68}$$

with solution $\ddot{x} = 1$ m/s^2 (see Figure 5.7). The Lagrangian function, however, yields a constrained min-max problem

$$\min_{\ddot{x}} \max_{\lambda_T \in C_D} L(\ddot{x}, \lambda_T), \qquad L(\ddot{x}, \lambda_T) = \frac{1}{2}m\ddot{x}^2 - f\ddot{x} + \frac{1}{2}mf^2 - \lambda_T\ddot{x}. \tag{5.69}$$

with solution $\ddot{x} = 1$ m/s^2, $\lambda_T = -1$ N. The saddle-point of L is $\ddot{x} = 0$ m/s^2, $\lambda_T = -2$ N, which corresponds to a point in the stick phase, but with a non-admissible friction force. The contour lines of the level sets of the Lagrangian $L(\ddot{x}, \lambda_T)$ are depicted in Figure 5.8. The horizontal lines $\lambda_T = \pm 1$ correspond to the borders of C_D. The inclined line corresponds to the condition $L_{\ddot{x}} = m\ddot{x} - f - \lambda_T = 0$, which is just the equation of motion. The vertical line $L_{\lambda_T} = \ddot{x} = 0$ corresponds to a stick condition which addresses only the actively constrained case of the set-valued friction law. The saddle-point has to fulfil the stationarity conditions $(L_{\ddot{x}}, L_{\lambda_T}) = (0, 0)$ and is located at the crossing of $m\ddot{x} - f - \lambda_T = 0$ and $\ddot{x} = 0$. Note that the lines $L_{\ddot{x}} = 0$ and $L_{\lambda_T} = 0$ are the 'ridges' of maxima and minima on the saddle surface L. The saddle-point, corresponding with the constraint $\ddot{x} = 0$, lies outside the admissible region C_D and is therefore not the solution we seek.

The augmented Lagrangian function

$$L_A(\ddot{x}, \lambda_T) = \frac{1}{2}m\ddot{x}^2 - f\ddot{x} + \frac{1}{2}mf^2 - \lambda_T\ddot{x} + \frac{r}{2}\ddot{x}^2 - \frac{1}{2r}\operatorname{dist}^2_{C_D}(\lambda_T - r\ddot{x}), \tag{5.70}$$

gives an unconstrained min-max problem

$$\min_{\ddot{x}} \max_{\lambda_T} L_A(\ddot{x}, \lambda_T), \tag{5.71}$$

with an (in this case) unique saddle-point $\ddot{x} = 1$ m/s^2, $\lambda_T = -1$ N. The contour lines of the level sets of the augmented Lagrangian $L_A(\ddot{x}, \lambda_T)$ are depicted in Figure 5.9 for $r = 0.2$ and $r = 0.5$. The level sets of the augmented Lagrangian have a much more complicated structure than the Lagrangian. The stationarity conditions for the saddle-point of the augmented Lagrangian are depicted as two non-smooth lines $L_{A,\ddot{x}} = 0$ and $L_{A,\lambda_T} = 0$. We recognize the graph of the friction law in the non-smooth line $L_{A,\lambda_T} = 0$. The saddle-point of the augmented Lagrangian coincides with the solution of the contact problem.

The difference between the Lagrangian and the augmented Lagrangian is now apparent: the Lagrangian has a saddle-point which corresponds with the constrained case ($\boldsymbol{s} = \mathbf{0}$) of the set-valued force law whereas the augmented Lagrangian has a saddle-point which corresponds with the valid case that is either constrained ($\boldsymbol{s} = \mathbf{0}$) or unconstrained ($\boldsymbol{s} \neq \mathbf{0}$).

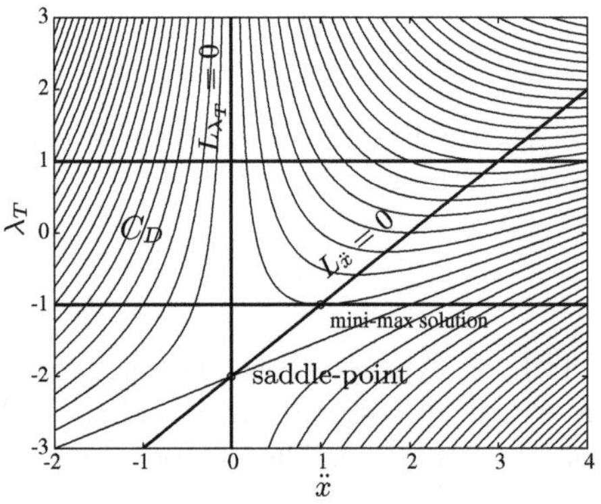

Fig. 5.8. Contour lines of the Lagrangian $L(\ddot{x}, \lambda_T)$.

5.5.4 Constrained Optimization Problems

The relation between the different constrained optimization problems is summarized in Table 5.1.

For bilateral constraints, one can pose a constrained minimization problem on the Zwang, or an unconstrained minimization problem on the extended Zwang. The extended Zwang $Z_C(\dot{\boldsymbol{u}}) = Z(\dot{\boldsymbol{u}}) + \Psi^*_{\mathbb{R}^m}(\boldsymbol{s}) = Z(\dot{\boldsymbol{u}}) + \Psi_{\{\mathbf{0}\}}(\boldsymbol{s})$ is however not locally bounded. More of use is to transform the problem to an *unconstrained* min-max problem of the Lagrangian or augmented Lagrangian (which are both locally bounded functions).

If the constraints are defined by cones, then the constrained minimization problem on the Zwang can be transformed to an unconstrained minimization problem on Z_C by making use of the indicator function on $C^\perp$, which is equal to the support function of C. Subsequently, we can transform the problem to a *constrained* min-max problem of the Lagrangian. A further transformation yields an *unconstrained* min-max problem on the augmented Lagrangian.

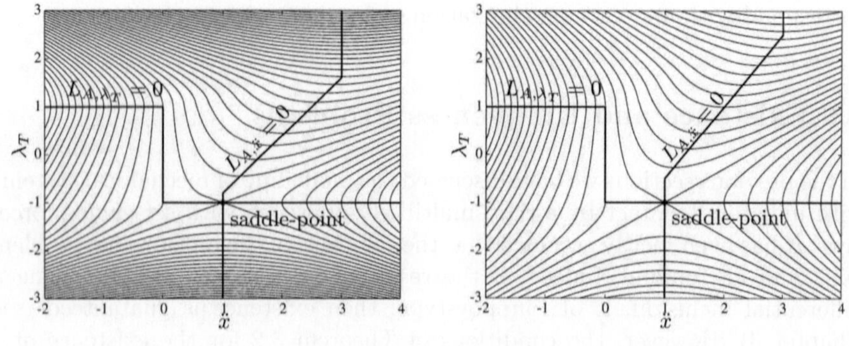

(a) $r = 0.2$ (b) $r = 0.5$

Fig. 5.9. Contour lines of the augmented Lagrangian $L_A(\ddot{x}, \lambda_T)$.

bilateral constraint $C = \mathbb{R}^m$	unilateral constraint $C =$ cone	general case $C =$ convex
$\min_{s=0} Z(\dot{u})$	$\min_{s \in C^{\perp}} Z(\dot{u})$	
↓	↓	
$\min Z_C(\dot{u})$ (not l.b.)	$\min Z_C(\dot{u})$ (not l.b.)	$\min Z_C(\dot{u})$ (not necess. l.b.)
↓	↓	↓
$\min_{\dot{u}} \max_{\lambda} L(\dot{u}, \lambda)$	$\min_{\dot{u}} \max_{\lambda \in C} L(\dot{u}, \lambda)$	$\min_{\dot{u}} \max_{\lambda \in C} L(\dot{u}, \lambda)$
↓	↓	↓
$\min_{\dot{u}} \max_{\lambda} L_A(\dot{u}, \lambda)$	$\min_{\dot{u}} \max_{\lambda} L_A(\dot{u}, \lambda)$	$\min_{\dot{u}} \max_{\lambda} L_A(\dot{u}, \lambda)$

Table 5.1. Relation between the different constrained optimization problems (l.b. = locally bounded). The functions are locally bounded unless stated otherwise.

In the more general case that the set-valued force laws are defined by convex sets, we can not pose a minimization problem on the Zwang. Instead, we have to add the support function of the force law to the Zwang, which gives an unconstrained minimization problem on Z_C (being not necessarily locally bounded). Subsequently, we can transform the problem to a *constrained* min-max problem of the Lagrangian or *unconstrained* min-max problem on the augmented Lagrangian.

The merit of the augmented Lagrangian is, that it always leads to an unconstrained min-max problem of a function L_A which is locally bounded. A (not necessarily unique) min-max solution of the augmented Lagrangian therefore corresponds with a saddle-point. The stationarity conditions for a saddle-point of the augmented Lagrangian form a set of non-smooth continuous algebraic equations, which can be solved with a root-finding algorithm.

The stationarity conditions for a saddle point of the augmented Lagrangian correspond with an exact regularization.

5.6 Existence and Uniqueness Problems

In the previous sections we have discussed the modelling of mechanical systems with unilateral contact by a rigid multibody approach with set-valued force-laws. It has been tacitly assumed that the solutions of the initial value problem exist and are unique. Indeed, if the resulting non-smooth system, being a differential inclusion, is of Filippov-type, then existence is guaranteed (see Chapter 3). However, the conditions of Theorem 3.2 for the existence of a solution of a Filippov system are not always met by mechanical systems with unilateral frictional contact. The loss of existence of solutions is sometimes named 'inconsistency' [12].

It has been known since the end of the 19th century that the combination of Amontons-Coulomb friction (non-associated Coulomb friction) together with the assumption of rigidity can cause inconsistency for high values of the friction coefficient [68, 109–111]. Painlevé [109–111] has considered the sliding motion of a rigid object in contact with the ground, which exhibits inconsistency for certain parameter values and initial conditions. The problem of a sliding rod studied by Painlevé has become the classical example of a non-smooth system with inconsistency. Lecornu [77] has proposed velocity jumps to escape from inconsistent states (i.e. states for which a solution in forward time does not exist within the used solution concept), often now addressed as impacts without collisions. A number of studies show that inconsistency occurs in the classical Painlevé example when the friction coefficient μ is large [12, 46, 89, 91]. The Painlevé problem has recently been studied in more detail in [41, 79].

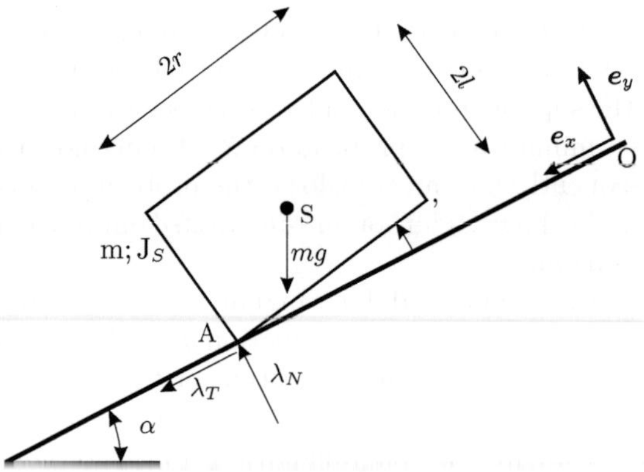

Fig. 5.10. The original Painlevé problem.

The original example of Painlevé is different from the classical example of Painlevé and is unfortunately almost forgotten. The original example of Painlevé studied in [110] (page 114-115) is that of a planar box sliding over an inclined plane (see Figure 5.10). In fact, Painlevé speaks of a three dimensional cylinder with its base in contact with the inclined plane, but this does not make any difference for the planar case. The box has the three degrees of freedom $\boldsymbol{q} = \begin{bmatrix} x \; y \; \varphi \end{bmatrix}^{\mathrm{T}}$, where x and y are the tangential and normal coordinates of the centre of mass S with respect to the frame $(\boldsymbol{e}_x, \boldsymbol{e}_y)$, which is fixed to the plane, and $\varphi = 0$ is the angle of rotation with respect to the plane. The inclined plane has a slope α. The mass of the box is m and its inertia around the centre of mass S is $J_S = k^2 m$, where k is the radius of gyration. The system matrices of the Painlevé problem are

$$\boldsymbol{M} = \begin{bmatrix} m & 0 & 0 \\ 0 & m & 0 \\ 0 & 0 & J_S \end{bmatrix}, \quad \boldsymbol{h} = \begin{bmatrix} mg\sin\alpha \\ -mg\cos\alpha \\ 0 \end{bmatrix}, \tag{5.72}$$

$$\boldsymbol{W}_N = \begin{bmatrix} 0 \\ 1 \\ -r\cos\varphi - l\sin\varphi \end{bmatrix}, \quad \boldsymbol{W}_T = \begin{bmatrix} 1 \\ 0 \\ -r\sin\varphi - l\cos\varphi \end{bmatrix}, \tag{5.73}$$

$$\tilde{\boldsymbol{w}}_N = \tilde{\boldsymbol{w}}_T = 0, \quad \bar{\boldsymbol{w}}_N = (r\sin\varphi + l\cos\varphi)\dot{\varphi}^2, \quad \bar{\boldsymbol{w}}_T = (-r\cos\varphi + l\sin\varphi)\dot{\varphi}^2, \tag{5.74}$$

with the normal contact distance and tangential contact velocity of contact point A

$$g_N = y - l\cos\varphi - r\sin\varphi, \quad \gamma_T = \dot{x} - (r\sin\varphi + l\cos\varphi)\dot{\varphi}. \tag{5.75}$$

Assume that at $t = 0$ the box is sliding downward, $\gamma_T > 0$, from which follows that $\lambda_T = -\mu\lambda_N$. Sliding of the box over the plane without rotation ($\varphi = \dot{\varphi} = 0$) is possible when $\mu < \frac{r}{l}$, which can be derived from a simple momentum analysis. If $\frac{r}{l} < \mu < \frac{r}{l} + \frac{k^2}{rl}$, then the box will rotate around the contact point A ($\ddot{\varphi} > 0$). If $\mu > \frac{r}{l} + \frac{k^2}{rl}$, then an inconsistency occurs which is now known as the Painlevé paradox. Assume that the box is slipping in forward direction, $\gamma_T > 0$, and the box begins to rotate, $\varphi = 0$. The equations of motion are in this mode ($g_N = 0$, $\gamma_T > 0$)

$$\boldsymbol{M}\dot{\boldsymbol{u}} = \boldsymbol{h} + \boldsymbol{W}_N\lambda_N + \boldsymbol{W}_T\lambda_T = \boldsymbol{h} + (\boldsymbol{W}_N - \mu\boldsymbol{W}_T)\lambda_N, \tag{5.76}$$

or

$$\begin{aligned} m\ddot{x} &= mg\sin\alpha + \lambda_T = mg\sin\alpha - \mu\lambda_N, \\ m\ddot{y} &= -mg\cos\alpha + \lambda_N, \\ J_S\ddot{\varphi} &= -r\lambda_N - l\lambda_T = (-r + \mu l)\lambda_N. \end{aligned} \tag{5.77}$$

The contact acceleration $\dot{\gamma}_N = \ddot{g}_N$ during forward slipping with $\varphi = 0$ is

$$\dot{\gamma}_N = \boldsymbol{W}_N^{\mathrm{T}}\dot{\boldsymbol{u}} + \bar{\boldsymbol{w}}_N = \ddot{y} - r\ddot{\varphi} + l\dot{\varphi}^2. \tag{5.78}$$

Substitution of the equations of motion (5.77) in the contact acceleration (5.78) gives a relation between the normal contact acceleration and the normal contact force

$$\begin{aligned} \dot{\gamma}_N &= \left(\frac{1}{m} - \frac{r}{k^2 m}(-r + \mu l)\right)\lambda_N - g\cos\alpha + l\dot{\varphi}^2 \\ &= A\lambda_N + b, \end{aligned} \tag{5.79}$$

with

$$\begin{aligned} A &= \boldsymbol{W}_N^{\mathrm{T}} \boldsymbol{M}^{-1}(\boldsymbol{W}_N - \mu \boldsymbol{W}_T)\big|_{\varphi=0} = \tfrac{1}{m} - \tfrac{r}{k^2 m}(-r + \mu l), \\ b &= \boldsymbol{W}_N^{\mathrm{T}} \boldsymbol{M}^{-1} \boldsymbol{h} + \bar{\boldsymbol{w}}_N\big|_{\varphi=0} = -g\cos\alpha + l\dot{\varphi}^2. \end{aligned} \tag{5.80}$$

Equation (5.79), which is linear in $\dot{\gamma}_N$ and λ_N, forms together with the complementarity condition $0 \leq \dot{\gamma}_N \perp \lambda_N \geq 0$ a scalar linear complementarity problem, describing the normal contact problem on acceleration level in the forward sliding mode. The scalar LCP with $\dot{\gamma}_N = A\lambda_N + b$, $0 \leq \dot{\gamma}_N \perp \lambda_N \geq 0$ has a unique solution for $A > 0$, two solutions for $A < 0 \wedge b > 0$, no solution for $A < 0 \wedge b < 0$ or infinitely many solutions for $A = b = 0$ (hyperstatic case). The solutions of the LCP are indicated in Table 2.1. The occurrence of non-existence (the LCP has no solution) and non-uniqueness (the LCP has more than one solution) in a mechanical system with friction is called the Painlevé paradox [41].

The Painlevé paradox will occur when A becomes negative, which happens for large enough values of μ. The critical value of μ for which $A = 0$ occurs at $\mu = \mu_c$ with

$$\mu_c = \frac{r}{l} + \frac{k^2}{rl}. \tag{5.81}$$

If $\mu > \mu_c$ and $b > 0$, then we do not know whether the block remains slipping or whether it detaches from the slope. The real physical system will in this case be extremely sensitive to initial conditions. If $\mu > \mu_c$ and $b < 0$, then we can not determine a solution at all. The non-existence of solution indicates that our mathematical solution concept can not describe the physical behaviour of the real physical system, being a frictional catastrophe. In reality, the relative tangential velocity of the block γ_T will drop extremely fast to zero and the block will stick or detach after the frictional catastrophe. The solution concept of Filippov does not allow velocity jumps. Solution concepts for systems with impact only allow velocity jumps at collisions. The existence problem of the Painlevé paradox requires an impact without collision, which is still a not completely solved problem in mechanics.

The critical value μ_c is dependent on the geometry and on the mass distribution of the block. If we assume the box to be homogeneous then we obtain $J_S = k^2 m = \frac{1}{3}(r^2 + l^2)m$. The critical friction coefficient for a homogeneous block becomes

$$\mu_c = \frac{r}{l} + \frac{1}{3}\left(\frac{r}{l} + \frac{l}{r}\right), \tag{5.82}$$

which is minimal for $\frac{r}{l} = \frac{1}{2}$. The minimal friction coefficient for a homogeneous block is therefore

$$\mu_{c\mathrm{min}} = \frac{4}{3}. \tag{5.83}$$

However, if all the mass is concentrated at S, giving a radius of gyration $k = 0$, then $\mu_c = \frac{r}{l}$ which can take any positive value.

Consequently, models of mechanical systems with set-valued force-laws do not always exhibit existence and uniqueness of the initial value problem. A frictional catastrophe can occur if Signorini's law (impenetrability) and a non-associated Coulomb's law are assumed and if the system has a coupling between the normal and tangential degree of freedom with respect to the contact surface. Apparently, the Painlevé paradox does not fulfil the requirements of the existence theorem of Filippov (Theorem 3.2). Filippov's theorem states that existence of solutions of a differential inclusion $\dot{\boldsymbol{x}} \in \boldsymbol{F}(\boldsymbol{x})$ is guaranteed if $\boldsymbol{F}(\boldsymbol{x})$ is upper semi-continuous, non-empty, closed, convex and bounded. Upper semi-continuity, non-emptyness, closedness and convexity are guaranteed by the nature of force-laws. Boundedness, however, is not always fulfilled. The set-valued function $\boldsymbol{F}(\boldsymbol{x})$ is in our example dependent on the normal contact force, which is during forward sliding given by

$$\lambda_N = -\frac{b}{A} \geq 0, \quad A \neq 0. \tag{5.84}$$

The normal contact force, and therefore the differential inclusion, becomes unbounded for $\mu \to \mu_c$ for which $A \to 0$. The example of Painlevé has therefore a kind of positive feedback of friction forces which ultimately leads to a 'frictional catastrophe' for which the normal contact force becomes unbounded.

5.7 Summary

The mathematical modelling of mechanical multibody systems with set-valued force-laws has been discussed in this chapter. Signorini's law expresses the impenetrability of the contact in normal direction and Coulomb's law gives a set-valued description of the friction forces. The force laws have been expressed on different kinematic levels in Section 5.3. The non-smooth potential theory (Section 5.4) provides the framework to define (augmented) Lagrangian functions for mechanical systems with frictional contact. The augmented Lagrangian method (Section 5.5) transforms algebraic inclusions to a set of non-smooth continuous algebraic equations, which can be solved with a root-finding algorithm. The theory and methods presented in this chapter form the basis for the numerical integration methods presented in Chapter 6.

6

Numerical Integration Methods

As explained in Chapter 3, solutions to differential inclusions can contain sliding modes. The occurrence of a sliding mode poses a difficulty for numerical integration. In this chapter we present numerical techniques for the integration of differential inclusions with sliding modes. First, the smoothing method will be briefly discussed in Section 6.1. The disadvantages of the smoothing method motivates to use the Switch Model (Section 6.2). Finally, more sophisticated integration methods such as the event-driven integration method (Section 6.3) and time-stepping method (Section 6.4) for the simulation of mechanical systems with set-valued force laws are presented.

6.1 Smoothing Method

Consider the differential inclusion (3.17)

$$\dot{\boldsymbol{x}}(t) \in \boldsymbol{F}(t, \boldsymbol{x}(t)) = \begin{cases} \boldsymbol{f}_-(t, \boldsymbol{x}(t)), & \boldsymbol{x} \in \mathcal{V}_-, \\ \overline{\text{co}}\{\boldsymbol{f}_-(t, \boldsymbol{x}(t)), \boldsymbol{f}_+(t, \boldsymbol{x}(t))\}, & \boldsymbol{x} \in \Sigma, \\ \boldsymbol{f}_+(t, \boldsymbol{x}(t)), & \boldsymbol{x} \in \mathcal{V}_+, \end{cases} \tag{6.1}$$

with

$$\begin{aligned} \mathcal{V}_- &= \{\boldsymbol{x} \in \mathbb{R}^n \mid h(\boldsymbol{x}) < 0\}, \\ \Sigma &= \{\boldsymbol{x} \in \mathbb{R}^n \mid h(\boldsymbol{x}) = 0\}, \\ \mathcal{V}_+ &= \{\boldsymbol{x} \in \mathbb{R}^n \mid h(\boldsymbol{x}) > 0\}. \end{aligned} \tag{6.2}$$

If we try to numerically integrate a discontinuous system of Filippov-type (or the associated differential inclusion) with an integration algorithm for ordinary differential equations, then we are faced with a difficulty. An ODE integrator will chatter around the attractive sliding mode, computing integration points alternating between $\mathcal{V}_-$ and $\mathcal{V}_+$. If the integration algorithm is equipped with a variable step size, then it will find a reasonable approximation to the exact solution but it will reduce the step size, taking a huge amount of integration points, within the attractive sliding mode. The discontinuous system can therefore not efficiently be integrated with integration methods for ordinary differential equations due to the presence of sliding modes. The discontinuous vector field is often approximated by a smoothened vector field (see for instance [58, 145]). A discontinuous right-hand side $\boldsymbol{f}(t, \boldsymbol{x})$ typically contains sign-functions. A common smoothening approximation of the sign-function is the arctangent approximation

$$\operatorname{sign}(x) \approx \frac{2}{\pi} \arctan(\varepsilon x), \qquad \varepsilon \gg 1. \tag{6.3}$$

With this arctangent approximation, the smoothened system becomes:

$$\dot{\boldsymbol{x}} = \frac{1}{2}\left(\boldsymbol{f}_+ + \boldsymbol{f}_- + \frac{2}{\pi}\arctan(\varepsilon h)(\boldsymbol{f}_+ - \boldsymbol{f}_-)\right), \tag{6.4}$$

with $\boldsymbol{f}_\pm := \boldsymbol{f}_\pm(t, \boldsymbol{x}(t))$ and $h := h(\boldsymbol{x})$. The smooth approximation normally yields a good approximation for large values of ε although difficulties can be expected at repulsive sliding modes. The smoothened system does not necessarily have the same equilibria as the convexified discontinuous system as follows from the following example.

Example 6.1
Consider the convexified discontinuous system

$$\dot{\boldsymbol{x}} \in \boldsymbol{F}(\boldsymbol{x}) = \begin{bmatrix} x_2 \\ a + b \operatorname{Sign} x_2 \end{bmatrix}, \quad \boldsymbol{x} = \begin{bmatrix} x_1 \\ x_2 \end{bmatrix}. \tag{6.5}$$

The differential inclusion has for $a \leq b$ a set of equilibria ($x_1 \in \mathbb{R}, x_2 = 0$). The smoothened system, with (6.3) as smoothing function, i.e.

$$\dot{\boldsymbol{x}} = \begin{bmatrix} x_2 \\ a + b\frac{2}{\pi}\arctan(\varepsilon x_2) \end{bmatrix}, \tag{6.6}$$

however, does not have an equilibrium for $a \neq 0$.

The advantage of the smoothing method is the easiness of use as standard integration techniques can be directly applied and no additional programming work has to be done. It should be noted that the smooth approximation always has existence and uniqueness of solutions whereas this is not the case for the discontinuous system. However, the main disadvantage of the smoothing method is the fact that it yields stiff differential equations which are numerically expensive to solve. The smoothened differential equation is especially stiff within a sliding mode.

6.2 Switch Model

In this section we will present the *Switch Model* [86], which is a numerical technique for integrating differential inclusions with sliding modes without suffering from stiff differential equations.

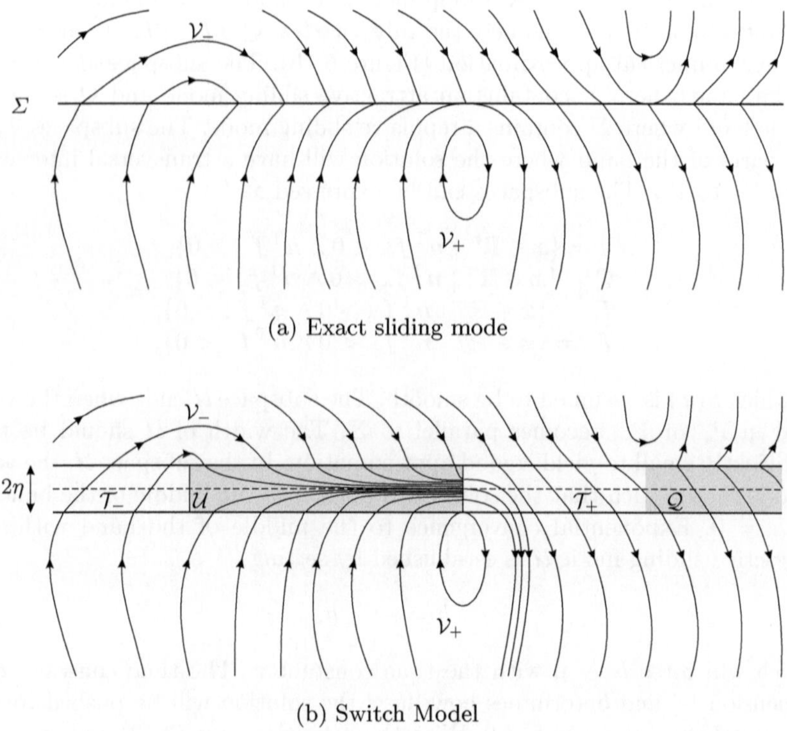

Fig. 6.1. Numerical approximation of a sliding mode with the Switch Model.

A remedy for the disadvantages of the smoothing method is found in the model presented by Karnopp [70]. The model of Karnopp introduces an interval of low relative velocity, i.e. $|v_{\mathrm{rel}}| \leq \eta$, also called 'stick band' which approximates the stick mode. For velocities within this interval, the relative acceleration is put to zero, $\dot{v}_{\mathrm{rel}} = 0$. The model of Karnopp leads therefore to a discontinuous system of Filippov-type, which is non-stiff within the stick mode. However, the model of Karnopp allows trajectories of the system within the stick band to remain close to the boundary of the stick band, which can cause numerical instabilities of the ODE integrator.

The Switch Model, presented in this section, is an improved (and generalized) version of the Karnopp model. The Switch Model introduces a vector field in the stick band, such, that the state of the system is pushed to the middle of the stick band, thereby avoiding numerical instabilities.

Consider the vector field of Figure 6.1a, which contains a switching boundary. The switching boundary contains (from left to right) a part with a transversal intersection from $\mathcal{V}_+$ to $\mathcal{V}_-$, an attractive sliding mode, a transversal intersection from $\mathcal{V}_-$ to $\mathcal{V}_+$ and a repulsive sliding mode. The Switch Model constructs, similar to the Karnopp model, a 'band' or 'boundary layer' with thickness 2η around Σ, namely the subspace $\mathcal{U} \cup \mathcal{Q} \cup \mathcal{T}_- \cup \mathcal{T}_+$, to allow for an efficient numerical approximation (Figure 6.1b). The subspace $\mathcal{U}$ is the part of the band where Σ contains an attractive sliding mode and $\mathcal{Q}$ is the part of the band where Σ contains a repulsive sliding mode. The subspaces $\mathcal{T}_\pm$ are the parts of the band where the solution will have a transversal intersection from $\mathcal{V}_\mp$ to $\mathcal{V}_\pm$. The subspaces can be expressed as

$$\begin{aligned} \mathcal{U} &= \{\boldsymbol{x} \in \mathbb{R}^n \mid \boldsymbol{n}^{\mathrm{T}}\boldsymbol{f}_+ < 0 \wedge \boldsymbol{n}^{\mathrm{T}}\boldsymbol{f}_- > 0\}, \\ \mathcal{Q} &= \{\boldsymbol{x} \in \mathbb{R}^n \mid \boldsymbol{n}^{\mathrm{T}}\boldsymbol{f}_+ > 0 \wedge \boldsymbol{n}^{\mathrm{T}}\boldsymbol{f}_- < 0\}, \\ \mathcal{T}_+ &= \{\boldsymbol{x} \in \mathbb{R}^n \mid \boldsymbol{n}^{\mathrm{T}}\boldsymbol{f}_+ > 0 \wedge \boldsymbol{n}^{\mathrm{T}}\boldsymbol{f}_- > 0\}, \\ \mathcal{T}_- &= \{\boldsymbol{x} \in \mathbb{R}^n \mid \boldsymbol{n}^{\mathrm{T}}\boldsymbol{f}_+ < 0 \wedge \boldsymbol{n}^{\mathrm{T}}\boldsymbol{f}_- < 0\}, \end{aligned} \tag{6.7}$$

in which $h(\boldsymbol{x})$ is assumed to be smooth. The subspace $\mathcal{U}$ ends when the vector field in $\mathcal{V}_+$ or $\mathcal{V}_-$ becomes parallel to Σ. The width of $\mathcal{U}$ should be taken sufficiently small to yield a good approximation. In the subspace $\mathcal{U}$, the vector field is chosen such that the solution is pushed to the middle of the band, i.e. to $h = 0$. Exponential convergence to the middle of the band within the attractive sliding mode $\mathcal{U}$ is established by setting

$$\dot{h} = -\tau^{-1}h, \tag{6.8}$$

which will force $h \to 0$ with the time constant τ. The time constant τ has dimension $[s]$ and determines how 'fast' the solution will be pushed towards the switching boundary $h = 0$. With the definition of $\boldsymbol{n}$ (3.12) and the vector field in the sliding mode $\dot{\boldsymbol{x}}(t) = \alpha\boldsymbol{f}_+ + (1-\alpha)\boldsymbol{f}_-$ (3.23) we obtain the relation

$$\begin{aligned} \dot{h}(\boldsymbol{x}) &= \frac{\partial h}{\partial \boldsymbol{x}}\frac{d\boldsymbol{x}}{dt} \\ &= \boldsymbol{n}^{\mathrm{T}}(\alpha\boldsymbol{f}_+ + (1-\alpha)\boldsymbol{f}_-). \end{aligned} \tag{6.9}$$

Substitution of (6.9) in (6.8) gives the value of α for $\boldsymbol{x} \in \mathcal{U}$

$$\alpha = \frac{\boldsymbol{n}^{\mathrm{T}}\boldsymbol{f}_- + \tau^{-1}h}{\boldsymbol{n}^{\mathrm{T}}(\boldsymbol{f}_- - \boldsymbol{f}_+)}. \tag{6.10}$$

The vector field in the transition subspaces $\mathcal{T}_\pm$ is $\boldsymbol{f}_\pm$ which implies that the solution drifts away from Σ. The solution on an attractive sliding mode and on transversal intersections is unique, and so is the solution in $\mathcal{U} \cup \mathcal{T}_- \cup \mathcal{T}_+$. The solution in the repulsive sliding mode is non-unique, which implies that the choice of the vector field in $\mathcal{Q}$ is arbitrary. We simply set the vector field in $\mathcal{Q}$ equal to $\boldsymbol{f}_+$, neglecting non-uniqueness of the differential inclusion. The pseudo-code for the function evaluation of the Switch Model is given in Table 6.1. The function can be passed to a standard ODE-integrator.

The thickness parameter η should be chosen *sufficiently small.* With *sufficiently small* it is meant that η is small enough to have no qualitative influence on the solution. The Switch Model maintains the continuity of the state vector

function $\dot{\boldsymbol{x}} = \mathit{odefunction}(t, \boldsymbol{x})$
calculate $h(\boldsymbol{x})$, $\boldsymbol{f}_{+}(t, \boldsymbol{x})$, $\boldsymbol{f}_{-}(t, \boldsymbol{x})$
if $|h| > \eta$ **then**
  if $h > \eta$ **then** — smooth motion
    $\dot{\boldsymbol{x}}(t) = \boldsymbol{f}_{+}, \quad \boldsymbol{x}(t) \in \mathcal{V}_{+}$
  else
    $\dot{\boldsymbol{x}}(t) = \boldsymbol{f}_{-}, \quad \boldsymbol{x}(t) \in \mathcal{V}_{-}$
  end
else
  if $\boldsymbol{n}^{\mathrm{T}}\boldsymbol{f}_{+} > 0 \wedge \boldsymbol{n}^{\mathrm{T}}\boldsymbol{f}_{-} > 0$ **then** — transition
    $\dot{\boldsymbol{x}}(t) = \boldsymbol{f}_{+}, \quad \boldsymbol{x}(t) \in \mathcal{T}_{+}$
  end
  if $\boldsymbol{n}^{\mathrm{T}}\boldsymbol{f}_{+} < 0 \wedge \boldsymbol{n}^{\mathrm{T}}\boldsymbol{f}_{-} < 0$ **then** — transition
    $\dot{\boldsymbol{x}}(t) = \boldsymbol{f}_{-}, \quad \boldsymbol{x}(t) \in \mathcal{T}_{-}$
  end
  if $\boldsymbol{n}^{\mathrm{T}}\boldsymbol{f}_{+} < 0 \wedge \boldsymbol{n}^{\mathrm{T}}\boldsymbol{f}_{-} > 0$ **then** — attractive sliding mode
    $\alpha = \dfrac{\boldsymbol{n}^{\mathrm{T}}\boldsymbol{f}_{-} + \tau^{-1}h}{\boldsymbol{n}^{\mathrm{T}}(\boldsymbol{f}_{-} - \boldsymbol{f}_{+})}$
    $\dot{\boldsymbol{x}}(t) = \alpha\boldsymbol{f}_{+} + (1-\alpha)\boldsymbol{f}_{-}, \quad \boldsymbol{x}(t) \in \mathcal{U}$
  end
  if $\boldsymbol{n}^{\mathrm{T}}\boldsymbol{f}_{+} > 0 \wedge \boldsymbol{n}^{\mathrm{T}}\boldsymbol{f}_{-} < 0$ **then** — repulsive sliding mode
    $\dot{\boldsymbol{x}}(t) = \boldsymbol{f}_{+}, \quad \boldsymbol{x}(t) \in \mathcal{Q}$
  end
end

Table 6.1. ODE-function for the Switch Model.

and yields a set of non-stiff ordinary differential equations. A disadvantage of the Switch Model is the rapidly increasing complexity of the logical structure with increasing number of switching boundaries.

Consider for instance the block-on-floor system of Figure 4.3. The block can slip in two directions or can stick to the floor. A Switch Model of this system will yield a model with the subspaces $\mathcal{V}_{\pm}$ for the two slip directions and the subspace $\mathcal{U}$ for the stick phase together, which gives 3 subspaces in total (+ the transition subspaces $\mathcal{T}_{\pm}$). A system of k blocks next to each-other will give $k \times 3$ subspaces (+ transition subspaces). However, if we stack the k blocks on top of each-other, we will obtain 3^k subspaces (+ transition subspaces).

The Switch Model is therefore not suitable for mechanical systems with many frictional contacts because the combinatorial contact problem becomes too large. A rigourous way to deal with mechanical systems with many unilateral contacts are integration techniques which solve the contact problem by transforming it to an (N)LCP or to a (quasi) optimization problem (e.g. the augmented Lagrangian method). An (N)LCP-based integration technique formulates the contact problem as an (N)LCP, which basically replaces the large combinatorial problem. Solving the (N)LCP then gives the next mode of the trajectory of the system. The augmented Lagrangian method uses an exact regularization of the set-valued contact laws, as explained in Section 5.5. The main (N)LCP or augmented Lagrangian-based integration techniques are the event-driven integration method and the time-stepping method [5, 12, 27, 46, 49, 56, 97, 118, 137, 139].

6.3 Event-driven Integration Method

The event-driven integration method uses a standard ODE-integrator (or DAE-integrator) in smooth phases of the system and an LCP, NCP or augmented Lagrangian method to determine the next mode at the switching boundaries. In this section we will briefly review an event-driven integration method for the dynamics of rigid bodies with Coulomb friction using an LCP approach. This method was developed in [46, 49, 118] for the dynamics of planar as well as spatial rigid bodies with impact and friction. Impact, however, leads to discontinuities in the states of the system and can therefore not be described by a differential inclusion of Filippov-type. We will confine the presentation of the method to planar rigid-body systems with stick-slip transitions and detachment. Coulomb's friction model for planar rigid bodies is unidirectional, which simplifies the analysis because the direction of the friction force is known in advance.

6.3.1 Index Sets

We introduce five index sets, which describe the kinematic state of the contact points:

$$\begin{aligned}
I_C &= \{1, 2, \ldots, n_C\}, \\
I_S &= \{i \in I_C \mid g_{Ni} = 0\} \text{ with } n_S \text{ elements}, \\
I_N &= \{i \in I_S \mid \gamma_{Ni} = 0\} \text{ with } n_N \text{ elements}, \\
I_H &= \{i \in I_N \mid \gamma_{Ti} = 0\} \text{ with } n_H \text{ elements}, \\
I_G &= I_N \backslash I_H \qquad\qquad\; \text{ with } n_G = n_N - n_H \text{ elements},
\end{aligned} \tag{6.11}$$

where g_{Ni}, γ_{Ni} and γ_{Ti} denote the normal contact distance, velocity and tangential relative velocity of contact point i. The set I_C consists of n_C indices of all (open and closed) contact points. The set I_S contains the n_S indices of the contact points with vanishing normal distance but arbitrary relative velocity. In the set I_N are the n_N indices of the potentially active normal constraints, which fulfil the necessary conditions for continuous contact (vanishing normal distance and no relative velocity in normal direction). The set I_N contains therefore all indices of slipping or sticking contact points. The n_H elements of the set I_H correspond to the potentially active constraints in tangential direction (sticking contact points). The remaining $n_G = n_N - n_H$ elements of the set I_G corresponds to the slipping contact points. The sets I_S, I_N, I_H and I_G are not constant, because the contact configuration of the dynamical system changes with time due to stick-slip transitions, impact and contact loss.

The dynamics of a multibody system can be expressed by the equation of motion (5.3). In the sequel we will make use of an abbreviated notation for (5.3)

$$\boldsymbol{M}\dot{\boldsymbol{u}} = \boldsymbol{h} + \boldsymbol{W}_N\boldsymbol{\lambda}_N + \boldsymbol{W}_T\boldsymbol{\lambda}_T. \tag{6.12}$$

The contact problem for stick-slip transitions and detachment will be formulated in Subsection 6.3.2 as an LCP on acceleration level. Signorini's law (5.28) and Coulomb's friction (5.30) law are taken as contact laws in normal and tangential direction.

6.3.2 Stick-slip Transitions and Detachment

For the description of stick-slip transitions and detachment we formulate the equation of motion (6.12) and the constraints on acceleration level as:

$$\boldsymbol{M}\dot{\boldsymbol{u}} - \boldsymbol{h} - \left[(\boldsymbol{W}_N + \boldsymbol{W}_G\bar{\bar{\boldsymbol{\mu}}}_G)\ \boldsymbol{W}_H\right]\begin{bmatrix}\boldsymbol{\lambda}_N\\ \boldsymbol{\lambda}_H\end{bmatrix} = \boldsymbol{0}, \tag{6.13}$$

$$\begin{bmatrix}\dot{\gamma}_N\\ \dot{\gamma}_H\end{bmatrix} = \begin{bmatrix}\boldsymbol{W}_N^{\mathrm{T}}\\ \boldsymbol{W}_H^{\mathrm{T}}\end{bmatrix}\dot{\boldsymbol{u}} + \begin{bmatrix}\bar{\boldsymbol{w}}_N\\ \bar{\boldsymbol{w}}_H\end{bmatrix} \qquad \in \mathbb{R}^{n_N+n_H}. \tag{6.14}$$

The vectors $\boldsymbol{\lambda}_N \in \mathbb{R}^{n_N}$ and $\dot{\gamma}_N \in \mathbb{R}^{n_N}$ contain the normal contact forces and normal contact accelerations of the closed contacts in I_N. Similarly, the vectors $\boldsymbol{\lambda}_H \in \mathbb{R}^{n_H}$ and $\dot{\gamma}_H \in \mathbb{R}^{n_H}$ consist of the tangential forces and accelerations of the potentially sticking contacts in I_H. The friction forces of the sliding constraints are already expressed by their corresponding normal forces $\boldsymbol{\lambda}_G = \bar{\bar{\boldsymbol{\mu}}}_G\boldsymbol{\lambda}_N \in \mathbb{R}^{n_G}$ and occur in (6.13) by the matrix expression $\boldsymbol{W}_G\bar{\bar{\boldsymbol{\mu}}}_G$, where $\bar{\bar{\boldsymbol{\mu}}}_G \in \mathbb{R}^{n_G \times n_N}$ is the matrix of friction coefficients of the sliding contacts with

$$\boldsymbol{\mu} = \mathrm{col}\{\mu_i\},\ i \in I_N, \tag{6.15}$$

$$\bar{\bar{\boldsymbol{\mu}}} = \mathrm{diag}\{\boldsymbol{\mu}\}, \tag{6.16}$$

$$\bar{\bar{\boldsymbol{\mu}}}_G = \mathrm{mat}\{-\bar{\bar{\boldsymbol{\mu}}}_i\,\mathrm{sign}(\gamma_{Ti})\}^{\mathrm{T}},\ i \in I_G. \tag{6.17}$$

The vector of tangential contact forces $\boldsymbol{\lambda}_T$ therefore contains potentially sticking contact forces $\boldsymbol{\lambda}_H$ and slipping contact forces $\boldsymbol{\lambda}_G$ and we have the relation

$$\boldsymbol{W}_T\boldsymbol{\lambda}_T = \boldsymbol{W}_H\boldsymbol{\lambda}_H + \boldsymbol{W}_G\boldsymbol{\lambda}_G, \tag{6.18}$$

where

$$\boldsymbol{W}_H = \mathrm{mat}\{\boldsymbol{w}_{Ti}\} \in \mathbb{R}^{n \times n_H},\ i \in I_H, \tag{6.19}$$

$$\boldsymbol{W}_G = \mathrm{mat}\{\boldsymbol{w}_{Ti}\} \in \mathbb{R}^{n \times n_G},\ i \in I_G. \tag{6.20}$$

The normal and tangential contact accelerations are expressed in the generalized coordinates in (6.14). The remaining elements of I_S correspond to collisions, which induce impulsive contact forces, and are not taken into account in (6.13).

Each closed contact $i \in I_N$ is characterized by a vanishing contact distance g_{Ni} and normal relative velocity γ_{Ni}. The normal contact law for a closed contact can be expressed on acceleration level (5.28)

$$-\dot{\gamma}_{Ni} \in N_{C_N}(\lambda_{Ni}), \quad g_{Ni} = 0, \; \gamma_{Ni} = 0. \tag{6.21}$$

The admissible set of normal contact forces C_N is the cone $\mathbb{R}^+$. The contact law on acceleration level can therefore directly be put in complementarity form

$$0 \leq \dot{\gamma}_{Ni} \perp \lambda_{Ni} \geq 0, \qquad i \in I_N. \tag{6.22}$$

The complementarity behaviour of the normal contact law is depicted in Figure 6.2a and shows a corner of admissible combinations of $\dot{\gamma}_{Ni}$ and λ_{Ni}. Unilateral laws of this type (showing one corner) lead, together with the equations of motion, to an LCP as we will see in the remainder of this section.

With respect to the tangential direction we will need to express Coulomb's friction law on acceleration level in similar unilateral primitives. Only non-associated Coulomb friction will be discussed (being more general than associated Coulomb friction). To determine the tangential contact force during sticking, the unilateral laws for sticking contacts have to be expressed on acceleration level (see (5.29) and (5.30))

$$-\dot{\gamma}_{Ti} \in N_{C_T(\boldsymbol{\lambda}_{Ni})}(\lambda_{Ti}), \qquad i \in I_H. \tag{6.23}$$

The tangential contact law is shown in the upper part of Figure 6.2b and shows two corners. The tangential contact law must therefore be decomposed into two separate unilateral primitives, which is shown in the lower part of Figure 6.2b. The decomposition involves the splitting of $\dot{\gamma}_{Ti}$ in right and left parts:

$$\dot{\gamma}_{Ri} = \frac{1}{2}(|\dot{\gamma}_{Ti}| + \dot{\gamma}_{Ti}), \quad \dot{\gamma}_{Li} = \frac{1}{2}(|\dot{\gamma}_{Ti}| - \dot{\gamma}_{Ti}), \quad \dot{\gamma}_{Ti} = \dot{\gamma}_{Ri} - \dot{\gamma}_{Li}. \tag{6.24}$$

The right and left parts of the accelerations are gathered in the vectors $\dot{\boldsymbol{\gamma}}_R = \mathrm{col}\{\dot{\gamma}_{Ri}\}$ and $\dot{\boldsymbol{\gamma}}_L = \mathrm{col}\{\dot{\gamma}_{Li}\}$, $i \in I_H$. Furthermore, we need to define the friction saturations $\boldsymbol{\lambda}_R$ and $\boldsymbol{\lambda}_L$, i.e. the differences in force on the horizontal axes relative to the origins of the two corners:

$$\boldsymbol{\lambda}_R = \bar{\bar{\boldsymbol{\mu}}}_H \boldsymbol{\lambda}_N + \boldsymbol{\lambda}_H, \tag{6.25}$$

$$\boldsymbol{\lambda}_L = \bar{\bar{\boldsymbol{\mu}}}_H \boldsymbol{\lambda}_N - \boldsymbol{\lambda}_H, \tag{6.26}$$

with $\boldsymbol{\mu}_H = \mathrm{col}\{\mu_i\}$, $i \in I_H$, and $\bar{\bar{\boldsymbol{\mu}}}_H = \mathrm{diag}\{\boldsymbol{\mu}_H\}$. Addition of equations (6.25) and (6.26) gives the relation

$$\boldsymbol{\lambda}_R = 2\bar{\bar{\boldsymbol{\mu}}}_H \boldsymbol{\lambda}_N - \boldsymbol{\lambda}_L, \tag{6.27}$$

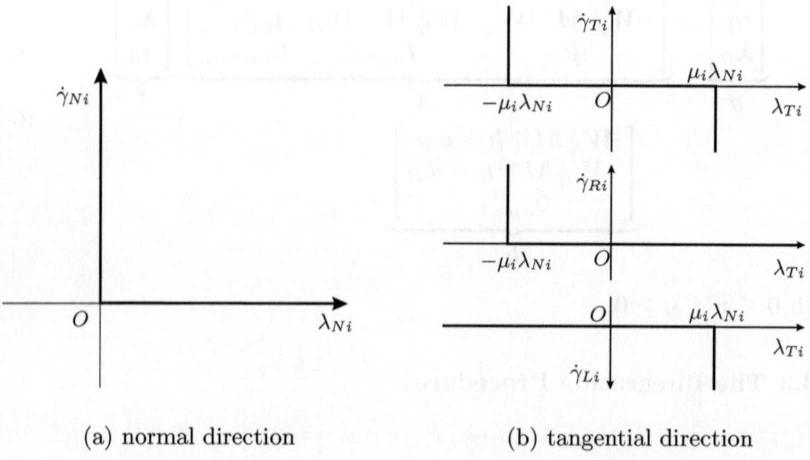

Fig. 6.2. Complementarity of contact laws on acceleration level.

which will be of use later on. The friction saturations $\boldsymbol{\lambda}_R$ and $\boldsymbol{\lambda}_L$ are complementary to the acceleration vectors $\dot{\boldsymbol{\gamma}}_R$ and $\dot{\boldsymbol{\gamma}}_L$, and can therefore be used to set up an LCP on acceleration level for the tangential contact problem. Equation (6.26) is substituted in the equation of motion (6.13)

$$\begin{aligned}\dot{\boldsymbol{u}} &= \boldsymbol{M}^{-1}\boldsymbol{h} + \boldsymbol{M}^{-1}(\boldsymbol{W}_N + \boldsymbol{W}_G\bar{\bar{\boldsymbol{\mu}}}_G)\boldsymbol{\lambda}_N + \boldsymbol{M}^{-1}\boldsymbol{W}_H\boldsymbol{\lambda}_H \\ &= \boldsymbol{M}^{-1}\boldsymbol{h} + \boldsymbol{M}^{-1}(\boldsymbol{W}_N + \boldsymbol{W}_G\bar{\bar{\boldsymbol{\mu}}}_G + \boldsymbol{W}_H\bar{\bar{\boldsymbol{\mu}}}_H)\boldsymbol{\lambda}_N - \boldsymbol{M}^{-1}\boldsymbol{W}_H\boldsymbol{\lambda}_L.\end{aligned} \tag{6.28}$$

The acceleration $\dot{\boldsymbol{u}}$ is subsequently substituted in the first line of the contact equations (6.14), which gives

$$\dot{\boldsymbol{\gamma}}_N = \boldsymbol{W}_N^{\mathrm{T}}\boldsymbol{M}^{-1}\boldsymbol{h} + \boldsymbol{W}_N^{\mathrm{T}}\boldsymbol{M}^{-1}\boldsymbol{W}_Q\boldsymbol{\lambda}_N - \boldsymbol{W}_N^{\mathrm{T}}\boldsymbol{M}^{-1}\boldsymbol{W}_H\boldsymbol{\lambda}_L + \bar{\boldsymbol{w}}_N, \tag{6.29}$$

with

$$\boldsymbol{W}_Q = \boldsymbol{W}_N + \boldsymbol{W}_G\bar{\bar{\boldsymbol{\mu}}}_G + \boldsymbol{W}_H\bar{\bar{\boldsymbol{\mu}}}_H. \tag{6.30}$$

Similarly, an expression for $\dot{\boldsymbol{\gamma}}_L$ is obtained, substituting the second line of (6.14)

$$\dot{\boldsymbol{\gamma}}_L = -\boldsymbol{W}_H^{\mathrm{T}}\boldsymbol{M}^{-1}\boldsymbol{h} - \boldsymbol{W}_H^{\mathrm{T}}\boldsymbol{M}^{-1}\boldsymbol{W}_Q\boldsymbol{\lambda}_N + \boldsymbol{W}_H^{\mathrm{T}}\boldsymbol{M}^{-1}\boldsymbol{W}_H\boldsymbol{\lambda}_L + \dot{\boldsymbol{\gamma}}_R - \bar{\boldsymbol{w}}_H, \tag{6.31}$$

where the relation $\dot{\boldsymbol{\gamma}}_H = \dot{\boldsymbol{\gamma}}_R - \dot{\boldsymbol{\gamma}}_L$ and the abbreviation $\boldsymbol{W}_Q$ (6.29) are used. Equations (6.29), (6.31) and (6.27) form together an LCP on acceleration level for the tangential contact problem:

$$\underbrace{\begin{bmatrix} \dot{\gamma}_N \\ \dot{\gamma}_L \\ \boldsymbol{\lambda}_R \end{bmatrix}}_{\boldsymbol{y}} = \underbrace{\begin{bmatrix} \boldsymbol{W}_N^{\mathrm{T}}\boldsymbol{M}^{-1}\boldsymbol{W}_Q & -\boldsymbol{W}_N^{\mathrm{T}}\boldsymbol{M}^{-1}\boldsymbol{W}_H & \mathbf{0}_{n_N \times n_H} \\ -\boldsymbol{W}_H^{\mathrm{T}}\boldsymbol{M}^{-1}\boldsymbol{W}_Q & \boldsymbol{W}_H^{\mathrm{T}}\boldsymbol{M}^{-1}\boldsymbol{W}_H & \boldsymbol{I}_{n_H \times n_H} \\ 2\bar{\bar{\boldsymbol{\mu}}}_H & -\boldsymbol{I}_{n_H \times n_H} & \mathbf{0}_{n_H \times n_H} \end{bmatrix}}_{\boldsymbol{A}} \underbrace{\begin{bmatrix} \boldsymbol{\lambda}_N \\ \boldsymbol{\lambda}_L \\ \dot{\gamma}_R \end{bmatrix}}_{\boldsymbol{x}} + \underbrace{\begin{bmatrix} \boldsymbol{W}_N^{\mathrm{T}}\boldsymbol{M}^{-1}\boldsymbol{h} + \bar{\boldsymbol{w}}_N \\ -\boldsymbol{W}_H^{\mathrm{T}}\boldsymbol{M}^{-1}\boldsymbol{h} - \bar{\boldsymbol{w}}_H \\ \mathbf{0}_{n_H \times 1} \end{bmatrix}}_{\boldsymbol{b}} \tag{6.32}$$

with $\mathbf{0} \leq \boldsymbol{x} \perp \boldsymbol{y} \geq \mathbf{0}$.

6.3.3 The Integration Procedure

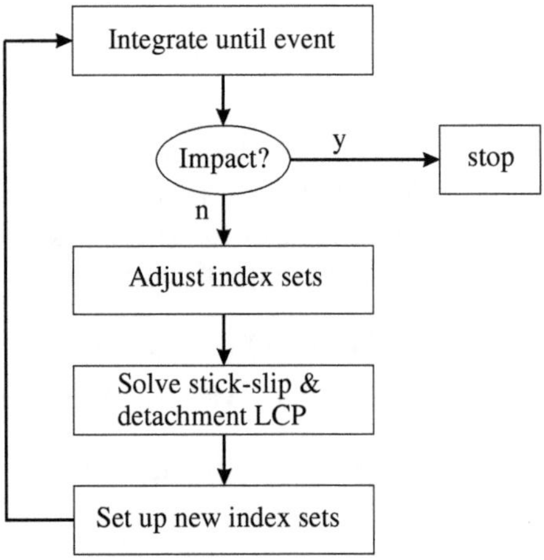

Fig. 6.3. Flowchart of the event-driven integration method.

Figure 6.3 shows the order of the different phases in the integration procedure. The equation of motion for given index sets is numerically integrated until an *event* occurs. An event occurs when the solution crosses a switching boundary. We have to distinguish between the following types of events:

1. *vanishing normal contact force*: $\lambda_{Ni} \leq 0$, contact i is likely to detach.
2. *vanishing contact distance*: $g_{Ni} \leq 0 \wedge \gamma_{Ni} < 0$, contact i is likely to make an impact.
3. *vanishing relative velocity*: γ_{Ti} changes sign, contact i might undergo a slip-to-stick transition or a transition in slip direction.
4. *vanishing friction saturation*: $\lambda_{Ri} \leq 0$ or $\lambda_{Li} \leq 0$, contact i is likely to undergo a stick-to-slip transition.

The words 'likely' and 'might' are used in the above list of event-types, because the event only is used to adjust the index sets, which set up the LCP. The actual new contact configuration is determined by the LCP. For instance, a vanishing relative velocity might cause a slipping contact to become sticking. The contact set I_H has therefore to be augmented with contact i when γ_{Ti} changes sign. Subsequently, the transition LCP (6.32) can be constructed with the adjusted index sets. The new contact set, describing the next mode, then follows from the solution of the LCP. A new smooth integration phase is started with the state at the old event-time as initial condition. Basically, any ODE-integrator can be used for the integration of the smooth phase as long as the ODE-integrator supports event detection. However, the algebraic constraint equation, defined by the index set, has to be maintained during integration. A drift-off of the constraint equation can occur during long integration intervals. The set of equations during a smooth phase is actually a differential algebraic equation for which dedicated integration methods exist [55]. The flowchart of the event-driven integration method is depicted in Figure 6.3. The algorithm is stopped when an impact event occurs, because we confine the analysis to Filippov systems. An event-driven integration method for systems with impact and friction is described in [46, 49, 81, 118].

6.4 Time-stepping Methods

The event-driven integration method (Section 6.3) integrates the system until an event occurs, calculates the next mode and proceeds integration. The event-driven integration method therefore clearly expresses the hybrid nature of systems with friction (or differential inclusions in general).

Time-stepping methods for mechanical rigid body systems are based on using a time-discretization of generalized positions $\boldsymbol{q}$ and velocities $\boldsymbol{u}$ ($\boldsymbol{u} = \dot{\boldsymbol{q}}$ almost everywhere), usually with a fixed step size. Integrals of forces over each time-step are used instead of the instantaneous values of the forces. The time-stepping method makes no distinction between impulsive forces (due to impacts) and finite forces. Only increments of the positions and velocities are computed. The acceleration $\dot{\boldsymbol{u}}$ is not computed by the algorithm, as it becomes infinite for impulsive forces. The positions and velocities at the end of the time-step are found by solving a (N)LCP which describes the contact problem or by means of an augmented Lagrangian approach. The time-stepping method does therefore not use the notion of 'events' and only one or no index sets are used. Multiple events might take place during one time-step, and the algorithm computes the overall integral of the forces over this time-step, which is finite. The time-stepping method is especially useful when one is interested in the global motion of systems with many contact points, leading to a large number of events. Each individual event is for those applications not of importance but the global motion is determined by the sum of all events. The benefit of time-stepping methods over event-driven integration methods is the fact that no event-detection and hardly any index sets are needed. This makes the algorithm less complex, more robust and will give a reduction in computation time when many contacts are involved.

The time-stepping method was introduced by Moreau [97] and has been subsequently developed by [5, 137, 139]. In [139] a time-stepping method is presented, formulating the normal contact problem on position level and the tangential contact problem on velocity level, leading to an algorithm without any index sets. In this section we will describe the time-stepping method of [97] with a formulation of both the normal contact problem and the tangential contact problem on velocity level, using one index set.

6.4.1 Discretization

The dynamics of a multibody system during an impact free part of the motion can be expressed by the equation of motion on acceleration level (5.3)

$$\boldsymbol{M}(t,\boldsymbol{q})\dot{\boldsymbol{u}} - \boldsymbol{h}(t,\boldsymbol{q},\boldsymbol{u}) - \boldsymbol{W}_N(t,\boldsymbol{q})\boldsymbol{\lambda}_N - \boldsymbol{W}_T(t,\boldsymbol{q})\boldsymbol{\lambda}_T = \boldsymbol{0}, \tag{6.33}$$

with the set-valued force laws

$$-\boldsymbol{g}_N \in N_{C_N}(\boldsymbol{\lambda}_N), \qquad -\boldsymbol{\gamma}_T \in N_{C_T(\boldsymbol{\lambda}_N)}(\boldsymbol{\lambda}_T). \tag{6.34}$$

Time-stepping methods do not make use of the usual equation of motion, which relates acceleration to forces, but replace the equation of motion on acceleration level by an equality of measures [51]

$$\boldsymbol{M}\mathrm{d}\boldsymbol{u} - \boldsymbol{h}\mathrm{d}t - \boldsymbol{W}_N\mathrm{d}\boldsymbol{\Lambda}_N - \boldsymbol{W}_T\mathrm{d}\boldsymbol{\Lambda}_T = \boldsymbol{0}, \tag{6.35}$$

or

$$\boldsymbol{M}\mathrm{d}\boldsymbol{u} - \boldsymbol{h}\mathrm{d}t - \boldsymbol{W}\mathrm{d}\boldsymbol{\Lambda} = \boldsymbol{0}, \tag{6.36}$$

where the dependence of the system matrices on $t, \boldsymbol{q}$ and $\boldsymbol{u}$ has been omitted for brevity. We denote with $\mathrm{d}t$ the Lebesgue-measure and with $\mathrm{d}\eta$ the sum of the dirac pulses at the impact times. The measure for the velocities $\mathrm{d}\boldsymbol{u} = \dot{\boldsymbol{u}}\mathrm{d}t + (\boldsymbol{u}^+ - \boldsymbol{u}^-)\mathrm{d}\eta$ is split in a Lebesgue-measurable part and an atomic part. The atomic part consists of the left and right limit of $\boldsymbol{u}$ at t. For impact free motion it holds that $\mathrm{d}\boldsymbol{u} = \dot{\boldsymbol{u}}\mathrm{d}t$. Similarly, the measure for the impulses is defined as $\mathrm{d}\boldsymbol{\Lambda} = \boldsymbol{\lambda}\mathrm{d}t + \boldsymbol{\Lambda}\mathrm{d}\eta$. The time-stepping method is generally applied to systems with impacts, for which $\boldsymbol{u}^+ \neq \boldsymbol{u}^-$ at the impact time. Here we focus on Filippov systems for which $\dot{\boldsymbol{q}} = \boldsymbol{u} = \boldsymbol{u}^+ = \boldsymbol{u}^-$.

The constraints on velocity level can be expressed in the left and right limits of $\boldsymbol{u}$

$$\boldsymbol{\gamma}^+ = \boldsymbol{W}^{\mathrm{T}}\boldsymbol{u}^+ + \tilde{\boldsymbol{w}}, \qquad \boldsymbol{\gamma}^- = \boldsymbol{W}^{\mathrm{T}}\boldsymbol{u}^- + \tilde{\boldsymbol{w}}. \tag{6.37}$$

For impact free motion it holds that $\boldsymbol{\gamma} = \boldsymbol{\gamma}^+ = \boldsymbol{\gamma}^-$.

At this point we discretize the measure differential equation by integrating over a small finite time interval Δt

$$\int_{\Delta t} \mathrm{d}\boldsymbol{u} = \Delta \boldsymbol{u}, \qquad \int_{\Delta t} \boldsymbol{h}\mathrm{d}t = \Delta \boldsymbol{h} \approx \boldsymbol{h}\Delta t, \qquad \int_{\Delta t} \mathrm{d}\boldsymbol{\Lambda} = \boldsymbol{\Lambda}, \tag{6.38}$$

and we obtain the difference-equation of motion

$$\boldsymbol{M}\Delta\boldsymbol{u} - \boldsymbol{h}\Delta t - \boldsymbol{W}_N\boldsymbol{\Lambda}_N - \boldsymbol{W}_T\boldsymbol{\Lambda}_T = \boldsymbol{0}. \tag{6.39}$$

The force laws for completely inelastic contact with friction can be put in the form

$$-\boldsymbol{\Lambda}_N \in \partial\Psi^*_{C_N}(\gamma_N^+), \qquad -\boldsymbol{\Lambda}_T \in \partial\Psi^*_{C_T(\boldsymbol{\Lambda}_N)}(\gamma_T^+), \tag{6.40}$$

with

$$\gamma_N^+ = \boldsymbol{W}_N^{\mathrm{T}}(\boldsymbol{u}^- + \Delta\boldsymbol{u}) + \tilde{\boldsymbol{w}}_N, \qquad \gamma_T^+ = \boldsymbol{W}_T^{\mathrm{T}}(\boldsymbol{u}^- + \Delta\boldsymbol{u}) + \tilde{\boldsymbol{w}}_T. \tag{6.41}$$

The conjugate of the indicator function of C_T is the velocity potential, i.e. $\Psi^*_{C_T(\boldsymbol{\Lambda}_N)}(\gamma_T^+) = \Phi(\gamma_T^+)$. For planar friction problems, the force law for the friction forces becomes

$$-\boldsymbol{\Lambda}_T \in \partial\Phi(\gamma_T^+) \iff -\boldsymbol{\Lambda}_T \in \mu\boldsymbol{\Lambda}_N \,\mathrm{Sign}(\gamma_T^+). \tag{6.42}$$

In each incremental step, say time-step, we have to solve $\Delta\boldsymbol{u}$ and $(\boldsymbol{\Lambda}_N, \boldsymbol{\Lambda}_T)$ from the equation of motion (6.39) and the force laws (6.40), which forms a set of algebraic inclusions. The set of algebraic inclusions can be solved with an (N)LCP or by means of an augmented Lagrangian method.

6.4.2 Moreau's Algorithm

The time-stepping method of Moreau is basically a special kind of midpoint DAE-integrator. Consider the positions $\boldsymbol{q}_A$ and velocities $\boldsymbol{u}_A$ known at the beginning of the time-step at time t_A. The algorithm takes first a half time-step for the positions and arrives at the midpoint $\boldsymbol{q}_M = \boldsymbol{q}_A + \frac{1}{2}\Delta t\boldsymbol{u}_A$. The midpoint is used to classify the status of the normal constraints, which allows for an index reduction. The contact set $I_N = \{i \mid g_{Ni}(t_M, \boldsymbol{q}_M) \leq 0\}$ is calculated at the midpoint and used to set-up the contact problem on velocity level for both the normal and tangential constraints. The set I_N contains all indices of the closed contact points. The velocity $\boldsymbol{u}_E$, at the end of time-step $t_E = t_A + \Delta t$, is subsequently calculated by a trapezoidal scheme

$$\boldsymbol{M}_M(\boldsymbol{u}_E - \boldsymbol{u}_A) = \boldsymbol{h}_M\Delta t + \boldsymbol{W}_{NM}\boldsymbol{\Lambda}_N + \boldsymbol{W}_{TM}\boldsymbol{\Lambda}_T, \tag{6.43}$$

and the set-valued force laws

$$-\boldsymbol{\Lambda}_N \in \partial\Psi_{\mathbb{R}^+}(\gamma_{NE}), \qquad -\boldsymbol{\Lambda}_T \in \partial\Psi^*_{C_T(\boldsymbol{\Lambda}_N)}(\gamma_{TE}), \tag{6.44}$$

where $\boldsymbol{M}_M$, $\boldsymbol{h}_M$, $\boldsymbol{W}_{NM}$ and $\boldsymbol{W}_{TM}$ are the system matrices evaluated at the midpoint. This set of algebraic inclusions can be solved with an LCP formulation or by augmented Lagrangian approach. Finally, the positions at the end of the time-step are calculated

$$q_E = q_M + \frac{1}{2}\Delta t u_E. \tag{6.45}$$

We will first discuss the time-stepping method using an LCP to solve the contact problem, and subsequently we will discuss the time-stepping method in combination with an augmented Lagrangian approach.

6.4.3 Formulation of the Contact Problem as an LCP

The formulation of the contact problem with planar Coulomb friction, i.e. a unidirectional friction force, requires a separation of the friction law into two unilateral complementarity conditions, similar to Section 6.3. We introduce therefore the right and left sliding velocities at $t = t_E$

$$\gamma_{RE} = \frac{1}{2}(|\gamma_{TE}| + \gamma_{TE}), \quad \gamma_{LE} = \frac{1}{2}(|\gamma_{TE}| - \gamma_{TE}), \quad \gamma_{TE} = \gamma_{RE} - \gamma_{LE}, \tag{6.46}$$

and the impulsive friction saturations

$$\begin{aligned} \Lambda_R &= \bar{\bar{\mu}}\Lambda_N + \Lambda_T, \\ \Lambda_L &= \bar{\bar{\mu}}\Lambda_N - \Lambda_T, \end{aligned} \tag{6.47}$$

with $\bar{\bar{\mu}} = \text{diag}\{\mu_i\}$, $i \in I_N$. Using (6.46) and (6.47), the force laws (6.42) can be written as the set of complementarity conditions

$$\begin{aligned} 0 \le \gamma_{NE} \perp \Lambda_N \ge 0, \\ 0 \le \gamma_{RE} \perp \Lambda_R \ge 0, \\ 0 \le \gamma_{LE} \perp \Lambda_L \ge 0. \end{aligned} \tag{6.48}$$

Substitution of $\gamma_{NE} = W_N^T u_E + \tilde{w}_N$ and $\gamma_{TE} = W_T^T u_E + \tilde{w}_T$ in (6.43) together with (6.46) and (6.47) gives the equations

$$\begin{aligned} \gamma_{NE} =& G_{NN}\Lambda_N - G_{NT}(\Lambda_L - \bar{\bar{\mu}}\Lambda_N) \\ &+ W_{NM}^T(u_A + M_M^{-1}h_M \Delta t) + \tilde{w}_{NM}, \end{aligned} \tag{6.49}$$

$$\begin{aligned} \gamma_{LE} =& - G_{TN}\Lambda_N + G_{TT}(\Lambda_L - \bar{\bar{\mu}}\Lambda_N) \\ &- W_{TM}^T M_M^{-1} h_M \Delta t - W_{TM}^T u_A + \gamma_{RE} - \tilde{w}_{TM}, \end{aligned} \tag{6.50}$$

with

$$\begin{aligned} &G_N = W_{NM}^T M_M^{-1}, \; G_T = W_{TM}^T M_M^{-1}, \\ &G_{NN} = G_N W_{NM}, \; G_{TN} = G_T W_{NM}, \\ &G_{NT} = G_N W_{TM}, \; G_{TT} = G_T W_{TM}. \end{aligned} \tag{6.51}$$

Equations (6.49) and (6.50) form together with the complementarity condition a linear complementarity problem

$$
\begin{bmatrix} \gamma_{NE} \\ \gamma_{LE} \\ \boldsymbol{\Lambda}_R \end{bmatrix} = \begin{bmatrix} \boldsymbol{G}_{NN} + \boldsymbol{G}_{NT}\bar{\bar{\boldsymbol{\mu}}} & -\boldsymbol{G}_{NT} & \boldsymbol{0} \\ -\boldsymbol{G}_{TN} - \boldsymbol{G}_{TT}\bar{\bar{\boldsymbol{\mu}}} & \boldsymbol{G}_{TT} & \boldsymbol{I} \\ 2\bar{\bar{\boldsymbol{\mu}}} & -\boldsymbol{I} & \boldsymbol{0} \end{bmatrix} \begin{bmatrix} \boldsymbol{\Lambda}_N \\ \boldsymbol{\Lambda}_L \\ \gamma_{RE} \end{bmatrix} + \begin{bmatrix} \boldsymbol{W}_{NM}^{\mathrm{T}}(\boldsymbol{u}_A + \boldsymbol{M}_M^{-1}\boldsymbol{h}_M\Delta t) + \tilde{\boldsymbol{w}}_{NM} \\ -\boldsymbol{W}_{TM}^{\mathrm{T}}(\boldsymbol{u}_A + \boldsymbol{M}_M^{-1}\boldsymbol{h}_M\Delta t) - \tilde{\boldsymbol{w}}_{TM} \\ \boldsymbol{0} \end{bmatrix}, \tag{6.52}
$$

with

$$
\boldsymbol{0} \leq \begin{bmatrix} \gamma_{NE} \\ \gamma_{LE} \\ \boldsymbol{\Lambda}_R \end{bmatrix} \perp \begin{bmatrix} \boldsymbol{\Lambda}_N \\ \boldsymbol{\Lambda}_L \\ \gamma_{RE} \end{bmatrix} \geq \boldsymbol{0}. \tag{6.53}
$$

The LCP (6.52) is solved in each integration step. The velocities $\boldsymbol{u}_E$ and positions $\boldsymbol{q}_E$ are subsequently found with (6.43) and (6.45). The pseudo-code of Moreau's time-stepping method with an LCP formulation of the contact problem is shown in Table 6.2.

6.4.4 Augmented Lagrangian Approach

The time-stepping method, introduced in the previous subsection, uses an LCP formulation to solve the contact problem. The contact problem can also be solved with an augmented Lagrangian approach. The augmented Lagrangian has been discussed in Section 5.5 for a system on acceleration level. The time-stepping method, however, makes use of a discretized version of a measure differential equation on velocity level. By making use of the Zwang in differences, $Z(\Delta\boldsymbol{u})$, one can also set up the augmented Lagrangian in differences $L_A(\Delta\boldsymbol{u}, \boldsymbol{\Lambda})$, similarly to (5.61). The (quasi) augmented Lagrangian for a frictional contact problem in differences can be expressed as (see (5.61))

$$
\begin{aligned}
L_A(\Delta\boldsymbol{u}, \boldsymbol{\Lambda}_N, \boldsymbol{\Lambda}_T) = & \frac{1}{2}\|\Delta\boldsymbol{u} - \boldsymbol{M}^{-1}\boldsymbol{h}\Delta t\|_{\boldsymbol{M}}^2 - \gamma_N^{\mathrm{T}}(\Delta\boldsymbol{u})\boldsymbol{\Lambda}_N - \gamma_T^{\mathrm{T}}(\Delta\boldsymbol{u})\boldsymbol{\Lambda}_T \\
& + \frac{r}{2}\|\gamma_N(\Delta\boldsymbol{u})\|^2 - \frac{1}{2r}\operatorname{dist}_{C_N}^2(\boldsymbol{\Lambda}_N - r\gamma_N(\Delta\boldsymbol{u})) \\
& + \frac{r}{2}\|\gamma_T(\Delta\boldsymbol{u})\|^2 - \frac{1}{2r}\operatorname{dist}_{C_T}^2(\boldsymbol{\Lambda}_T - r\gamma_T(\Delta\boldsymbol{u})),
\end{aligned} \tag{6.54}
$$

where $C_T := C_T(\operatorname{prox}_{\mathbb{R}^+}(\boldsymbol{\Lambda}_N - r\gamma_N(\Delta\boldsymbol{u}))$ and $r > 0$. The augmented Lagrangian L_A is dependent on a parameter $r > 0$, which (loosely speaking) controls the steepness of L_A in the region where L is constrained as well as an additional penalty term in the direction $\Delta\boldsymbol{u}$. The value of r should be taken large enough to make the problem well conditioned in the constrained region, but not too high in order to prevent ill-conditioning due to the penalty term $\frac{r}{2}\|\gamma(\Delta\boldsymbol{u})\|^2$.

Evaluating the stationarity conditions of the saddle-point of L_A gives the equations

function $[t, \boldsymbol{q}, \boldsymbol{u}] = \mathit{TimeSteppingMethodMoreauLCP}(t_0, t_e, \boldsymbol{q}_0, \boldsymbol{u}_0, N, \mathrm{tol})$

$\Delta t = \frac{t_e - t_0}{N-1}$ Initialize

$t^1 = t_0, \quad \boldsymbol{q}^1 = \boldsymbol{q}_0, \quad \boldsymbol{u}^1 = \boldsymbol{u}_0$

for $j = 1$ **to** N Do all time-steps

$t_A = t^j, \quad \boldsymbol{q}_A = \boldsymbol{q}^j, \quad \boldsymbol{u}_A = \boldsymbol{u}^j$

$t_M = t_A + \frac{1}{2}\Delta t, \quad \boldsymbol{q}_M = \boldsymbol{q}_A + \frac{1}{2}\Delta t \boldsymbol{u}_A$ Do half a time-step

$\boldsymbol{M}_M = \boldsymbol{M}(t_M, \boldsymbol{q}_M), \quad \boldsymbol{h}_M = \boldsymbol{h}(t_M, \boldsymbol{q}_M, \boldsymbol{u}_A)$

$I_N = \{i \mid g_{Ni}(t_M, \boldsymbol{q}_M) \leq 0\}, \quad n_N = \mathit{length}(I_N)$

$\boldsymbol{W}_{NM} = \mathrm{mat}\{\boldsymbol{w}_{Ni}(t_M, \boldsymbol{q}_M)\}, \ \tilde{\boldsymbol{w}}_{NM} = \mathrm{col}\{\tilde{w}_{Ni}(t_M, \boldsymbol{q}_M)\}, \quad i \in I_N$

$\boldsymbol{W}_{TM} = \mathrm{mat}\{\boldsymbol{w}_{Ti}(t_M, \boldsymbol{q}_M)\}, \ \tilde{\boldsymbol{w}}_{TM} = \mathrm{col}\{\tilde{w}_{Ti}(t_M, \boldsymbol{q}_M)\}, \quad i \in I_N$

$\bar{\bar{\boldsymbol{\mu}}} = \mathrm{diag}\{\mu_i\}, \quad i \in I_N$

$\boldsymbol{G}_N = \boldsymbol{W}_{NM}^{\mathrm{T}} \boldsymbol{M}_M^{-1}, \ \boldsymbol{G}_T = \boldsymbol{W}_{TM}^{\mathrm{T}} \boldsymbol{M}_M^{-1}$

$\boldsymbol{G}_{NN} = \boldsymbol{G}_N \boldsymbol{W}_{NM}, \ \boldsymbol{G}_{TN} = \boldsymbol{G}_T \boldsymbol{W}_{NM}$

$\boldsymbol{G}_{NT} = \boldsymbol{G}_N \boldsymbol{W}_{TM}, \ \boldsymbol{G}_{TT} = \boldsymbol{G}_T \boldsymbol{W}_{TM}$

$$\boldsymbol{A} = \begin{bmatrix} \boldsymbol{G}_{NN} + \boldsymbol{G}_{NT}\bar{\bar{\boldsymbol{\mu}}} & -\boldsymbol{G}_{NT} & \boldsymbol{0} \\ -\boldsymbol{G}_{TN} - \boldsymbol{G}_{TT}\bar{\bar{\boldsymbol{\mu}}} & \boldsymbol{G}_{TT} & \boldsymbol{I} \\ 2\bar{\bar{\boldsymbol{\mu}}} & -\boldsymbol{I} & \boldsymbol{0} \end{bmatrix}$$

$$\boldsymbol{b} = \begin{bmatrix} \boldsymbol{W}_{NM}^{\mathrm{T}}(\boldsymbol{u}_A + \boldsymbol{M}_M^{-1}\boldsymbol{h}_M \Delta t) + \tilde{\boldsymbol{w}}_{NM} \\ -\boldsymbol{W}_{TM}^{\mathrm{T}}(\boldsymbol{u}_A + \boldsymbol{M}_M^{-1}\boldsymbol{h}_M \Delta t) - \tilde{\boldsymbol{w}}_{TM} \\ \boldsymbol{0} \end{bmatrix}$$

$(\boldsymbol{x}, \boldsymbol{y}) = \mathit{LCP}(\boldsymbol{A}, \boldsymbol{b})$ Call LCP solver

$\boldsymbol{\Lambda}_N = \mathrm{col}\{x_i\}, \quad i = 1 \cdots n_N$

$\boldsymbol{\Lambda}_L = \mathrm{col}\{x_i\}, \quad i = n_N + 1 \cdots 2n_N$

$\boldsymbol{\Lambda}_R = \mathrm{col}\{y_i\}, \quad i = 2n_N + 1 \cdots 3n_N$

$\boldsymbol{\Lambda}_T = \frac{1}{2}(\boldsymbol{\Lambda}_R - \boldsymbol{\Lambda}_L)$

$\boldsymbol{u}_E = \boldsymbol{u}_A + \boldsymbol{M}_M^{-1}(\boldsymbol{h}_M \Delta t + \boldsymbol{W}_{NM}\boldsymbol{\Lambda}_N + \boldsymbol{W}_{TM}\boldsymbol{\Lambda}_T)$

$\boldsymbol{q}_E = \boldsymbol{q}_M + \frac{1}{2}\Delta t \boldsymbol{u}_E$

$t_E = t_M + \frac{1}{2}\Delta t$ Do the second half time-step

$t^{j+1} = t_E, \quad \boldsymbol{q}^{j+1} = \boldsymbol{q}_E, \quad \boldsymbol{u}^{j+1} = \boldsymbol{u}_E$

end

end

Table 6.2. Moreau's time-stepping method with an LCP formulation.

$$\begin{aligned} \nabla_{\Delta \boldsymbol{u}} L_A &= \boldsymbol{M}\Delta\boldsymbol{u} - \boldsymbol{h}\Delta t - \boldsymbol{W}_N \Pi_N(\boldsymbol{\Lambda}_N, \gamma_N) - \boldsymbol{W}_T \Pi_T(\boldsymbol{\Lambda}_N, \boldsymbol{\Lambda}_T, \gamma_T) = \boldsymbol{0}, \\ \nabla_{\boldsymbol{\Lambda}_N} L_A &= -\tfrac{1}{r}\big(\boldsymbol{\Lambda}_N - \Pi_N(\boldsymbol{\Lambda}_N, \gamma_N)\big) = \boldsymbol{0}, \\ \nabla_{\boldsymbol{\Lambda}_T} L_A &= -\tfrac{1}{r}\big(\boldsymbol{\Lambda}_T - \Pi_T(\boldsymbol{\Lambda}_N, \boldsymbol{\Lambda}_T, \gamma_T)\big) = \boldsymbol{0}, \end{aligned} \tag{6.55}$$

where use has been made of (2.16), (2.20), (2.21) and (2.26) and the abbreviations

$$\begin{aligned} \Pi_N(\boldsymbol{\Lambda}_N, \gamma_N) &:= \mathrm{prox}_{C_N}(\boldsymbol{\Lambda}_N - r\gamma_N), \\ \Pi_T(\boldsymbol{\Lambda}_N, \boldsymbol{\Lambda}_T, \gamma_T) &:= \mathrm{prox}_{C_T(\boldsymbol{\Lambda}_N)}(\boldsymbol{\Lambda}_T - r\gamma_T). \end{aligned} \tag{6.56}$$

Note that $L_A := L_A(\Delta\boldsymbol{u}, \boldsymbol{\Lambda}_N, \boldsymbol{\Lambda}_T)$, $\gamma_N := \gamma_N(\Delta\boldsymbol{u})$ and $\gamma_T := \gamma_T(\Delta\boldsymbol{u})$. We therefore obtain the set of algebraic equations

$$
\begin{aligned}
&\boldsymbol{M}\Delta\boldsymbol{u} - \boldsymbol{h}\Delta t - \boldsymbol{W}_N \Pi_N(\boldsymbol{\Lambda}_N, \gamma_N) - \boldsymbol{W}_T \Pi_T(\boldsymbol{\Lambda}_N, \boldsymbol{\Lambda}_T, \gamma_T) = \boldsymbol{0},\\
&\boldsymbol{\Lambda}_N = \Pi_N(\boldsymbol{\Lambda}_N, \gamma_N), \qquad (6.57)\\
&\boldsymbol{\Lambda}_T = \Pi_T(\boldsymbol{\Lambda}_N, \boldsymbol{\Lambda}_T, \gamma_T).
\end{aligned}
$$

The saddle point of the augmented Lagrangian, being the solution to the set of algebraic equations, can be found with a Modified Newton algorithm [2]. More elaborate methods also exist but we will discuss the Modified Newton Method because of+ its simplicity.

The Modified Newton Method uses initial guesses for the impulses, $\boldsymbol{\Lambda}_N^k$ and $\boldsymbol{\Lambda}_T^k$ for $k = 1$, and iterates the scheme

1. Solve $\Delta\boldsymbol{u}^{k+1}$ from $\boldsymbol{M}\Delta\boldsymbol{u}^{k+1} = \boldsymbol{h}\Delta t + \boldsymbol{W}_N\boldsymbol{\Lambda}_N^k + \boldsymbol{W}_T\boldsymbol{\Lambda}_T^k$
2. Project $\boldsymbol{\Lambda}_N^{k+1} = \Pi_N(\boldsymbol{\Lambda}_N^k, \gamma_N^{k+1})$, $\boldsymbol{\Lambda}_T^{k+1} = \Pi_T(\boldsymbol{\Lambda}_N^{k+1}, \boldsymbol{\Lambda}_T^k, \gamma_T^{k+1})$
   with $\gamma_N^{k+1} = \boldsymbol{W}_N^{\mathrm{T}}(\Delta\boldsymbol{u}^{k+1} + \boldsymbol{u}^k) + \boldsymbol{w}_N$, $\gamma_T^{k+1} = \boldsymbol{W}_T^{\mathrm{T}}(\Delta\boldsymbol{u}^{k+1} + \boldsymbol{u}^k) + \boldsymbol{w}_T$

until the error $\|\boldsymbol{\Lambda}_N^{k+1} - \boldsymbol{\Lambda}_N^k\| + \|\boldsymbol{\Lambda}_T^{k+1} - \boldsymbol{\Lambda}_T^k\|$ is within some tolerance.

The algorithm is summarized in Table 6.3, which shows how the augmented Lagrangian approach is combined with the time-stepping method of Moreau. The algorithm is implemented such that the calculated impulses $\boldsymbol{\Lambda}_N$ and $\boldsymbol{\Lambda}_T$ are used as initial guesses for the augmented Lagrangian calculation in the next time-step.

The time-stepping method was introduced in this section to simulate mechanical systems with unidirectional friction. Detachment and completely inelastic impact are, however, also described by the present formulation. If a contact distance vanishes during the time-evolution, i.e. $g_{Ni} = 0$ and $\gamma_{Ni} < 0$, then a collision occurs. The time-stepping method does not distinguish between finite forces and impulsive forces due to impacts. The method simply computes the total normal impulse Λ_{Ni} over the time-step, such that $g_{Ni} = 0$ and $\gamma_{Ni} = 0$ after the time-step. This corresponds with a completely inelastic impact law, and such impulses are named *standard inelastic shocks* by Moreau [97].

We have presented the event-driven integration method and time-stepping method to simulate mechanical systems with unidirectional friction or Filippov systems in general. Event-driven integration methods and time-stepping methods can easily be extended to systems with impact and spatial friction [12, 46, 50, 118]. Newton's or Poisson's impact law is used to describe partially elastic impact. The restriction to unidirectional friction allows bodies to stick to each other, or slip over each other in forward or backward direction. The description of the contact problem becomes more complicated when we deal with spatial contact problems. The friction force lies on a plane for spatial contact problems, whereas the friction force lies on a line for planar problems. The calculation of the slip direction in spatial contact problems is therefore much more complicated because there exists an infinite number of slip directions. Spatial frictional contact can be formulated as a *nonlinear* complementarity problem (2.31), see [49, 51, 139]. Formulation of the contact

function $[t, \boldsymbol{q}, \boldsymbol{u}] = \mathit{TimeSteppingMethodMoreauAL}(t_0, t_e, \boldsymbol{q}_0, \boldsymbol{u}_0, N, \mathrm{tol})$
  $\Delta t = \frac{t_e - t_0}{N-1}$ Initialize
  $t^1 = t_0, \quad \boldsymbol{q}^1 = \boldsymbol{q}_0, \quad \boldsymbol{u}^1 = \boldsymbol{u}_0$
  $\boldsymbol{\Lambda}_N = \boldsymbol{0}, \quad \boldsymbol{\Lambda}_T = \boldsymbol{0}$
  for $j = 1$ **to** N Do all time-steps
    $t_A = t^j, \quad \boldsymbol{q}_A = \boldsymbol{q}^j, \quad \boldsymbol{u}_A = \boldsymbol{u}^j$
    $t_M = t_A + \frac{1}{2}\Delta t, \quad \boldsymbol{q}_M = \boldsymbol{q}_A + \frac{1}{2}\Delta t \boldsymbol{u}_A$ Do half a time-step
    $\boldsymbol{M}_M = \boldsymbol{M}(t_M, \boldsymbol{q}_M), \quad \boldsymbol{h}_M = \boldsymbol{h}(t_M, \boldsymbol{q}_M, \boldsymbol{u}_A)$
    $I_N = \{i \mid g_{Ni}(t_M, \boldsymbol{q}_M) \leq 0\}$
    $\boldsymbol{W}_{NM} = \mathrm{mat}\{\boldsymbol{w}_{Ni}(t_M, \boldsymbol{q}_M)\}, \; \tilde{\boldsymbol{w}}_{NM} = \mathrm{col}\{\tilde{w}_{Ni}(t_M, \boldsymbol{q}_M)\}, \quad i \in I_N$
    $\boldsymbol{W}_{TM} = \mathrm{mat}\{\boldsymbol{w}_{Ti}(t_M, \boldsymbol{q}_M)\}, \quad \tilde{\boldsymbol{w}}_{TM} = \mathrm{col}\{\tilde{w}_{Ti}(t_M, \boldsymbol{q}_M)\}, \quad i \in I_N$
    $k = 1$, converged = false, $\boldsymbol{\Lambda}_N^k = \boldsymbol{\Lambda}_N(I_N), \, \boldsymbol{\Lambda}_T^k = \boldsymbol{\Lambda}_T(I_N)$
    while converged = false augmented Lagrangian method
      $\boldsymbol{u}_E = \boldsymbol{u}_A + \boldsymbol{M}_M^{-1}(\boldsymbol{h}_M \Delta t + \boldsymbol{W}_{NM}\boldsymbol{\Lambda}_N^k + \boldsymbol{W}_{TM}\boldsymbol{\Lambda}_T^k)$
      $\boldsymbol{q}_E = \boldsymbol{q}_M + \frac{1}{2}\Delta t \boldsymbol{u}_E$
      $\boldsymbol{\gamma}_{NE} = \boldsymbol{W}_{NM}^{\mathrm{T}}\boldsymbol{u}_E + \tilde{\boldsymbol{w}}_{NM}, \quad \boldsymbol{\gamma}_{TE} = \boldsymbol{W}_{TM}^{\mathrm{T}}\boldsymbol{u}_E + \tilde{\boldsymbol{w}}_{TM}$
      $\boldsymbol{\Lambda}_N^{k+1} = \Pi_N(\boldsymbol{\Lambda}_N^k, \boldsymbol{\gamma}_{NE}), \quad \boldsymbol{\Lambda}_T^{k+1} = \Pi_T(\boldsymbol{\Lambda}_N^{k+1}, \boldsymbol{\Lambda}_T^k, \boldsymbol{\gamma}_{TE})$
      error $= \|\boldsymbol{\Lambda}_N^{k+1} - \boldsymbol{\Lambda}_N^k\| + \|\boldsymbol{\Lambda}_T^{k+1} - \boldsymbol{\Lambda}_T^k\|$
      converged = (error < tol)
      $k = k + 1$
    end
    $t_E = t_M + \frac{1}{2}\Delta t$ Do the second half time-step
    $t^{j+1} = t_E, \quad \boldsymbol{q}^{j+1} = \boldsymbol{q}_E, \quad \boldsymbol{u}^{j+1} = \boldsymbol{u}_E$
    $\boldsymbol{\Lambda}_N(I_N) = \boldsymbol{\Lambda}_N^k, \quad \boldsymbol{\Lambda}_T(I_N) = \boldsymbol{\Lambda}_T^k$
  end
end

Table 6.3. Moreau's time-stepping method with an augmented Lagrangian approach.

problem as an NCP has a number of drawbacks. First of all, the numerical solution of the resulting type of NCPs is cumbersome. Secondly, the formulation of spatial anisotropic Coulomb fiction or more elaborate friction models is complicated and not suitable from a practical point of view. The augmented Lagrangian method has the advantage that complicated sets of admissible friction forces (e.g. C_T) can directly be incorporated without altering the method itself.

6.5 Simulation of the Stick-slip System

A single-degree-of-freedom system will be used to introduce and evaluate the numerical aspects of the methods described in this chapter. A mass-spring-damper system can, under the influence of dry friction with Stribeck effect, undergo stick-slip vibrations and is called a 'stick-slip system'. In Section 6.5.1, a stick-slip system with a Coulomb-Stribeck friction model (i.e. a Coulomb friction law with a velocity dependent friction coefficient) is studied. Section 6.5.2 studies a stick-slip system with stiction friction model. The stiction friction model exposes a similar phenomenon as the Stribeck effect due to the difference between the static friction coefficient and the dynamic friction coefficient. The same system treated in [42] and [145] is used to facilitate the comparison of results.

6.5.1 Coulomb–Stribeck Friction Model

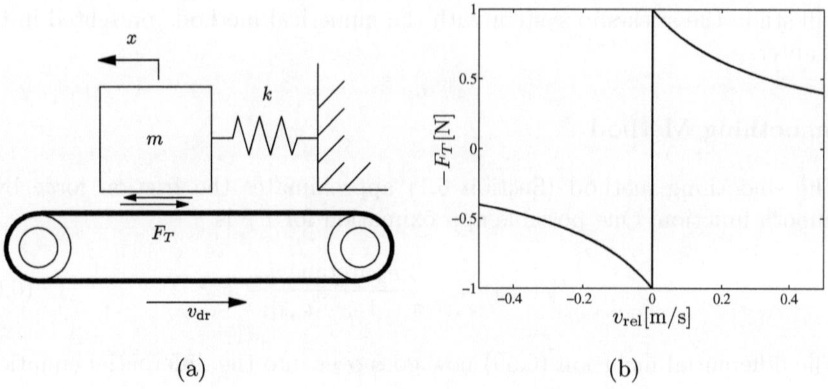

Fig. 6.4. 1–DOF system with dry friction.

Consider a mass m attached to inertial space by a spring k. The mass is riding on a belt, that is moving at constant velocity v_{dr} (Figure 6.4a). The relative velocity of the mass m with respect to the belt is denoted by $v_{rel} = \dot{x} - v_{dr}$. Dry friction occurs between the mass and the belt, with a friction force F_T. The friction force $F_T(v_{rel})$ is in the slip phase a function of the relative velocity v_{rel}. The friction model reads as

$$F_T(v_{rel}) = -\mu(v_{rel})F_N \operatorname{sign}(v_{rel}), \qquad \mu(v_{rel}) = \frac{\mu_s}{1+\delta|v_{rel}|}, \qquad v_{rel} \neq 0, \tag{6.58}$$

where $F_N = mg$. The friction force in the stick phase is limited by the maximum static friction force, i.e. $|F_T| \leq F_s = \mu_s F_N$. The friction curve is drawn in Figure 6.4b for $F_s = 1$ [N] and $\delta = 3$ [s/m]. The differential inclusion, describing this system, reads as

$$\dot{\boldsymbol{x}}(t) \in \boldsymbol{F}(\boldsymbol{x}(t)) = \begin{cases} \boldsymbol{f}_-(\boldsymbol{x}(t)), & \boldsymbol{x} \in \mathcal{V}_-, \\ \overline{\text{co}}\{\boldsymbol{f}_-(\boldsymbol{x}(t)), \boldsymbol{f}_+(\boldsymbol{x}(t))\}, & \boldsymbol{x} \in \Sigma, \\ \boldsymbol{f}_+(\boldsymbol{x}(t)), & \boldsymbol{x} \in \mathcal{V}_+, \end{cases} \tag{6.59}$$

where $\boldsymbol{x} = \begin{bmatrix} x \ \dot{x} \end{bmatrix}^{\mathrm{T}}$, $h(\boldsymbol{x}) = v_{\mathrm{rel}}$ and

$$\boldsymbol{f}_-(\boldsymbol{x}) = \begin{bmatrix} \dot{x} \\ -\dfrac{k}{m}x + \dfrac{1}{m}\dfrac{F_{\mathrm{s}}}{1+\delta|v_{\mathrm{rel}}|} \end{bmatrix}, \quad \boldsymbol{f}_+(\boldsymbol{x}) = \begin{bmatrix} \dot{x} \\ -\dfrac{k}{m}x - \dfrac{1}{m}\dfrac{F_{\mathrm{s}}}{1+\delta|v_{\mathrm{rel}}|} \end{bmatrix}, \tag{6.60}$$

$$\begin{aligned} \mathcal{V}_- &= \{\boldsymbol{x} \in \mathbb{R}^n \mid h(\boldsymbol{x}) < 0\}, \\ \Sigma &= \{\boldsymbol{x} \in \mathbb{R}^n \mid h(\boldsymbol{x}) = 0\}, \\ \mathcal{V}_+ &= \{\boldsymbol{x} \in \mathbb{R}^n \mid h(\boldsymbol{x}) > 0\}. \end{aligned} \tag{6.61}$$

The parameter values are $m = 1$ kg, $k = 1$ N/m, $v_{\mathrm{dr}} = 0.2$ m/s, $\mu_{\mathrm{s}} = 0.1$, $g = 10$ m/s^2, $F_{\mathrm{s}} = 1$ N and $\delta = 3$ [s/m]. In the remainder of this section we will study the stick-slip system with the numerical methods presented in this chapter.

Smoothing Method

The smoothing method (Section 6.1) approximates the friction force by a smooth function. One possible approximation for F_T is

$$\tilde{F}_T(v_{\mathrm{rel}}) = -\frac{2}{\pi}\frac{F_{\mathrm{s}} \arctan \varepsilon v_{\mathrm{rel}}}{1+\delta|v_{\mathrm{rel}}|}. \tag{6.62}$$

The differential inclusion (6.59) now goes over into the differential equation

$$\dot{\boldsymbol{x}} = \boldsymbol{f}(\boldsymbol{x}) = \begin{bmatrix} \dot{x} \\ -\dfrac{k}{m}x + \dfrac{\tilde{F}_T(v_{\mathrm{rel}})}{m} \end{bmatrix}. \tag{6.63}$$

The smoothened friction model (6.62) is identical to the friction model depicted in Figure 4.2a with $F_{\mathrm{d}} = 0$. Clearly, increasing the steepness parameter ε improves the approximation, especially for v_{rel} close to 0. The friction curve will almost be the same as in Figure 6.4b for large values of the steepness parameter ($\varepsilon = 10^6$). However, a steep slope in $\tilde{F}$ arises at $v_{\mathrm{rel}} = 0$, given by $-(2/\pi)\varepsilon F_{\mathrm{s}}$. This causes a stiff differential equation, which is numerically costly to integrate.

The stick-slip system contains an unstable equilibrium and a stable stick-slip limit cycle around the equilibrium. To prove that this system has only one stable periodic solution is beyond the scope of this section. Existence of at least one periodic solution can be proven by using a non-smooth version of the Poincaré-Bendixson theorem [37, 74] and has been observed before [3, 42, 74, 78, 86, 145]. In Figure 6.6a the numerical integration along the limit cycle is obtained with the smoothing method for $\varepsilon = 10^4$. A Runge-Kutta integrator (ode45 of MATLAB) was used for the numerical integration with a relative tolerance of 10^{-8} and an absolute tolerance of 10^{-8}. As can be seen, small time-steps are taken during the whole stick phase, resulting in 3921 integration points. A drift during the stick phase can be observed, which is due to the fact that the smoothing method can not properly describe stiction. A better result for this extremely stiff differential equation can be obtained using a Backward Differentiation Formula (the Runge-Kutta method has been used for comparison).

Switch Model

The Switch Model (see Section 6.2) approximates a differential inclusion by a number of sets of ordinary differential equations. The switching boundaries split the state space in smooth subspaces $\mathcal{V}_i$. Subsequently, subspaces are defined around each switching boundary: $\mathcal{U}$ around the attractive sliding mode part of the switching boundary, $\mathcal{Q}$ around the repulsive sliding mode and $\mathcal{T}_\pm$ for the transition modes around the transversal intersections on the switching boundary.

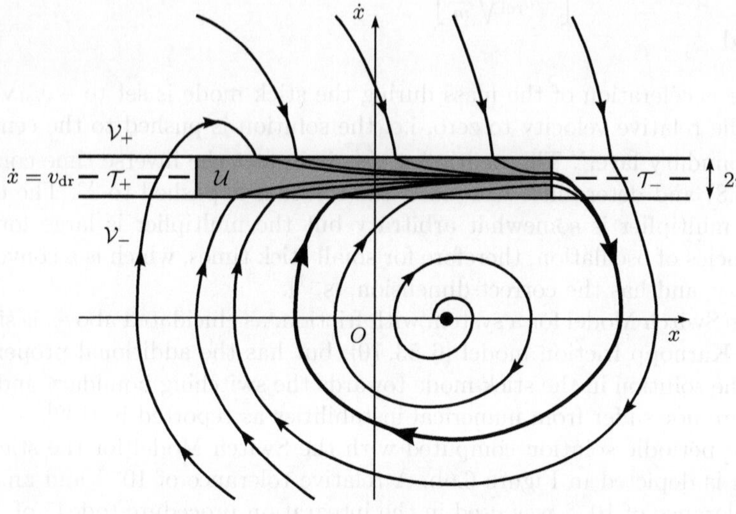

Fig. 6.5. Phase spaces of the stick-slip system.

The stick-slip system has a single switching boundary Σ defined by $h = v_{\mathrm{rel}} = 0$. The switching boundary Σ splits the state space $(x, \dot{x}) \in \mathbb{R}^2$ in two smooth subspaces: $\mathcal{V}_+$ for $v_{\mathrm{rel}} > 0$ and $\mathcal{V}_-$ for $v_{\mathrm{rel}} < 0$ defining the forward respectively backward slip mode. The system contains an attractive sliding mode, which coincides with the stick phase $\mathcal{U}$. A repulsive sliding mode is not present. The stick mode $\mathcal{U}$ is defined by

$$\boldsymbol{n}^{\mathrm{T}} \boldsymbol{f}_+ < 0 \wedge \boldsymbol{n}^{\mathrm{T}} \boldsymbol{f}_- > 0, \tag{6.64}$$

from which follows with $\boldsymbol{n}^{\mathrm{T}} = [0\ 1]$

$$-kx - F_s < 0 \wedge -kx + F_s > 0 \Longrightarrow |kx| < F_s. \tag{6.65}$$

The subspaces $\mathcal{T}_+$ and $\mathcal{T}_-$, on either side of $\mathcal{U}$, are the transition phases from forward slip ($v_{\mathrm{rel}} > 0$) to backward slip ($v_{\mathrm{rel}} < 0$) and vice versa. The Switch Model can be directly implemented with the pseudo-code given in Section 6.2. However, the pseudo-code can for this problem be simplified to:

if $|v_{\mathrm{rel}}| > \eta$ **then**

$$\dot{\boldsymbol{x}} = \boldsymbol{f}_{\mathcal{V}\pm}(\boldsymbol{x}) = \begin{bmatrix} \dot{x} \\ -\frac{k}{m}x + \frac{F(v_{\mathrm{rel}})}{m} \end{bmatrix} \quad \text{slip, } \boldsymbol{x} \in \mathcal{V}_+ \vee \boldsymbol{x} \in \mathcal{V}_-$$

elseif $|kx| > F_s$

$$\dot{\boldsymbol{x}} = \boldsymbol{f}_{\mathcal{T}\pm}(\boldsymbol{x}) = \begin{bmatrix} \dot{x} \\ -\frac{k}{m}x + \frac{F_s}{m}\,\mathrm{sign}\,kx \end{bmatrix} \quad \text{transition, } \boldsymbol{x} \in \mathcal{T}_+ \vee \boldsymbol{x} \in \mathcal{T}_-$$

else

$$\dot{\boldsymbol{x}} = \boldsymbol{f}_{\mathcal{U}}(\boldsymbol{x}) = \begin{bmatrix} v_{\mathrm{dr}} \\ -v_{\mathrm{rel}}\sqrt{\frac{k}{m}} \end{bmatrix} \quad \text{stick, } \boldsymbol{x} \in \mathcal{U}$$

end

The acceleration of the mass during the stick mode is set to $-v_{\mathrm{rel}}\sqrt{\frac{k}{m}}$ to force the relative velocity to zero, i.e. the solution is pushed to the centre of the 'boundary layer'. The multiplier $\sqrt{\frac{k}{m}}$ is in fact the inverse time constant τ^{-1} (6.8) and determines how 'fast' the solution is pushed to Σ. The choice of the multiplier is somewhat arbitrary but the multiplier is large for high frequencies of oscillation, therefore for small stick times, which is a convenient property, and has the correct dimension, $[\mathrm{s}^{-1}]$.

The Switch Model for a system with friction, as elucidated above, is similar to the Karnopp friction model [6, 53, 70], but has the additional property to push the solution in the stick mode towards the switching boundary and does therefore not suffer from numerical instabilities as reported in [132].

The periodic solution computed with the Switch Model for the stick-slip system is depicted in Figure 6.6b. A relative tolerance of 10^{-8} and an absolute tolerance of 10^{-8} was used in the integration procedure (ode45 of MATLAB). Small time-steps are taken only near the transitions between slip and stick phase, resulting in 341 integration points. In fact, the adaptive time-step control of the integration algorithm determines the switching point with the desired accuracy. The constant η was taken to be 10^{-6}. The Switch Model was introduced in Section 6.2 as an efficient numerical method for integrating a system with dry friction or Filippov systems in general. The importance of the Switch Model for theoretical considerations will become evident in Sections 7.3. The approximation of the stick phase by a finite narrow band, used by the Switch Model, will turn out to be essential to understand bifurcations occurring in systems with dry friction (Sections 9.8.2 and 9.9.1). The use of the Switch Model is however limited to systems with a small number of switching boundaries as the logical structure becomes increasingly complex

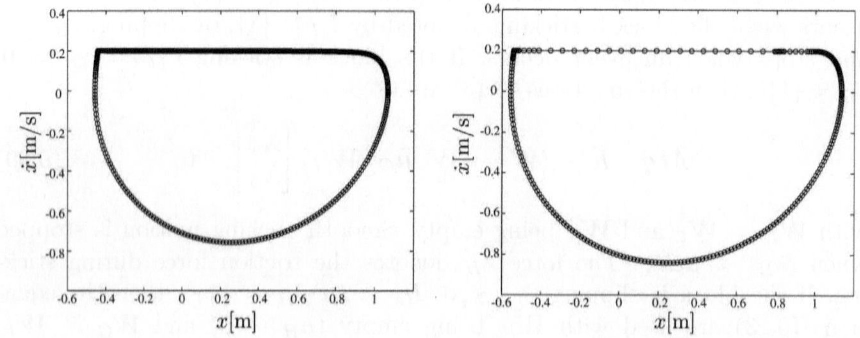

(a) Smoothing method (b) Switch Model

Fig. 6.6. Periodic solutions obtained with the smoothing method and with the Switch Model.

for multiple switching boundaries. The event-driven integration method and the time-stepping method are useful for more complex systems, but will also be demonstrated on the relatively simple stick-slip system.

Event-driven Integration Method

Normal and tangential contact are treated simultaneously by the event-driven integration method, as presented in Section 6.3. We therefore model the system with normal as well as tangential degrees of freedom, which allows us to directly use the results obtained in Section 6.3. The vector of generalized coordinates becomes $\boldsymbol{q} = \begin{bmatrix} x \ y \end{bmatrix}^{\mathrm{T}}$, where y is the vertical displacement of the block relative to the belt. The block is in contact with the belt if $y = 0$. The normal and tangential contact force are denoted by λ_N and λ_T. The event-driven integration method describes the dynamics of the system with the equation of motion (5.3)

$$\boldsymbol{M}\ddot{\boldsymbol{q}} = \boldsymbol{h} + \boldsymbol{W}_N \lambda_N + \boldsymbol{W}_T \lambda_T. \tag{6.66}$$

The system matrices for the stick-slip system become

$$\begin{gathered} \boldsymbol{M} = \begin{bmatrix} m & 0 \\ 0 & m \end{bmatrix}, \quad \boldsymbol{h} = \begin{bmatrix} -kx \\ -mg \end{bmatrix}, \quad \boldsymbol{W}_N = \begin{bmatrix} 0 \\ 1 \end{bmatrix}, \quad \boldsymbol{W}_T = \begin{bmatrix} 1 \\ 0 \end{bmatrix}, \\ \tilde{w}_N = \bar{w}_N = 0, \quad \tilde{w}_T = -v_{\mathrm{dr}}, \quad \bar{w}_T = 0, \end{gathered} \tag{6.67}$$

with the contact equations

$$g_N = y, \qquad \gamma_T = \dot{x} - v_{\mathrm{dr}}. \tag{6.68}$$

The contact configuration of the system is expressed by index sets (6.11). The stick-slip system has only a single contact point, $I_C = \{1\}$. The block is assumed to be always in contact with the belt, $I_N = \{1\}$. Smooth motion occurs when the block is sticking, denoted by $I_H = \{1\}$, or slipping, $I_H = \emptyset$, and stops when an event occurs. If the block is sticking ($\gamma_T = \dot{\gamma}_T = 0$, $I_H = \{1\}$), then the equations (6.13) are used

$$\boldsymbol{M}\ddot{\boldsymbol{q}} - \boldsymbol{h} - \left[(\boldsymbol{W}_N + \boldsymbol{W}_G\bar{\boldsymbol{\mu}}_G)\ \boldsymbol{W}_H\right] \begin{bmatrix} \boldsymbol{\lambda}_N \\ \boldsymbol{\lambda}_H \end{bmatrix} = \boldsymbol{0}, \tag{6.69}$$

with $\boldsymbol{W}_H = \boldsymbol{W}_T$ and $\boldsymbol{W}_G$ being empty. Smooth sticking motion is stopped when $|\lambda_H| \not< \mu_s \lambda_N$. The force λ_H denotes the friction force during sticking. If the block is slipping ($\gamma_T \neq 0$, $I_G = I_N \backslash I_H = \{1\}$), then the equations (6.13) are used with $\boldsymbol{W}_H$ being empty ($n_H = 0$) and $\boldsymbol{W}_G = \boldsymbol{W}_T$, $\bar{\boldsymbol{\mu}}_G = -\mu(v_{\mathrm{rel}})\,\mathrm{sign}(v_{\mathrm{rel}})$. Smooth slipping motion is stopped when $\gamma_T = v_{\mathrm{rel}}$ changes sign.

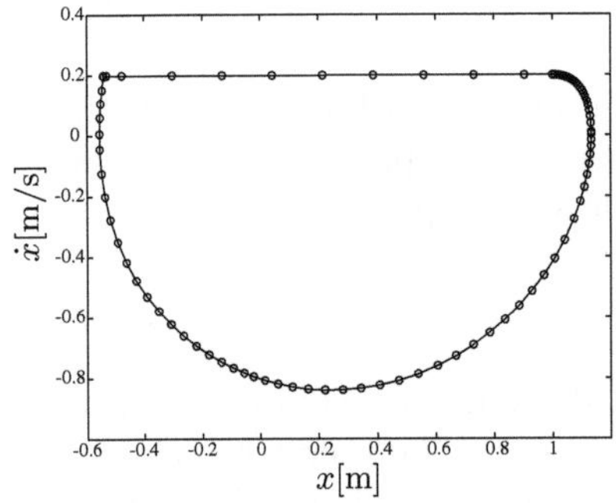

(a) Event-driven integration

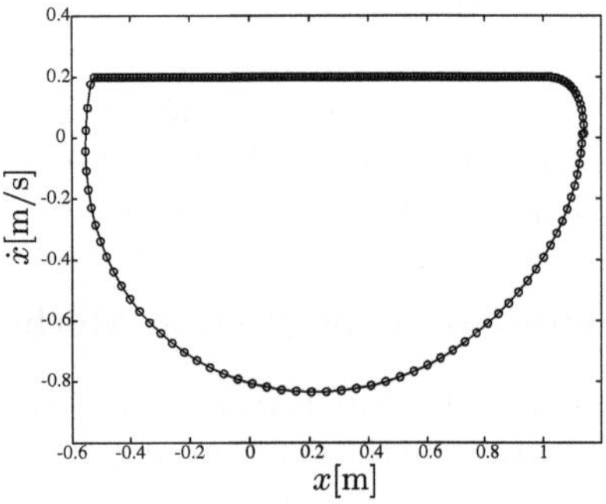

(b) Time-stepping

Fig. 6.7. Periodic solutions obtained with the LCP-based integration methods.

The new contact configuration has to be found after the occurrence of an event. The index set is adjusted to $I_H = \{1\}$ and the LCP (6.32) is solved after which the new contact configuration can be determined. The LCP (6.32) evaluated for the stick-slip system yields the set of equations

$$\underbrace{\begin{bmatrix} \dot{\gamma}_N \\ \dot{\gamma}_L \\ \lambda_R \end{bmatrix}}_{\boldsymbol{y}} = \underbrace{\begin{bmatrix} \frac{1}{m} & 0 & 0 \\ -\frac{\mu_s}{m} & \frac{1}{m} & 1 \\ 2\mu_s & -1 & 0 \end{bmatrix}}_{\boldsymbol{A}} \underbrace{\begin{bmatrix} \lambda_N \\ \lambda_L \\ \dot{\gamma}_R \end{bmatrix}}_{\boldsymbol{x}} + \underbrace{\begin{bmatrix} -g \\ \frac{k}{m}x \\ 0 \end{bmatrix}}_{\boldsymbol{b}}, \tag{6.70}$$

with the complementarity conditions $\boldsymbol{0} \leq \boldsymbol{x} \perp \boldsymbol{y} \geq \boldsymbol{0}$. It follows from the first equation in (6.70) that $\dot{\gamma}_N = 0$ and $\lambda_N = mg$. Consequently, the block remains in contact with the belt. We therefore arrive at the two-dimensional linear complementarity problem

$$\begin{bmatrix}\dot{\gamma}_L\\ \lambda_R\end{bmatrix} = \begin{bmatrix}\frac{1}{m} & 1\\ -1 & 0\end{bmatrix}\begin{bmatrix}\lambda_L\\ \dot{\gamma}_R\end{bmatrix} + \begin{bmatrix}\frac{k}{m}x - \mu_s g\\ 2\mu_s mg\end{bmatrix}, \qquad \mathbf{0} \le \begin{bmatrix}\dot{\gamma}_L\\ \lambda_R\end{bmatrix} \perp \begin{bmatrix}\lambda_L\\ \dot{\gamma}_R\end{bmatrix} \ge \mathbf{0}. \tag{6.71}$$

The two-dimensional LCP (6.71) has $2^2 = 4$ possible solutions:

1. $\dot{\gamma}_L > 0 \wedge \lambda_R > 0 \wedge \dot{\gamma}_R = 0 \wedge \lambda_L = 0 \Longrightarrow \dot{\gamma}_L = \frac{k}{m}x - \mu_s g \wedge \lambda_R = 2\mu_s mg$. It follows that the block will slip in backward direction if $kx > \mu_s mg$.
2. $\dot{\gamma}_R > 0 \wedge \lambda_L > 0 \wedge \dot{\gamma}_L = 0 \wedge \lambda_R = 0 \Longrightarrow \dot{\gamma}_R = -\frac{k}{m}x - \mu_s g \wedge \lambda_L = 2\mu_s mg$. It follows that the block will slip in forward direction if $kx < -\mu_s mg$.
3. $\lambda_R > 0 \wedge \lambda_L > 0 \wedge \dot{\gamma}_R = 0 \wedge \dot{\gamma}_L = 0 \Longrightarrow \lambda_R = kx - \mu_s mg \wedge \lambda_L = -kx + \mu_s mg$. It follows that the block will stick if $-\mu_s mg < kx < \mu_s mg$.
4. $\dot{\gamma}_R > 0 \wedge \dot{\gamma}_L > 0 \wedge \lambda_R = 0 \wedge \lambda_L = 0 \Longrightarrow 2\mu_s mg = 0$. This solution is not possible for $\mu_s > 0$. Physically, this solution is impossible because the block can not slip simultaneously in forward and backward direction.

The solutions of the transition LCP show that the border of the stick mode is given by $x = \pm\mu_s \frac{mg}{k}$, which yields in our numerical example $x = \pm 1$.

The numerical results for the calculation of the periodic solution of the stick-slip system with the event-driven integration method are depicted in Figure 6.7a. An ODE integration method (with event-detection) was used for the integration during the smooth phases of motion (ode45 with a relative tolerance of 10^{-8} and absolute tolerance of 10^{-13}). The event-driven integration method took 95 integration points along the periodic solution. The Switch Model, as used in the example, does not use event-detection and therefore uses more integration points than the event-driven integration method.

Time-stepping Method

The time-stepping method (Section 6.4) does also treat normal and tangential contact simultaneously. We can therefore use the same system matrices (6.67) as used for the event-driven integration method. The time-stepping method solves in each time-step the LCP (6.52). Let $x_A = x^l$, $y_A = y^l$, $\dot{x}_A = \dot{x}^l$ and $\dot{y}_A = \dot{y}^l$ be given at time-step l. The mid-point step yields $x_M = x_A + \dot{x}_A \Delta t$ and $y_M = y_A + \dot{y}_A \Delta t$. It holds that $y_M = 0$ because the block is assumed to be in contact ($y_A = \dot{y}_A = 0$). The contact LCP of the stick-slip system becomes

$$\underbrace{\begin{bmatrix}\gamma_{NE}\\ \gamma_{LE}\\ \Lambda_R\end{bmatrix}}_{\boldsymbol{y}} = \underbrace{\begin{bmatrix}\frac{1}{m} & 0 & 0\\ -\frac{\mu_s}{m} & \frac{1}{m} & 1\\ 2\mu_s & -1 & 0\end{bmatrix}}_{\boldsymbol{A}}\underbrace{\begin{bmatrix}\Lambda_N\\ \Lambda_L\\ \gamma_{RE}\end{bmatrix}}_{\boldsymbol{x}} + \underbrace{\begin{bmatrix}\dot{y}_A - g\Delta t\\ -\dot{x}_A + \frac{k}{m}x_M \Delta t\\ 0\end{bmatrix}}_{\boldsymbol{b}}, \tag{6.72}$$

with $\mathbf{0} \le \boldsymbol{x} \perp \boldsymbol{y} \ge \mathbf{0}$. It follows that if $\dot{y}_A - g\Delta t < 0$ in time-step l, then the LCP will give the solution $\gamma_{NE} = 0 \wedge \Lambda_N = mg\Delta t$ for the next time-step. The block will therefore always remain in contact with the belt if we initialize the stick-slip system with $y = \dot{y} = 0$. The subproblem of the LCP in tangential direction is similar to the two-dimensional LCP (6.71) and determines whether the block will stick or slip during the next time-step.

The numerical results for the calculation of the periodic solution of the stick-slip system with the time-stepping method are depicted in Figure 6.7b. An amount of 200 time-steps were sufficient to obtain an accuracy comparable to the event-driven integration method.

6.5.2 Stiction Friction Model

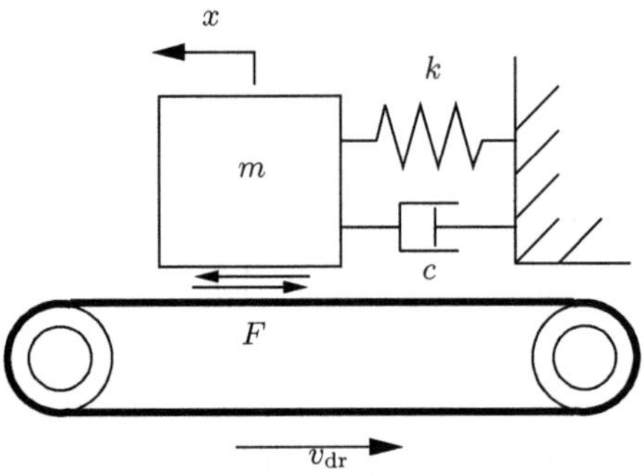

Fig. 6.8. 1–DOF system with dry friction.

In this section we study a similar stick-slip system as studied in the previous section, but with a friction model with a static friction coefficient which is higher than the dynamic friction coefficient (see Figure 4.2c). A system with the stiction friction model can not be regarded as a differential inclusion as explained in Section 4.2. This system can therefore not be numerically analyzed with the event-driven integration method or time-stepping method, as presented in Sections 6.3 and 6.4[1]. However, the Switch Model is able to approximate a system with stiction friction model with a differential inclusion and can therefore still be applied. An approximation with the smoothing method is also possible but will result in two steep parts in the friction graph. We will only show numerical results for this example using the Switch Model.

The model of the stick-slip system is shown in Figure 6.8 and differs slightly from the system in the previous section as it also contains a viscous damper with constant c. The parameter values are $k = 1$ N/m, $c = 0.1$ Ns/m, $m = 1$ kg, $v_{dr} = 0.2$ m/s, $\mu_s = 0.2$, $\mu_d = 0.1$, $g = 10$ m/s^2, $F_s = 2$ N and $F_d = 1$ N.

[1] The event-driven and time-stepping integration methods as presented in this book are only applicable to friction models that can be derived from a potential or dissipation function. The resulting differential inclusion has a minimal, upper-semi continuous, closed, convex (and bounded) image. An attempt to apply stiction friction models, for which the image is not a minimal set, to a time-stepping integration method is reported in [98].

The pseudo-code can for the stick-slip system with stiction friction model be expressed as:

if $|v_{\mathrm{rel}}| > \eta$ **then**

$$\dot{\boldsymbol{x}} = \boldsymbol{f}_{\mathcal{V}\pm}(\boldsymbol{x}) = \begin{bmatrix} \dot{x} \\ -\frac{k}{m}x - \frac{c}{m}\dot{x} - \frac{F_{\mathrm{d}}\,\mathrm{sign}(v_{\mathrm{rel}})}{m} \end{bmatrix} \qquad \text{slip, } \boldsymbol{x} \in \mathcal{V}_+ \vee \boldsymbol{x} \in \mathcal{V}_-$$

elseif $|kx + c\dot{x}| > F_{\mathrm{s}}$

$$\dot{\boldsymbol{x}} = \boldsymbol{f}_{\mathcal{T}\pm}(\boldsymbol{x}) = \begin{bmatrix} \dot{x} \\ -\frac{k}{m}x - \frac{c}{m}\dot{x} + \frac{F_{\mathrm{s}}}{m}\,\mathrm{sign}(kx + c\dot{x}) \end{bmatrix} \qquad \text{transition, } \boldsymbol{x} \in \mathcal{T}_+ \vee \boldsymbol{x} \in \mathcal{T}_-$$

else

$$\dot{\boldsymbol{x}} = \boldsymbol{f}_{\mathcal{U}}(\boldsymbol{x}) = \begin{bmatrix} v_{\mathrm{dr}} \\ -v_{\mathrm{rel}}\sqrt{\frac{k}{m}} \end{bmatrix} \qquad \text{stick, } \boldsymbol{x} \in \mathcal{U}$$

end;

Note how the presence of a viscous damper and static friction alter the pseudo-code for the Switch Model.

The equilibrium ($x = F_{\mathrm{d}}/k, \dot{x} = 0$) is stable due to the presence of the viscous damper in the system. A coexisting stable and unstable limit cycle exist. The stable limit cycle, computed with the Switch Model, is depicted in Figure 6.9. A relative tolerance of 10^{-8} and an absolute tolerance of 10^{-8} was used in the integration procedure (ode45, MATLAB). The constant η was taken to be 10^{-6} and 389 integration points were taken by the integration procedure. The existence of an unstable limit cycle, located between the equilibrium and the stable limit cycle, will be studied in Subsection 9.9.1.

The transition from stick to slip in Figure 6.9 is non-smooth due to the jump of the friction curve from F_{s} to F_{d}. Figure 6.6b shows a smooth transition from stick to slip because the friction force, depicted in Figure 6.4b, remains continuous during this transition.

The stick-slip system with stiction friction model will be studied again in Sections 7.3 and 9.9.1.

6.6 Summary

In this chapter, we have presented different methods to solve differential inclusions and paid special attention to mechanical systems with friction. The numerical methods have been applied to a stick-slip system in Section 6.5.1 and 6.5.2. The smoothing method is still used for its simplicity but has a number of drawbacks. Stiction can not be properly described (a block on a slope will never stand still) and the computation time can become excessively large. The Switch Model partly solves these problems, by switching between a set of non-stiff differential equations and the ability to describe stiction. It has been shown in Section 6.5.2 that the Switch Model is also capable to approximate

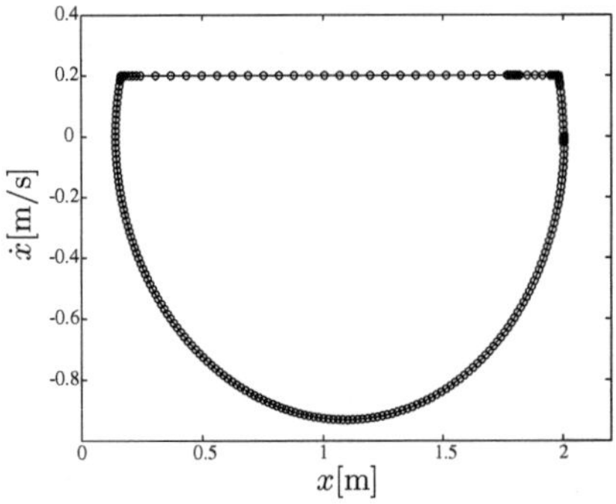

Fig. 6.9. Stick-slip limit cycle of system with stiction friction model, calculated with the Switch Model.

a system with a stiction type of friction model. However, the logical structure of the Switch Model becomes too complex when the system has intersecting switching boundaries, which occurs for mechanical systems with multiple frictional contacts. Event-driven integration methods and time-stepping methods are able to deal with many contacts in a rigorous way. Event-driven integration methods formulate the contact problem on acceleration level and use standard integration methods for the smooth phases of motion. Time-stepping methods formulate the contact problem on velocity level and integrate the system with fixed time-steps. Time-stepping methods can become advantageous over event-driven methods when many contact points are involved or when the number of events is excessively large. The event-driven integration method and the time-stepping method both have to solve for the contact problem. A standard approach is to transform the contact problem to an LCP or NCP. Such a transformation is not always possible and might be very cumbersome. The augmented Lagrangian method gives a more general approach to solve the contact problem of complicated set-valued force laws.

7

Fundamental Solution Matrix

Trajectories of a dynamical system, starting from a particular initial state, might evolve towards a steady state of the system. A steady state can be an equilibrium of the system but can also be a (quasi-)periodic motion. The stability of equilibria is (for the hyperbolic case) determined by the eigenvalues of the local linearization of the system. The stability of periodic solutions, however, is determined by so-called *Floquet multipliers*, which are eigenvalues of the *fundamental solution matrix*. The fundamental solution matrix is, in fact, a local linearization of the system along the periodic solution. In this chapter, the discontinuous behaviour of fundamental solution matrices of Filippov systems is discussed. The definition of a fundamental solution matrix for smooth systems is extensively reviewed in Section 7.1. The differences in the properties of the fundamental solution matrix between linear and nonlinear systems, as well as between autonomous and non-autonomous systems are discussed. Subsequently, the stability of periodic solutions is addressed by means of the concept of Floquet multipliers. The special behaviour of fundamental solution matrices of Filippov systems is treated in Section 7.2. The theory is applied on a system with stick-slip motion in Section 7.3 and on a system with a discontinuous support in Section 7.4.

7.1 Fundamental Solution Matrix of Smooth Systems

We first review some periodicity properties of autonomous and non-autonomous smooth systems. Stability of periodic solutions of smooth systems will be addressed in Subsection 7.1.3.

7.1.1 Nonlinear Autonomous Systems

An autonomous system does not explicitly depend on time. Consider an nth-order, autonomous, nonlinear dynamical system represented by the differential equation

$$\dot{\boldsymbol{x}}(t) = \boldsymbol{f}(\boldsymbol{x}(t)), \tag{7.1}$$

with the initial condition

$$\boldsymbol{x}(t_0) = \boldsymbol{x}_0. \tag{7.2}$$

Let $\varphi_{t_d}(\boldsymbol{x}_0) := \varphi(t_0 + t_d, t_0, \boldsymbol{x}_0)$ denote the solution of an autonomous system after a time lapse t_d starting from the initial condition $\boldsymbol{x}_0$. With the notation $\varphi(t_0 + t_d, t_0, \boldsymbol{x}_0)$ we stress the fact that it is the solution of the initial value problem (7.1) from a certain specific initial condition in space and time. However, we will in the sequel sometimes be less strict and simply write $\boldsymbol{x}(t)$ to denote a solution $\varphi(t, t_0, \boldsymbol{x}_0)$ in order to simplify the notation.

A solution of an autonomous system can be shifted over the time axis, because the vector field is not explicitly dependent on time, i.e.

$$\varphi(t + \Delta t, t_0 + \Delta t, \boldsymbol{x}_0) = \varphi(t, t_0, \boldsymbol{x}_0), \forall \Delta t. \tag{7.3}$$

Only the time lapse $t_d = t + \Delta t - (t_0 + \Delta t) = t - t_0$ is of interest for a solution of an autonomous system, not the initial time t_0.

A periodic solution of an autonomous system is denoted by $\varphi^p(t_0 + t_d, t_0, \boldsymbol{x}_0)$, which has the property

$$\varphi^p(t_0 + \alpha T, t_0, \boldsymbol{x}_0) = \boldsymbol{x}_0, \quad \alpha = 0, 1, 2, ... \tag{7.4}$$

where $T > 0$ is the minimal period time of the periodic solution.

In this section we will assume $\boldsymbol{f}(\boldsymbol{x})$ to be Lipschitz continuous and to be continuously differentiable with respect to $\boldsymbol{x}$. The initial value problem (7.1) has therefore locally existence and uniqueness of solutions in forward and backward time.

7.1.2 Nonlinear Non-autonomous Systems

We consider an nth-order, non-autonomous, nonlinear dynamical system

$$\dot{\boldsymbol{x}}(t) = \boldsymbol{f}(t, \boldsymbol{x}(t)), \tag{7.5}$$

with the initial condition

$$\boldsymbol{x}(t_0) = \boldsymbol{x}_0, \tag{7.6}$$

for which the right-hand side $\boldsymbol{f}(t, \boldsymbol{x}(t))$ depends explicitly on time. Note that we can not translate the solution over the time-axis (like we did for autonomous systems) because $\boldsymbol{f}$ is explicitly dependent on t. Let $\varphi_{t_d}(t_0, \boldsymbol{x}_0) := \varphi(t_0 + t_d, t_0, \boldsymbol{x}_0)$ denote the solution of a non-autonomous system after a time lapse t_d starting from the initial condition $\boldsymbol{x}_0$ at $t = t_0$.

We define a time-periodic non-autonomous system by (7.5) with the additional property

$$\boldsymbol{f}(t, \boldsymbol{y}) = \boldsymbol{f}(t + T, \boldsymbol{y}), \qquad \forall \boldsymbol{y} \in \mathbb{R}^n, \ \forall t, \tag{7.7}$$

where $T > 0$ is the minimal period time of the system.

A periodic solution of a non-autonomous system is denoted by $\varphi^p(t_0 + t_d, t_0, \boldsymbol{x}_0)$, which has the property

$$\varphi(t_0 + \alpha T, t_0, \boldsymbol{x}_0) = \boldsymbol{x}_0, \quad \alpha = 0, 1, 2, ... \tag{7.8}$$

where T is the minimal period time of the periodic solution. If a non-autonomous system has a periodic solution then it must hold that $\dot{\boldsymbol{x}}(t+T) = \dot{\boldsymbol{x}}(t)$ from which property (7.7) follows. Consequently, if a non-autonomous system has a periodic solution of period T, then it follows that the system is time-periodic of the same period T (or a fraction thereof, e.g. $\frac{1}{2}T$, $\frac{1}{3}T$ etc.).

With respect to periodic solutions of non-autonomous systems we have to make the following important remark: when specifying a point on a periodic solution, we do not only need to specify the state, say $\boldsymbol{x}_a$, but also at which time it is on the periodic solution, e.g. at $t = t_a$. An initial condition of a non-autonomous system must specify the state of the system together with a time, e.g. $(t_0, \boldsymbol{x}_0)$. We therefore have in general for a non-autonomous system that

$$\varphi(t + \Delta t, t_0 + \Delta t, \boldsymbol{x}_0) \neq \varphi(t, t_0, \boldsymbol{x}_0). \tag{7.9}$$

which turns into the equality (7.8) for a time-periodic system with property (7.7) when $\Delta t = T$.

7.1.3 Periodic Solutions and their Stability

In this subsection we will discuss the stability of periodic solutions of autonomous as well as non-autonomous systems. We will consider the more general form (7.5) with periodicity property (7.7) for our discussion and we have to bear in mind that we do not have the explicit time dependence for autonomous systems, which are of the form (7.1). Consider a nonlinear system of the form (7.5) and a trajectory $\boldsymbol{x}^p(t) = \varphi^p(t, t_0, \boldsymbol{x}_0)$, which is a periodic solution of the system. We linearize (7.5) around $\boldsymbol{x}^p(t)$ in order to access the stability properties of the periodic solution. A perturbed solution $\boldsymbol{x}(t) = \boldsymbol{x}^p(t) + \Delta\boldsymbol{x}(t)$ of (7.5) is considered. The perturbed solution must satisfy (7.5):

$$\begin{aligned} \dot{\boldsymbol{x}}^p(t) + \Delta\dot{\boldsymbol{x}}(t) &= \boldsymbol{f}(t, \boldsymbol{x}^p(t) + \Delta\boldsymbol{x}(t)) \\ &= \boldsymbol{f}(t, \boldsymbol{x}^p(t)) + \left.\frac{\partial \boldsymbol{f}(t, \boldsymbol{x})}{\partial \boldsymbol{x}}\right|_{\boldsymbol{x}^p(t)} \Delta\boldsymbol{x}(t) + H.O.T., \end{aligned} \tag{7.10}$$

where $H.O.T.$ denotes the higher-order terms in $\Delta\boldsymbol{x}$ and $\Delta\dot{\boldsymbol{x}} = d(\Delta\boldsymbol{x})/dt$. Using $\dot{\boldsymbol{x}}^p(t) = \boldsymbol{f}(t, \boldsymbol{x}^p(t))$ in (7.10), the following differential equation for the perturbation $\Delta\boldsymbol{x}$ with respect to the periodic solution holds:

$$\Delta\dot{\boldsymbol{x}} = \left.\frac{\partial \boldsymbol{f}(t, \boldsymbol{x})}{\partial \boldsymbol{x}}\right|_{\boldsymbol{x}^p(t)} \Delta\boldsymbol{x} + H.O.T. \tag{7.11}$$

Since $\boldsymbol{x}^p(t)$ is time-dependent (and periodic), $\frac{\partial \boldsymbol{f}(t,\boldsymbol{x})}{\partial \boldsymbol{x}}|_{\boldsymbol{x}^p(t)}$ is a time-dependent (and periodic) Jacobian matrix, which we will denote by $\boldsymbol{A}(t, \boldsymbol{x}^p(t))$:

$$\Delta\dot{\boldsymbol{x}}(t) = \boldsymbol{A}(t, \boldsymbol{x}^p(t))\Delta\boldsymbol{x}(t) + H.O.T. \tag{7.12}$$

If the system is autonomous, then the Jacobian at $\boldsymbol{x}$ is not explicitly dependent on time, i.e. $\boldsymbol{A}(\boldsymbol{x})$. However, if we consider a periodic solution $\boldsymbol{x}^p(t)$, then the Jacobian along this periodic solution is implicity dependent on time, i.e. $\boldsymbol{A}(\boldsymbol{x}^p(t))$.

We will now study the specific properties of fundamental solution matrices of Linear Time-Invariant, Linear Time-Variant and Linear Periodic Time-Variant Systems. This will provide a good stepping stone for studying the stability of periodic motion of non-autonomous and autonomous nonlinear systems.

Linear Time-Invariant Systems

First we consider a linear time-invariant (LTI) system given by

$$\dot{\boldsymbol{x}}(t) = \boldsymbol{A}\boldsymbol{x}(t) + \boldsymbol{b}. \tag{7.13}$$

Note that $\boldsymbol{A}$ is a constant matrix for an LTI system. It follows from (7.13) that the perturbation equation (7.12) becomes

$$\Delta\dot{\boldsymbol{x}}(t) = \boldsymbol{A}\Delta\boldsymbol{x}(t), \tag{7.14}$$

without higher-order terms. The perturbation equation (7.14) is also linear time-invariant and always admits a set of n independent fundamental solutions $\boldsymbol{\phi}_1(t), \dots, \boldsymbol{\phi}_n(t)$. In particular, if $\boldsymbol{A}$ has n distinct eigenvalues one can write:

$$\boldsymbol{\phi}_i(t) = c_{1_i}\boldsymbol{u}_1 e^{\lambda_1 t} + \ldots + c_{n_i}\boldsymbol{u}_n e^{\lambda_n t}, \tag{7.15}$$

where $\boldsymbol{u}_1, \dots, \boldsymbol{u}_n$ and $\lambda_1, \dots, \lambda_n$ are the eigenvectors and eigenvalues of $\boldsymbol{A}$, respectively. We gather the fundamental solutions $\boldsymbol{\phi}_i(t)$, $i = 1, \dots, n$ in the matrix $\boldsymbol{\Phi}(t) = \left[\boldsymbol{\phi}_1(t) \ \dots \ \boldsymbol{\phi}_n(t)\right]$, which is called the *fundamental solution matrix* (or state transition matrix). Now, the constants $c_{1_i}, \dots, c_{n_i}$, $i = 1, \dots, n$, are chosen such that $\boldsymbol{\phi}_i(0) = \boldsymbol{e}_i$, where $\boldsymbol{e}_i$ is a unit column vector with zeros at every row except at the i-th row where the entry equals unity (for example $e_2 = \left[0\ 1\ 0 \dots 0\right]^T$).

Without going into detail, also if the eigenvalues of $\boldsymbol{A}$ are not all distinct, the equation (7.14) admits n independent solutions, but in addition to (7.15) also solutions of the form $te^{\lambda_i t}$ may appear if λ_i is an eigenvalue with algebraic multiplicity larger than 1 (and which has a geometric multiplicity 1).

Since the columns of $\boldsymbol{\Phi}$ are solutions of the differential equation (7.14), the following relation holds:

$$\dot{\boldsymbol{\Phi}}(t) = \boldsymbol{A\Phi}(t), \quad \text{with} \quad \boldsymbol{\Phi}(0) = \boldsymbol{I}, \tag{7.16}$$

which has the solution

$$\boldsymbol{\Phi}(t) = e^{\boldsymbol{A}t}\boldsymbol{\Phi}(0) = e^{\boldsymbol{A}t}. \tag{7.17}$$

Since, firstly, $\boldsymbol{\Phi}(t)$ contains the fundamental solutions of (7.14) for specific initial conditions $\boldsymbol{x}_i(0) = \boldsymbol{e}_i$ and, secondly, (7.14) is linear and thus the superposition principle with respect to initial conditions holds, we can write

$$\Delta\boldsymbol{x}(t) = \boldsymbol{\Phi}(t)\Delta\boldsymbol{x}(0), \tag{7.18}$$

for arbitrary $\Delta\boldsymbol{x}(0)$. Note that for $\Delta\boldsymbol{x}(0) = \boldsymbol{e}_i$, $i = 1, \ldots, n$, $\Delta\boldsymbol{x}_i(t) = \boldsymbol{\Phi}(t)\boldsymbol{e}_i = \boldsymbol{\phi}_i(t)$ is satisfied by the equation. So, the fundamental solution matrix takes an initial state (perturbation) $\Delta\boldsymbol{x}(0)$ to the state (perturbation) $\Delta\boldsymbol{x}(t)$ at time t.

Linear Time-Variant Systems

We now consider a linear time-variant (LTV) system

$$\dot{\boldsymbol{x}}(t) = \boldsymbol{A}(t)\boldsymbol{x}(t) + \boldsymbol{b}(t), \tag{7.19}$$

with $\boldsymbol{A}(t)$ time-dependent but not necessarily periodic. The perturbation equation (7.12) becomes

$$\Delta\dot{\boldsymbol{x}}(t) = \boldsymbol{A}(t)\Delta\boldsymbol{x}(t), \tag{7.20}$$

without the presence of higher-order terms. We can write the related fundamental solution matrix as

$$\boldsymbol{\Phi}(t, t_0) = \left[\boldsymbol{\phi}_1(t, t_0) \ldots \boldsymbol{\phi}_n(t, t_0)\right], \tag{7.21}$$

with $\boldsymbol{\Phi}(t_0, t_0) = \boldsymbol{I}$. Note that the fundamental solutions (and thus the fundamental solution matrix $\boldsymbol{\Phi}(t, t_0)$) depend not only on time but also on the initial time instant t_0 due to the time-variant nature of $\boldsymbol{A}(t)$. So, $\boldsymbol{\phi}_i(t, t_0)$ are solutions of (7.20) with $\boldsymbol{\phi}_i(t_0, t_0) = \boldsymbol{e}_i$. Consequently,

$$\dot{\boldsymbol{\Phi}}(t, t_0) = \boldsymbol{A}(t)\boldsymbol{\Phi}(t, t_0), \quad \text{with} \quad \boldsymbol{\Phi}(t_0, t_0) = \boldsymbol{I}. \tag{7.22}$$

Unfortunately, for LTV systems we can not write the fundamental solution matrix (solution of (7.22)) as an exponential function of the system matrix $\boldsymbol{A}(t)$ as was done for LTI system in (7.17). In fact, in general we are not able to give an analytic solution for (7.22). Therefore, solutions of (7.22) should in general be calculated by integrating (7.22) numerically, taking the appropriate initial condition into account.

Equivalent to (7.18) for LTI systems, we can now write

$$\Delta\boldsymbol{x}(t) = \boldsymbol{\Phi}(t, t_0)\Delta\boldsymbol{x}(t_0), \tag{7.23}$$

for LTV systems. Clearly, this equation expresses the fact (as (7.18) did for LTI systems) that the fundamental solution matrix determines the way in which a solution depends on the initial condition.

An important property of the fundamental solution matrix, which we need later on, can be derived from (7.23). Namely, using (7.23), we can write

$$\begin{aligned}\Delta \boldsymbol{x}(t_1) &= \boldsymbol{\Phi}(t_1, t_0)\Delta \boldsymbol{x}(t_0),\\ \Delta \boldsymbol{x}(t_2) &= \boldsymbol{\Phi}(t_2, t_1)\Delta \boldsymbol{x}(t_1),\end{aligned}$$

for $t_0 \leq t_1 \leq t_2$. Consequently, it holds that

$$\Delta \boldsymbol{x}(t_2) = \boldsymbol{\Phi}(t_2, t_1)\boldsymbol{\Phi}(t_1, t_0)\Delta \boldsymbol{x}(t_0) = \boldsymbol{\Phi}(t_2, t_0)\Delta \boldsymbol{x}(t_0)$$

and thus

$$\boldsymbol{\Phi}(t_2, t_0) = \boldsymbol{\Phi}(t_2, t_1)\boldsymbol{\Phi}(t_1, t_0), \quad t_0 \leq t_1 \leq t_2, \tag{7.24}$$

which is called the transition property of the fundamental solution matrix.

Another property of the fundamental solution matrix of smooth systems is non-singularity.

Lemma 7.1. *The fundamental solution matrix $\boldsymbol{\Phi}(t, t_0)$ of a smooth system is non-singular for all t.*

Proof: We will prove that singularity is impossible by *reductio ad absurdum*. If $\boldsymbol{\Phi}(t, t_0)$ is singular, then there exists an initial perturbation $\Delta \boldsymbol{x}(t_0)$ such that $\Delta \boldsymbol{x}(t) = \boldsymbol{\Phi}(t, t_0)\Delta \boldsymbol{x}(t_0) = \boldsymbol{0}$. This would imply that there exist an infinite number of initial conditions $\boldsymbol{x}^p(t_0) + \alpha\Delta \boldsymbol{x}(t_0)$, $\alpha \in \mathbb{R}$ which evolve in finite time to $\boldsymbol{x}^p(t)$. This contradicts uniqueness of solutions in backward time of smooth systems. □

Note that the above lemma holds for linear as well as smooth nonlinear systems.

Linear Periodic Time-Variant Systems

We now consider the case that $\boldsymbol{A}(t)$ depends on time in a periodic fashion: $\boldsymbol{A}(t + T) = \boldsymbol{A}(t)$, where T is the period time of the periodically varying coefficients of $\boldsymbol{A}(t)$. The following lemma will be useful in understanding the behaviour of solutions of Linear Periodic Time-Variant Systems:

Lemma 7.2 (Sastry [131]). *The fundamental solution matrix of (7.19) with $\boldsymbol{A}(t + T) = \boldsymbol{A}(t)$ may be written as*

$$\boldsymbol{\Phi}(t, t_0) = \boldsymbol{K}(t)e^{\boldsymbol{B}(t-t_0)}, \tag{7.25}$$

where $\boldsymbol{K}(t) = \boldsymbol{K}(t+T) \in \mathbb{R}^{n\times n}$ and $\boldsymbol{K}(t_0) = \boldsymbol{I}$, and $\boldsymbol{B} = \frac{1}{T}\log\boldsymbol{\Phi}(t_0 + T, t_0)$.

Proof: Using the definition of $\boldsymbol{B}$, define

$$\boldsymbol{K}(t) = \boldsymbol{\Phi}(t, t_0)e^{-\boldsymbol{B}(t-t_0)}. \tag{7.26}$$

Equation (7.25) follows from the definitions. To establish the periodicity of $\boldsymbol{K}(t)$ note that

$$\begin{aligned}\boldsymbol{K}(t+T) &= \boldsymbol{\Phi}(t+T, t_0)e^{-\boldsymbol{B}(t+T-t_0)} \\ &= \boldsymbol{\Phi}(t+T, t_0+T)\boldsymbol{\Phi}(t_0+T, t_0)e^{-\boldsymbol{B}(t+T-t_0)} \\ &= \boldsymbol{\Phi}(t+T, t_0+T)e^{-\boldsymbol{B}(t-t_0)} \\ &= \boldsymbol{\Phi}(t, t_0)e^{-\boldsymbol{B}(t-t_0)} \\ &= \boldsymbol{K}(t).\end{aligned}$$

□

It follows from the above lemma that the long-term behaviour of the fundamental solution matrix $\boldsymbol{\Phi}(t, t_0)$ is determined by the constant matrix $\boldsymbol{B}$ and hence by $\boldsymbol{\Phi}(t_0 + T, t_0)$. We call

$$\boldsymbol{\Phi}_T = \boldsymbol{\Phi}(t_0 + T, t_0) \tag{7.27}$$

the *monodromy matrix*. From the periodicity of $\boldsymbol{A}(t)$ it follows that

$$\boldsymbol{\Phi}(t, t_0) = \boldsymbol{\Phi}(t+T, t_0+T) = \boldsymbol{\Phi}(t+kT, t_0+kT), \quad k = 0, 1, 2, \ldots \tag{7.28}$$

and

$$\boldsymbol{\Phi}_T = \boldsymbol{\Phi}(t_0+T, t_0) = \boldsymbol{\Phi}(t_0+2T, t_0+T) = \boldsymbol{\Phi}(t_0+(k+1)T, t_0+kT), \tag{7.29}$$

for $k = 0, 1, 2, \ldots$.

Using the fact that $\Delta\boldsymbol{x}(t_0+T) = \boldsymbol{\Phi}(t_0+T, t_0)\Delta\boldsymbol{x}(t_0)$ we can conclude that $\boldsymbol{\Phi}_T = \boldsymbol{\Phi}(t_0+T, t_0)$ maps an initial condition $\Delta\boldsymbol{x}(t_0)$ to the response $\Delta\boldsymbol{x}(t_0+T)$ at time T (one period) later. The monodromy matrix plays an important role in analyzing whether or not the fundamental solution $\boldsymbol{\Phi}(t, t_0)$ matrix grows as time proceeds. Namely, repeated application of (7.28) and (7.29) gives

$$\boldsymbol{\Phi}(t+2T, t_0) = \boldsymbol{\Phi}(t+T, t_0)\ \boldsymbol{\Phi}_T = \boldsymbol{\Phi}(t, t_0)\ (\boldsymbol{\Phi}_T)^2 \tag{7.30}$$

and

$$\boldsymbol{\Phi}(t+kT, t_0) = \boldsymbol{\Phi}(t, t_0)\ (\boldsymbol{\Phi}_T)^k, \quad k = 0, 1, 2, \ldots. \tag{7.31}$$

This equation expresses again that the long-term behaviour of the fundamental solution matrix is determined by the monodromy matrix. A consequence of (7.31) is that the fundamental solution matrix $\boldsymbol{\Phi}(t, t_0)$ is known for all time t if it is known for $t_0 \leq t \leq t_0 + T$ (for example through numerical integration of (7.22)).

We use the spectral decomposition (or eigenvalue decomposition) of the monodromy matrix:

$$\boldsymbol{\Phi}_T = \boldsymbol{U}\boldsymbol{\Lambda}\boldsymbol{U}^{-1}. \tag{7.32}$$

The matrix $\boldsymbol{\Lambda}$ is a diagonal matrix with the eigenvalues on the diagonal and $\boldsymbol{U}$ contains the eigencolumns. Using (7.32), we can write

$$(\boldsymbol{\Phi}_T)^k = \boldsymbol{U}\boldsymbol{\Lambda}^k\boldsymbol{U}^{-1}. \tag{7.33}$$

As a consequence, (7.31) transforms to

$$\boldsymbol{\Phi}(t + kT, t_0) = \boldsymbol{\Phi}(t, t_0)\, \boldsymbol{U}\boldsymbol{\Lambda}^k\boldsymbol{U}^{-1}, \quad k = 0, 1, 2, \ldots, \tag{7.34}$$

which leads to the conclusion that the long-term behaviour of the fundamental solution matrix (for a periodic LTV system) is determined by the eigenvalues of the monodromy matrix. These eigenvalues are usually called *Floquet multipliers* or *characteristic multipliers.* The growth or decay of a perturbation follows from (7.23) and (7.33)

$$\Delta\boldsymbol{x}(t_0 + kT) = \boldsymbol{U}\boldsymbol{\Lambda}^k\boldsymbol{U}^{-1}\Delta\boldsymbol{x}(t_0), \quad k = 0, 1, 2, \ldots \tag{7.35}$$

If the magnitude of the k-th Floquet multiplier λ_k is larger than unity, $|\lambda_k| > 1$, then a perturbation grows in the direction $\boldsymbol{u}_k$, while if $|\lambda_k| < 1$ then a perturbation decays in the direction $\boldsymbol{u}_k$. The Floquet multipliers therefore determine the exponential growth or decay of a perturbation (except if $|\lambda_k| = 1$). The Floquet multipliers are particulary useful for the stability analysis of (hyperbolic) periodic solutions as we will discuss in the remainder of the section.

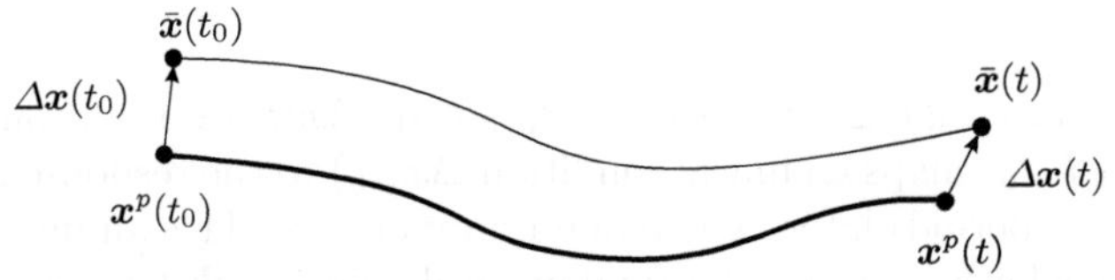

Fig. 7.1. Evolution of a solution and its perturbed solution.

Nonlinear Systems

We finally consider the case that the system (7.5) is nonlinear. When we consider $\Delta\boldsymbol{x}(t)$ in (7.20) to be a perturbation around a trajectory of the nonlinear system (7.5) (or (7.1) for the autonomous case), then the fundamental solution matrix relates how an infinitely small perturbation evolves to an infinitely small perturbation after a time lapse $t - t_0$. A solution of the initial value problem starting at $\boldsymbol{x}^p(t_0)$ (see Figure 7.1) evolves during a time

$t_d = t - t_0$ to $\boldsymbol{x}^p(t)$. If we perturb the initial condition with a small perturbation $\Delta\boldsymbol{x}(t_0) = \bar{\boldsymbol{x}}(t_0) - \boldsymbol{x}^p(t_0)$, then the perturbed solution will evolve during a time t_d to $\bar{\boldsymbol{x}}(t)$. The perturbation on the initial condition $\Delta\boldsymbol{x}(t_0)$ yields a perturbation $\Delta\boldsymbol{x}(t) = \bar{\boldsymbol{x}}(t) - \boldsymbol{x}^p(t)$ after a time t, which is different in direction and magnitude. The fundamental solution matrix relates $\Delta\boldsymbol{x}(t_0)$ to $\Delta\boldsymbol{x}(t)$ as

$$\Delta\boldsymbol{x}(t) = \boldsymbol{\Phi}(t, t_0, \boldsymbol{x}^p(t_0))\Delta\boldsymbol{x}(t_0) + H.O.T., \tag{7.36}$$

where the fundamental solution matrix is now not only dependent on time but also on the initial condition $\boldsymbol{x}^p(t_0)$. It should be noted that the higher-order terms are negligible when infinitely small perturbations are considered. The fundamental solution matrix for nonlinear systems can be obtained by integrating the IVP

$$\dot{\boldsymbol{\Phi}}(t, t_0, \boldsymbol{x}^p(t_0)) = \boldsymbol{A}(t, \boldsymbol{x}^p(t))\boldsymbol{\Phi}(t, t_0, \boldsymbol{x}^p(t_0)), \quad \boldsymbol{\Phi}(t_0, t_0, \boldsymbol{x}^p(t_0)) = \boldsymbol{I}. \tag{7.37}$$

For an autonomous system we have the property that the fundamental solution matrix only depends on the time span, i.e.

$$\boldsymbol{\Phi}(t, t_0, \boldsymbol{x}_0) = \boldsymbol{\Phi}(t + t_a, t_0 + t_a, \boldsymbol{x}_0), \quad \forall \boldsymbol{x}_0 \in \mathbb{R}^n, \tag{7.38}$$

where $\boldsymbol{x}_0$ is not necessarily on a periodic solution.

For a non-autonomous system we can write down a similar property if the fundamental solution is along a periodic solution $\boldsymbol{x}^p(t)$ and if the time shift is exactly the period time T

$$\boldsymbol{\Phi}(t, t_0, \boldsymbol{x}_0^p) = \boldsymbol{\Phi}(t + T, t_0 + T, \boldsymbol{x}_0^p). \tag{7.39}$$

Stability of Periodic Solutions of Nonlinear Systems

With the notion of the fundamental solution in mind, we are now able to consider the stability of a periodic solution of a nonlinear system. In the sequel we will assume that $\boldsymbol{x}^p(t) = \boldsymbol{x}^p(t + T)$ is periodic with period time T. Alternatively, we use the notation $\varphi^p(t, t_0, \boldsymbol{x}_0) := \boldsymbol{x}^p(t)$ with $\boldsymbol{x}^p(t_0) = \boldsymbol{x}_0$ to explicitly state the initial condition of the solution. We will use the following definition of local asymptotical stability of a periodic solution:

Definition 7.3 (Local asymptotical stability of periodic solutions). *A periodic solution $\varphi^p(t, t_0, \boldsymbol{x}_0)$ is locally asymptotically stable if and only if there exists an $\epsilon > 0$ and a function $t_1(t)$ such that*

$$\|\boldsymbol{x}_0^* - \boldsymbol{x}_0\| < \epsilon \Longrightarrow \lim_{t\to\infty} \|\varphi(t, t_0, \boldsymbol{x}_0^*) - \varphi^p(t_1(t), t_0, \boldsymbol{x}_0)\| = 0.$$

Hence, a solution which starts in an ϵ-tube around a locally asymptotically stable periodic solution will converge towards the periodic solution, apart from a possible time-shift expressed by $t_1(t)$.

We recall that the nonlinear system (7.11) can be linearized around the periodic solution of this system $\boldsymbol{x}^p(t)$, which leads to an LTV system as in (7.12). So, given the fact that the monodromy matrix $\boldsymbol{\Phi}_T$ and more specifically its eigenvalues (the Floquet multipliers $\lambda_1, \lambda_2, \ldots, \lambda_n$) can, for a periodic LTV system, give information on the stability of the equilibrium point of this system, we can now use the Floquet multipliers to assess the local stability of a periodic solution of a nonlinear system. For a graphical representation of the role that the monodromy matrix plays in the evolution of perturbations around a periodic solution, see Figure 7.2.

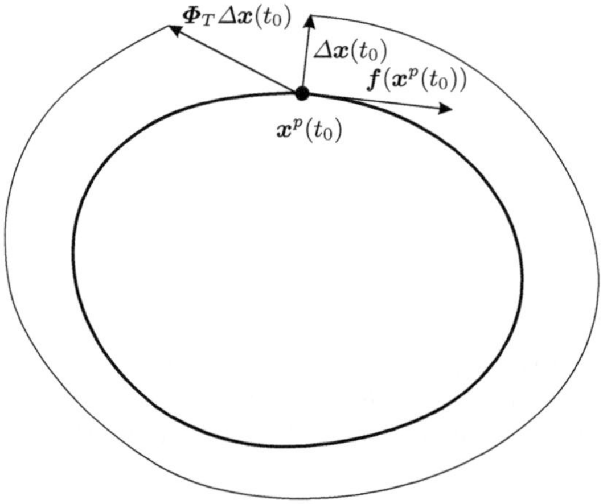

Fig. 7.2. Periodic solution with fundamental solution matrix.

The Floquet multipliers determine the exponential growth or decay of small perturbations in the eigendirections of the monodromy matrix $\boldsymbol{\Phi}_T$ and hence the local stability of the periodic solution. If a Floquet multiplier has a magnitude larger (smaller) than one, then a perturbation of the initial condition in the corresponding eigendirection will grow (decay) after one period of oscillation.

Theorem 7.4 (Local asymptotical stability of periodic solutions of non-autonomous systems). *Let $\boldsymbol{\varphi}^p(t, t_0, \boldsymbol{x}_0)$ be a periodic solution of a non-autonomous system and $\boldsymbol{\Phi}_T$ be the monodromy matrix along this periodic solution. Let λ_i, $i = 1 \ldots n$, denote the Floquet multipliers belonging to $\boldsymbol{\Phi}_T$. If $|\lambda_i| < 1$, $i = 1 \ldots n$, then the periodic solution is locally asymptotically stable.*

Proof: Let $\lambda_{\max}$ denote the Floquet multiplier with the largest magnitude. It holds that

$$\|\Delta\boldsymbol{x}(t_0 + kT)\| \leq |\lambda_{\max}|^k \|\Delta\boldsymbol{x}_0\| + H.O.T.$$

If $\|\Delta\boldsymbol{x}_0\| < \epsilon$, for some small $\epsilon > 0$, and if $|\lambda_{\max}| < 1$ then it must hold, taking the limit of $k \to \infty$, that

$$\lim_{t \to \infty} \Delta \boldsymbol{x}(t) = 0. \tag{7.40}$$

Consequently, a solution $\varphi(t, t_0, \boldsymbol{x}_0^*)$ with $\boldsymbol{x}_0^* = \boldsymbol{x}_0 + \Delta \boldsymbol{x}_0$ converges to the periodic solution $\varphi^p(t, t_0, \boldsymbol{x}_0)$. □

A periodic solution of a *non-autonomous* system is therefore locally asymptotically stable if *all* Floquet multipliers are within the unit circle, i.e. $|\lambda_i| < 1$ $\forall i$. The periodic solution is unstable if one or more Floquet multipliers are located outside the unit circle. The non-hyperbolic case for a non-autonomous system occurs when one or more Floquet multipliers are on the unit circle, i.e. $|\lambda_i| = 1$. If the the Floquet multiplier with the largest magnitude is on the unit circle, i.e. $|\lambda_{\max}| = 1$, then the stability or instability of the periodic solution is determined by higher order terms and can not be assessed by the Floquet multipliers.

The stability analysis of autonomous systems is similar to that of non-autonomous systems but has a minor complication. If the infinitely small perturbation on the initial condition is in the direction of the vector field in $\boldsymbol{x}^p(t)$, i.e. $\Delta \boldsymbol{x}(t_0) = \alpha \boldsymbol{f}(\boldsymbol{x}^p(t_0))$ ($\alpha \ll 1$), then the resulting perturbation will, provided that the system is autonomous, be identical to the initial perturbation $\Delta \boldsymbol{x}(T + t_0) = \Delta \boldsymbol{x}(t_0)$, due to the periodicity of the periodic solution (and thus of the vector field on this solution) and due to the time invariance (7.3):

$$\begin{aligned} \Delta \boldsymbol{x}(T + t_0) &= \boldsymbol{\Phi}(T + t_0, t_0, \boldsymbol{x}^p(t_0)) \Delta \boldsymbol{x}(t_0) \\ \Rightarrow \boldsymbol{f}(\boldsymbol{x}^p(t_0)) &= \boldsymbol{\Phi}_T \boldsymbol{f}(\boldsymbol{x}^p(t_0)). \end{aligned} \tag{7.41}$$

It therefore holds for autonomous systems that always one eigenvector is $\boldsymbol{f}(\boldsymbol{x}^p(t_0))$ with eigenvalue $\lambda_1 = 1$. The above equality hinges on the fact that if $(t_0, \boldsymbol{x}_0)$ is on the periodic solution, then also $(t_0, \boldsymbol{x}_0 + \Delta \boldsymbol{x}(t_0))$ is on the periodic solution provided that $\Delta \boldsymbol{x}(t_0)$ is in the direction of $\boldsymbol{f}(\boldsymbol{x}^p(t_0))$ and infinitely small.

Theorem 7.5 (Local asymptotical stability of periodic solutions of autonomous systems). *Let $\varphi^p(t, t_0, \boldsymbol{x}_0)$ be a periodic solution of an autonomous system and $\boldsymbol{\Phi}_T$ be the monodromy matrix along this periodic solution. Let $\lambda_1 = 1$ denote the Floquet multiplier corresponding to the eigendirection $\boldsymbol{f}(\boldsymbol{x}^p(t_0))$. Let λ_i, $i = 2 \dots n$ denote the other $(n-1)$ Floquet multipliers belonging to $\boldsymbol{\Phi}_T$. If $|\lambda_i| < 1$, for $i = 2 \dots n$ then the periodic solution is locally asymptotically stable.*

Proof: An initial perturbation $\Delta \boldsymbol{x}_0$ in the direction of $\boldsymbol{f}(\boldsymbol{x}^p(t_0))$ gives a perturbation $\Delta \boldsymbol{x}(T + t_0) = \Delta \boldsymbol{x}_0$. The perturbed solution therefore stays within the ϵ-tube. The proof for the other eigendirections is similar to the proof of Theorem 7.4. Consequently, a solution $\varphi(t, t_0, \boldsymbol{x}_0^*)$ with $\boldsymbol{x}_0^* = \boldsymbol{x}_0 + \Delta \boldsymbol{x}_0$ converges to the periodic solution $\varphi^p(t, t_0, \boldsymbol{x}_0)$ apart from a possible time-shift due to perturbation in the direction of $\boldsymbol{f}(\boldsymbol{x}^p(t_0))$. We therefore have that

$$\|\boldsymbol{x}_0^* - \boldsymbol{x}_0\| < \epsilon \Longrightarrow \lim_{t \to \infty} \|\varphi(t, t_0, \boldsymbol{x}_0^*) - \varphi^p(t_1(t), t_0, \boldsymbol{x}_0)\| = 0,$$

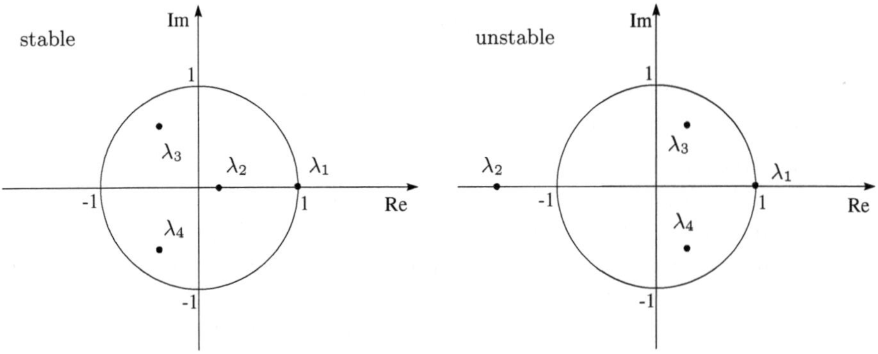

Fig. 7.3. Floquet multipliers in the complex plane.

for some function $t_1(t)$ which expresses the time-shift. □

Two different sets of Floquet multipliers of two different periodic solutions of an autonomous system are depicted in the complex plane in Figure 7.3. One Floquet multiplier in both figures is $\lambda_1 = 1$, because of the freedom of phase of an autonomous system. The other Floquet multipliers in the left figure, $\lambda_2, \lambda_3, \lambda_4$, are within the unit circle and the corresponding periodic solution is therefore locally asymptotically stable. The Floquet multipliers in the right figure indicate an unstable periodic solution because one Floquet multiplier lies outside the unit circle, $\lambda_2 < -1$.

As has been said before, the fundamental solution matrix $\boldsymbol{\Phi}_T$, for a periodic solution $\boldsymbol{x}^p(t)$, is dependent on the initial condition $\boldsymbol{x}^p(t_0)$. The Floquet multipliers, being the eigenvalues of $\boldsymbol{\Phi}_T$, are not dependent on the choice of $\boldsymbol{x}^p(t_0)$. This holds for autonomous as well as non-autonomous systems. Consider the following lemmas:

Lemma 7.6. *Let* $\boldsymbol{A}$ *and* $\boldsymbol{B} \in \mathbb{R}^{n\times n}$ *be square matrices. Then, the matrices* $\boldsymbol{C}_I = \boldsymbol{AB}$ *and* $\boldsymbol{C}_{II} = \boldsymbol{BA}$ *have an identical set of eigenvalues.*

Proof: Let λ_{Ii}, $\boldsymbol{u}_{Ii}$, $i = 1, \ldots, n$, be the eigenvalues and eigenvectors of $\boldsymbol{C}_I$, such that

$$\lambda_{Ii}\boldsymbol{u}_{Ii} = \boldsymbol{AB}\boldsymbol{u}_{Ii}.$$

Multiplication with $\boldsymbol{B}$ gives

$$\lambda_{Ii}\boldsymbol{B}\boldsymbol{u}_{Ii} = \boldsymbol{BAB}\boldsymbol{u}_{Ii}.$$

Hence, λ_{Ii} is also an eigenvalue of $\boldsymbol{C}_{II} = \boldsymbol{BA}$ with eigenvector $\boldsymbol{u}_{IIi} = \boldsymbol{B}\boldsymbol{u}_{Ii}$. □

Lemma 7.7. *Let* $\boldsymbol{x}^p(t)$ *be a periodic solution of a system being autonomous or non-autonomous with the fundamental solution matrices*

$$\boldsymbol{\Phi}_a = \boldsymbol{\Phi}(t_a + T, t_a, \boldsymbol{x}^p(t_a))$$

and

$$\boldsymbol{\Phi}_b = \boldsymbol{\Phi}(t_b + T, t_b, \boldsymbol{x}^p(t_b)),$$

with $\boldsymbol{x}^p(t_a) \neq \boldsymbol{x}^p(t_b)$ *and* $t_b > t_a$. *Then,* $\boldsymbol{\Phi}_a$ *and* $\boldsymbol{\Phi}_b$ *have an identical set of eigenvalues.*

Proof: Splitting of the fundamental solution matrices using (7.24) and (7.28) gives

$$\boldsymbol{\Phi}_a = \boldsymbol{\Phi}(t_a + T, t_b, \boldsymbol{x}^p(t_b))\boldsymbol{\Phi}(t_b, t_a, \boldsymbol{x}^p(t_a))$$

and

$$\boldsymbol{\Phi}_b = \boldsymbol{\Phi}(t_b + T, t_a + T, \boldsymbol{x}^p(t_a + T))\boldsymbol{\Phi}(t_a + T, t_b, \boldsymbol{x}^p(t_b)).$$

With the property (7.38) for autonomous systems and (7.39) for non-autonomous systems it holds that $\boldsymbol{\Phi}(t_b, t_a, \boldsymbol{x}^p(t_a)) = \boldsymbol{\Phi}(t_b + T, t_a + T, \boldsymbol{x}^p(t_a + T))$. It follows from Lemma 7.6 that $\boldsymbol{\Phi}_a$ and $\boldsymbol{\Phi}_b$ have an identical set of eigenvalues. □

We can conclude from the above lemma that if we choose another initial condition on the periodic solution, then we will find a different fundamental solution matrix, but with the same Floquet multipliers (only the eigendirections are different). For the stability analysis it does therefore not matter which initial condition we choose (as long as it is on the periodic solution). This is true for autonomous as well as non-autonomous systems. But for an autonomous system we only have to specify a point on the periodic solution $\boldsymbol{x}_0^p$ as initial condition and for a non-autonomous system we have to specify $(t_0, \boldsymbol{x}_0^p)$.

Fundamental solution matrices are furthermore used in shooting methods for finding periodic solutions, in continuation methods to follow branches of periodic solutions, and they are used in the determination of Lyapunov exponents.

7.2 Fundamental Solution Matrix of Filippov Systems

For smooth systems the fundamental solution matrix can be obtained in an elegant manner by integrating the IVP (7.37). Discontinuous systems however, exhibit discontinuities (or 'saltations' / 'jumps') in the time evolution of the fundamental solution matrix. The jumps in the fundamental solution matrix can be computed by means of the theory of Aizerman and Gantmakher [1], which will be discussed in Sections 7.2.1 and 7.2.2.

The theory of Aizerman and Gantmakher was used by [11] and [99] to calculate Lyapunov exponents of discontinuous systems.

7.2.1 Jumping Conditions: A Single Discontinuity

In this section we will derive how the fundamental solution matrix $\boldsymbol{\Phi}$ jumps if the solution $\boldsymbol{x}(t)$ crosses a switching boundary Σ, on which the vector field is discontinuous. Consider the nonlinear system (3.16) with discontinuous right-hand side as described in Chapter 3 and its set-valued extension (3.17)

$$\dot{\boldsymbol{x}}(t) \in \boldsymbol{F}(t, \boldsymbol{x}(t)) = \begin{cases} \boldsymbol{f}_-(t, \boldsymbol{x}(t)), & \boldsymbol{x}(t) \in \mathcal{V}_-, \\ \overline{\text{co}}\{\boldsymbol{f}_-(t, \boldsymbol{x}(t)), \boldsymbol{f}_+(t, \boldsymbol{x}(t))\}, & \boldsymbol{x}(t) \in \Sigma, \\ \boldsymbol{f}_+(t, \boldsymbol{x}(t)), & \boldsymbol{x}(t) \in \mathcal{V}_+, \end{cases} \tag{7.42}$$

with the initial condition

$$\boldsymbol{x}(t=0) = \boldsymbol{x}_0. \tag{7.43}$$

Assume that at a certain point in time, say t_p, the solution $\boldsymbol{x}(t)$ will cross Σ. With the definition of the switching boundary function (3.10), $h(\boldsymbol{x}) = 0 \Longleftrightarrow \boldsymbol{x} \in \Sigma$ we obtain

$$h(\boldsymbol{x}(t_p)) = 0.$$

At this switching boundary there are two derivatives $\boldsymbol{f}_{p-}$ and $\boldsymbol{f}_{p+}$, which lie in the direction of the solution as denoted in Figure 7.4. The derivatives have components $\boldsymbol{f}_{p-}^{\mathrm{N}}$ and $\boldsymbol{f}_{p+}^{\mathrm{N}}$ perpendicular to the switching boundary with magnitudes $\boldsymbol{n}^{\mathrm{T}}\boldsymbol{f}_{p-}$ and $\boldsymbol{n}^{\mathrm{T}}\boldsymbol{f}_{p+}$. We first consider only transversal intersections. Uniqueness of the solution is therefore assured (this is not the case for a repulsive sliding mode problem). In order to assure a transversal intersection, we assume that the projections of the derivatives $\boldsymbol{f}_{p-}$ and $\boldsymbol{f}_{p+}$ on the normal $\boldsymbol{n}$ have the same sign

$$\boldsymbol{n}^{\mathrm{T}}\boldsymbol{f}_{p-}\boldsymbol{n}^{\mathrm{T}}\boldsymbol{f}_{p+} > 0. \tag{7.44}$$

Equation (7.44) assures that the solution leaves the switching boundary and stays on the switching boundary at one point of time and not on an interval of time (i.e. the solution curve *crosses* the switching boundary).

An infinitesimal disturbance $\delta\boldsymbol{x}_0$ on the initial condition will cause a disturbance $\delta\boldsymbol{x}(t)$ on the state $\boldsymbol{x}(t)$. The fundamental solution matrix $\boldsymbol{\Phi}(t, t_0)$ relates $\delta\boldsymbol{x}(t)$ to $\delta\boldsymbol{x}_0$,

$$\delta\boldsymbol{x}(t) = \boldsymbol{\Phi}(t, t_0, \boldsymbol{x}_0)\delta\boldsymbol{x}_0 + O(\|\delta\boldsymbol{x}_0\|^2). \tag{7.45}$$

Let the solution start in the subspace $\mathcal{V}_-$, that is $\boldsymbol{x}_0 \in \mathcal{V}_-$. Suppose the solution curve crosses the switching boundary Σ at $t = t_p$, i.e. $h(\boldsymbol{x}(t_p)) = 0$. The system is continuous on the interval $D = \{t \in \mathbb{R} \mid t_0 \le t \le t_p\}$. The fundamental solution matrix will also be continuous on the interior of D. The time evolution of the fundamental solution matrix on the *interior* of D can be obtained from the IVP (7.37). The Jacobian $\partial\boldsymbol{f}/\partial\boldsymbol{x}$ is not uniquely defined on the border of D at $t = t_p$ where $\boldsymbol{x}(t_p)$ is located on the switching boundary Σ. This causes a jump (or discontinuity) in the fundamental solution matrix. We will derive an expression for the jump in Section 7.2.2. For the moment

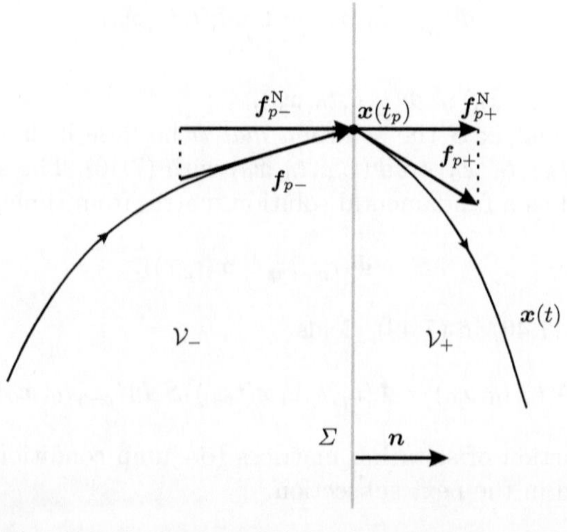

Fig. 7.4. Projection of derivatives on the normal.

we will assume that we know how the fundamental solution matrix jumps and we assume that we can express the jump with a matrix $\boldsymbol{S}$, which maps the fundamental solution matrix just before the jump, $\boldsymbol{\Phi}(t_{p-}, t_0, \boldsymbol{x}_0)$, to the fundamental solution matrix just after the jump, $\boldsymbol{\Phi}(t_{p+}, t_0, \boldsymbol{x}_0)$, as

$$\boldsymbol{\Phi}(t_{p+}, t_0, \boldsymbol{x}_0) = \boldsymbol{S}\,\boldsymbol{\Phi}(t_{p-}, t_0, \boldsymbol{x}_0), \tag{7.46}$$

where

$$\boldsymbol{\Phi}(t_{p-}, t_0, \boldsymbol{x}_0) = \lim_{t \uparrow t_p} \boldsymbol{\Phi}(t, t_0, \boldsymbol{x}_0). \tag{7.47}$$

On D the fundamental solution matrix can be obtained from integrating the variational equation (7.37) which gives for $\boldsymbol{\Phi}(t_{p-}, t_0, \boldsymbol{x}_0)$

$$\boldsymbol{\Phi}(t_{p-}, t_0, \boldsymbol{x}_0) = \int_{t_0}^{t_p} \dot{\boldsymbol{\Phi}}(t, t_0, \boldsymbol{x}_0)dt + \boldsymbol{I}. \tag{7.48}$$

The fundamental solution matrix after the jump can then be obtained by (7.46) where $\boldsymbol{S}$ should of course be known. The solution enters the subspace $\mathcal{V}_+$ (as transversality was assumed) at $t = t_p$, and traverses $\mathcal{V}_+$ during the interval $G = \{t \in \mathbb{R} \mid t_p < t \leq t_q\}$. We can now construct the fundamental solution matrix on G after the jump as

$$\begin{aligned} \boldsymbol{\Phi}(t_q, t_0, \boldsymbol{x}_0) &= \boldsymbol{\Phi}(t_q, t_{p+}, \boldsymbol{x}(t_{p+}))\boldsymbol{\Phi}(t_{p+}, t_0, \boldsymbol{x}_0) \\ &= \left(\int_{t_p}^{t_q} \dot{\boldsymbol{\Phi}}(t, t_0, \boldsymbol{x}_0)dt + \boldsymbol{I} \right) \boldsymbol{\Phi}(t_{p+}, t_0, \boldsymbol{x}_0). \end{aligned} \tag{7.49}$$

If the fundamental solution matrix is known on G, then we can express $\boldsymbol{\Phi}(t_{p+}, t_0, \boldsymbol{x}_0)$ by the right time limit to the jump as

$$\boldsymbol{\Phi}(t_{p+}, t_0, \boldsymbol{x}_0) = \lim_{t \downarrow t_p} \boldsymbol{\Phi}(t, t_0, \boldsymbol{x}_0). \tag{7.50}$$

In general $\boldsymbol{\Phi}(t_{p-}, t_0, \boldsymbol{x}_0) \neq \boldsymbol{\Phi}(t_{p+}, t_0, \boldsymbol{x}_0)$.

We call the matrix $\boldsymbol{S}$ the *saltation matrix* because it describes the jump by mapping $\boldsymbol{\Phi}(t_{p-}, t_0, \boldsymbol{x}_0)$ to $\boldsymbol{\Phi}(t_{p+}, t_0, \boldsymbol{x}_0)$ with (7.46). The saltation matrix can be regarded as a fundamental solution matrix from time t_{p-} to t_{p+}

$$\boldsymbol{S} = \boldsymbol{\Phi}(t_{p+}, t_{p-}, \boldsymbol{x}(t_{p-})). \tag{7.51}$$

Substitution of (7.46) in (7.49) yields

$$\boldsymbol{\Phi}(t_q, t_0, \boldsymbol{x}_0) = \boldsymbol{\Phi}(t_q, t_{p+}, \boldsymbol{x}(t_{p+}))\boldsymbol{S}\,\boldsymbol{\Phi}(t_{p-}, t_0, \boldsymbol{x}_0). \tag{7.52}$$

The construction of saltation matrices (or jump conditions) is due to [1] and is explained in the next subsection.

7.2.2 Construction of Saltation Matrices

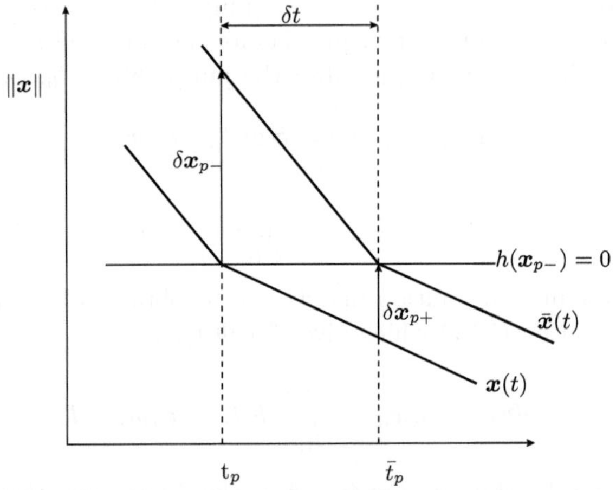

Fig. 7.5. Disturbed and undisturbed solution.

One question has not been answered up to now: how do we obtain the saltation matrix $\boldsymbol{S}$? The saltation matrix will be derived by inspecting the nonlinear dynamical system in the neighbourhood of switching boundary. Consider the disturbed and undisturbed solutions depicted in Figure 7.5. Time is on the horizontal axis of Figure 7.5 and an arbitrary norm of $\boldsymbol{x}$ is on the vertical axis. The disturbed solution $\bar{\boldsymbol{x}}(t)$ is due to an initial disturbance

$$\bar{\boldsymbol{x}}_0 = \boldsymbol{x}_0 + \delta\boldsymbol{x}_0. \tag{7.53}$$

The disturbed solution stays $\delta t = \bar{t}_p - t_p$ longer (if $\delta t > 0$) or shorter (if $\delta t < 0$) in $\mathcal{V}_-$ before hitting the switching boundary Σ. The differences between the disturbed and undisturbed solutions at the crossings are denoted by

$$\delta \boldsymbol{x}_{p-} = \bar{\boldsymbol{x}}(t_p) - \boldsymbol{x}(t_p), \tag{7.54}$$

$$\delta \boldsymbol{x}_{p+} = \bar{\boldsymbol{x}}(\bar{t}_p) - \boldsymbol{x}(\bar{t}_p). \tag{7.55}$$

We can express the undisturbed and disturbed solutions in a first-order Taylor expansion

$$\boldsymbol{x}(\bar{t}_p) \approx \boldsymbol{x}(t_p) + \boldsymbol{f}_{p+}\delta t, \tag{7.56}$$

$$\bar{\boldsymbol{x}}(\bar{t}_p) \approx \boldsymbol{x}(t_p) + \delta \boldsymbol{x}_{p-} + \boldsymbol{f}_{p-}\delta t, \tag{7.57}$$

with the abbreviations

$$\begin{aligned} \boldsymbol{f}_{p+} &:= \boldsymbol{f}(t_{p+}, \boldsymbol{x}(t_{p+})), \\ \boldsymbol{f}_{p-} &:= \boldsymbol{f}(t_{p-}, \boldsymbol{x}(t_{p-})). \end{aligned} \tag{7.58}$$

The equations (7.56) and (7.57) are inserted into (7.55)

$$\begin{aligned} \delta \boldsymbol{x}_{p+} &= \bar{\boldsymbol{x}}(\bar{t}_p) - \boldsymbol{x}(\bar{t}_p) \\ &\approx \boldsymbol{x}(t_p) + \delta \boldsymbol{x}_{p-} + \boldsymbol{f}_{p-}\delta t - (\boldsymbol{x}(t_p) + \boldsymbol{f}_{p+}\delta t) \\ &\approx \delta \boldsymbol{x}_{p-} + \boldsymbol{f}_{p-}\delta t - \boldsymbol{f}_{p+}\delta t, \end{aligned} \tag{7.59}$$

The disturbed solution satisfies the switching boundary function (3.10), $0 = h(\bar{\boldsymbol{x}}(\bar{t}_p))$. We apply a Taylor series expansion up to the first-order terms [99]:

$$\begin{aligned} 0 &= h(\bar{\boldsymbol{x}}(\bar{t}_p)) \\ &\approx h(\boldsymbol{x}(t_p) + \delta \boldsymbol{x}_{p-} + \boldsymbol{f}_{p-}\delta t) \\ &\approx \underbrace{h(\boldsymbol{x}(t_p))}_{=0} + \boldsymbol{n}^{\mathrm{T}}(\delta \boldsymbol{x}_{p-} + \boldsymbol{f}_{p-}\delta t) \\ &\approx \boldsymbol{n}^{\mathrm{T}}(\delta \boldsymbol{x}_{p-} + \boldsymbol{f}_{p-}\delta t), \end{aligned} \tag{7.60}$$

where the normal $\boldsymbol{n}$ is defined by $\boldsymbol{n}(\boldsymbol{x}) = \nabla h(\boldsymbol{x})$ (3.12).

From (7.60) we can express the variation δt in terms of $\delta \boldsymbol{x}_{p-}$

$$\boldsymbol{n}^{\mathrm{T}} \boldsymbol{f}_{p-}\delta t = -\boldsymbol{n}^{\mathrm{T}} \delta \boldsymbol{x}_{p-}. \tag{7.61}$$

The dependence between the variation δt and $\delta \boldsymbol{x}_{p-}$ can be envisaged from Figure 7.6. Due to the variation $\delta \boldsymbol{x}_{p-}$, the disturbed solution after a time t_p does not lie exactly on the (fixed) surface Σ. The disturbed solution has to stay a time δt longer/shorter in $\mathcal{V}_-$, covering an additional distance $\boldsymbol{f}_{p-}\delta t$, to reach Σ. We infer from Figure 7.6 that the vectors $\delta \boldsymbol{x}_{p-}$ and $\boldsymbol{f}_{p-}\delta t$ are related by (7.61).

Combination of (7.59) and (7.61) gives

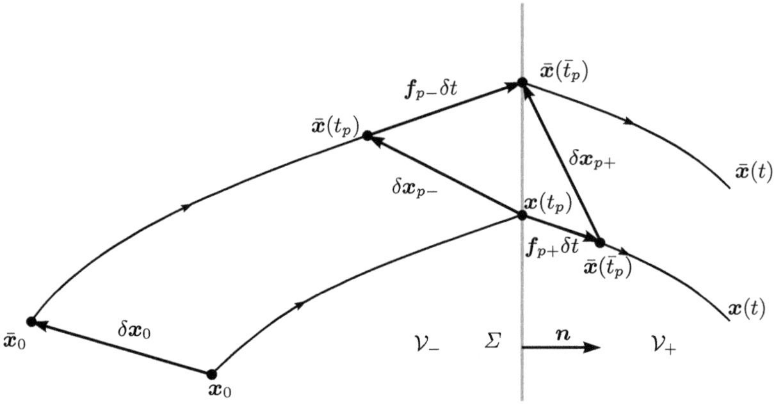

Fig. 7.6. Construction of δt for autonomous Σ.

$$\delta \boldsymbol{x}_{p+} = \delta \boldsymbol{x}_{p-} + (\boldsymbol{f}_{p+} - \boldsymbol{f}_{p-}) \frac{\boldsymbol{n}^{\mathrm{T}} \delta \boldsymbol{x}_{p-}}{\boldsymbol{n}^{\mathrm{T}} \boldsymbol{f}_{p-}}. \tag{7.62}$$

We have now expressed the variation $\delta \boldsymbol{x}_{p+}$ in terms of the variation $\delta \boldsymbol{x}_{p-}$. The saltation matrix relates $\delta \boldsymbol{x}_{p+}$ to $\delta \boldsymbol{x}_{p-}$

$$\delta \boldsymbol{x}_{p+} = \boldsymbol{S} \delta \boldsymbol{x}_{p-}. \tag{7.63}$$

We obtain the saltation matrix $\boldsymbol{S} = \boldsymbol{\Phi}(t_{p+}, t_{p-})$ from (7.62) and (7.63) as

$$\boldsymbol{S} = \boldsymbol{I} + \frac{(\boldsymbol{f}_{p+} - \boldsymbol{f}_{p-}) \boldsymbol{n}^{\mathrm{T}}}{\boldsymbol{n}^{\mathrm{T}} \boldsymbol{f}_{p-}}. \tag{7.64}$$

The inverse of the saltation matrix $\boldsymbol{S}^{-1} = \boldsymbol{\Phi}(t_{p-}, t_{p+}, \boldsymbol{x}(t_{p+}))$ is given by (for non-singular $\boldsymbol{S}$)

$$\boldsymbol{S}^{-1} = \boldsymbol{I} + \frac{(\boldsymbol{f}_{p-} - \boldsymbol{f}_{p+}) \boldsymbol{n}^{\mathrm{T}}}{\boldsymbol{n}^{\mathrm{T}} \boldsymbol{f}_{p+}}. \tag{7.65}$$

The saltation matrix $\boldsymbol{S}$ becomes singular if $\boldsymbol{n}^{\mathrm{T}} \boldsymbol{f}_{p+} = 0$ which will not happen if the transversality condition (7.44) is fulfilled.

The saltation matrix was derived in this section for an autonomous switching boundary function $h(\boldsymbol{x}(t))$. Non-autonomous systems can give rise to non-autonomous switching boundary functions. However, non-autonomous time periodic systems can be transformed into autonomous systems, having autonomous switching boundary functions. Alternatively, one can derive the saltation matrix for a non-autonomous switching boundary function $h(t, \boldsymbol{x})$ [1,37,99]. The saltation matrix for a non-autonomous switching boundary function is given by

$$\boldsymbol{S} = \boldsymbol{I} + \frac{(\boldsymbol{f}_{p+} - \boldsymbol{f}_{p-}) \boldsymbol{n}^{\mathrm{T}}}{\boldsymbol{n}^{\mathrm{T}} \boldsymbol{f}_{p-} + \frac{\partial h}{\partial t}(t_p, \boldsymbol{x}(t_p))}. \tag{7.66}$$

7.3 Example I: The Stick-slip System

To demonstrate the above theory we will study the stick-slip system with stiction friction model, as studied in Section 6.5.2, which possesses a locally asymptotically stable stick-slip limit cycle. We will use the same parameters as in Section 6.5.2 but with $c = 0$.

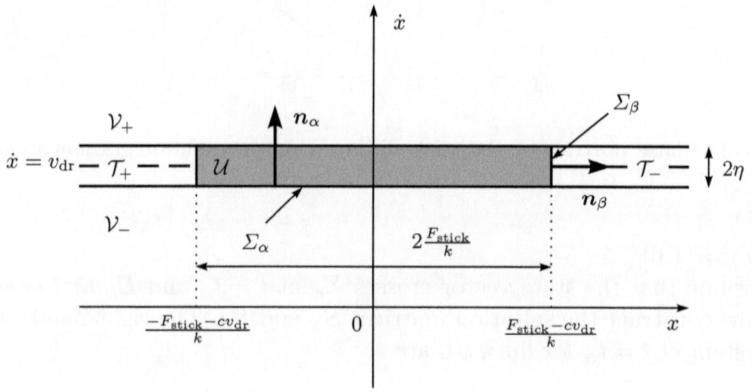

Fig. 7.7. Definition of subspaces $\mathcal{V}_-$, $\mathcal{V}_+$ and $\mathcal{U}$.

The Switch Model was used in the previous chapters as a numerical method. In this section we will use the Switch Model as a theoretical concept to determine the jumps in the fundamental solution matrix during the time-evolution on the stable limit cycle of the stick-slip system with stiction friction model. The pseudo-code of the Switch Model for the stick-slip system with stiction friction model has been given in Section 6.5.2. The subspaces $\mathcal{U}$, $\mathcal{V}_+$ and $\mathcal{V}_-$ for the Switch Model are depicted schematically in Figure 7.7. The small parameter η is enlarged in Figure 7.7 to make the stick subspace $\mathcal{U}$ visible.

A locally asymptotically stable stick-slip limit cycle of this system exists and is depicted in Figure 7.8 together with the equilibrium point $(x, \dot{x}) = (1, 0)$. As this system is autonomous, the switching boundaries are not dependent on time. It can be seen that the state traverses $\mathcal{V}_-$ (the backward slip phase) and $\mathcal{U}$ (the stick phase). If the state leaves $\mathcal{V}_-$ and enters $\mathcal{U}$, the switching boundary Σ_α is crossed with normal $\boldsymbol{n}_\alpha$ where

$$h_\alpha(x, \dot{x}) = \dot{x} - v_{dr}, \tag{7.67}$$

and $\boldsymbol{n}_\alpha = \begin{bmatrix} 0 & 1 \end{bmatrix}^T$. Likewise, if the state leaves $\mathcal{U}$ and enters $\mathcal{V}_-$ again, the switching boundary Σ_β is crossed with normal $\boldsymbol{n}_\beta$ where

$$h_\beta(x, \dot{x}) = kx + cv_{dr} - F_{stick}, \tag{7.68}$$

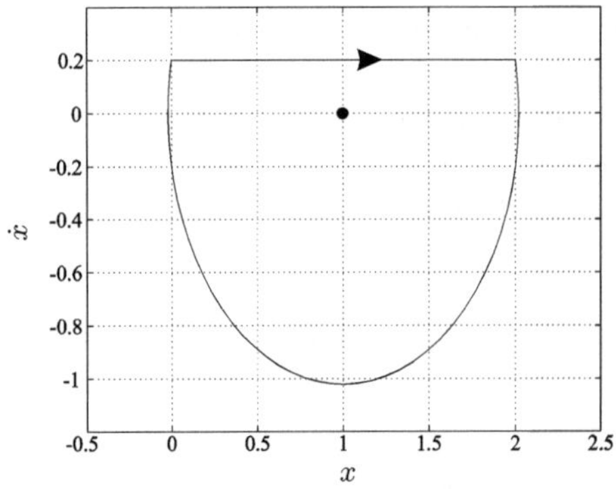

Fig. 7.8. Phase portrait of the stick-slip system with stiction friction model for $c = 0$.

and $\boldsymbol{n}_\beta = \begin{bmatrix}1 & 0\end{bmatrix}^{\mathrm{T}}$.

Assume that the state vector crosses Σ_α at $t = t_\alpha$ and Σ_β at $t = t_\beta$. We can now construct the saltation matrices $\boldsymbol{S}_\alpha$ and $\boldsymbol{S}_\beta$. The right-hand sides of the system at $t = t_\alpha$ for $\lim \eta \downarrow 0$ are

$$\boldsymbol{f}_{\alpha-} = \begin{bmatrix} v_{\mathrm{dr}} \\ \ddot{x}_{\alpha-} \end{bmatrix}, \qquad \boldsymbol{f}_{\alpha+} = \begin{bmatrix} v_{\mathrm{dr}} \\ 0 \end{bmatrix}. \tag{7.69}$$

The saltation matrix $\boldsymbol{S}_\alpha$ becomes

$$\boldsymbol{S}_\alpha = \boldsymbol{I} + \frac{(\boldsymbol{f}_{\alpha+} - \boldsymbol{f}_{\alpha-})\boldsymbol{n}_\alpha^{\mathrm{T}}}{\boldsymbol{n}_\alpha^{\mathrm{T}} \boldsymbol{f}_{\alpha-}} = \begin{bmatrix} 1 & 0 \\ 0 & 0 \end{bmatrix}, \tag{7.70}$$

which is independent of any system parameter. Conducting the same for $\boldsymbol{S}_\beta$ yields

$$\boldsymbol{f}_{\beta-} = \begin{bmatrix} v_{\mathrm{dr}} \\ 0 \end{bmatrix}, \qquad \boldsymbol{f}_{\beta+} = \begin{bmatrix} v_{\mathrm{dr}} \\ -\frac{\Delta F}{m} \end{bmatrix}, \tag{7.71}$$

with $\Delta F = F_{\mathrm{s}} - F_{\mathrm{d}}$. Substitution yields

$$\boldsymbol{S}_\beta = \boldsymbol{I} + \frac{(\boldsymbol{f}_{\beta+} - \boldsymbol{f}_{\beta-})\boldsymbol{n}_\beta^{\mathrm{T}}}{\boldsymbol{n}_\beta^{\mathrm{T}} \boldsymbol{f}_{\beta-}} = \begin{bmatrix} 1 & 0 \\ -\frac{\Delta F}{m v_{\mathrm{dr}}} & 1 \end{bmatrix}. \tag{7.72}$$

Note that the saltation matrix $\boldsymbol{S}_\alpha$ is singular causing the fundamental solution matrix to be singular. The physical meaning of the singularity is that the solution of the state vector is uniquely mapped from $\boldsymbol{x}_0$ in $\mathcal{V}_-$ to $\boldsymbol{x}(t)$ in $\mathcal{U}$ but the inverse mapping does not exist. If different solutions enter the stick phase, they all pass the same states on the stick phase and leave the stick phase from the same state $\boldsymbol{x}_\beta$. So, if the solution enters the stick phase, knowledge about its initial state is lost. The elements Φ_{ij} of the fundamental solution matrix for the stable limit cycle of the stick-slip system are plotted in Figure 7.9. Jumps at $t = t_\alpha$ and $t = t_\beta$ in the fundamental solution matrix can be clearly distinguished.

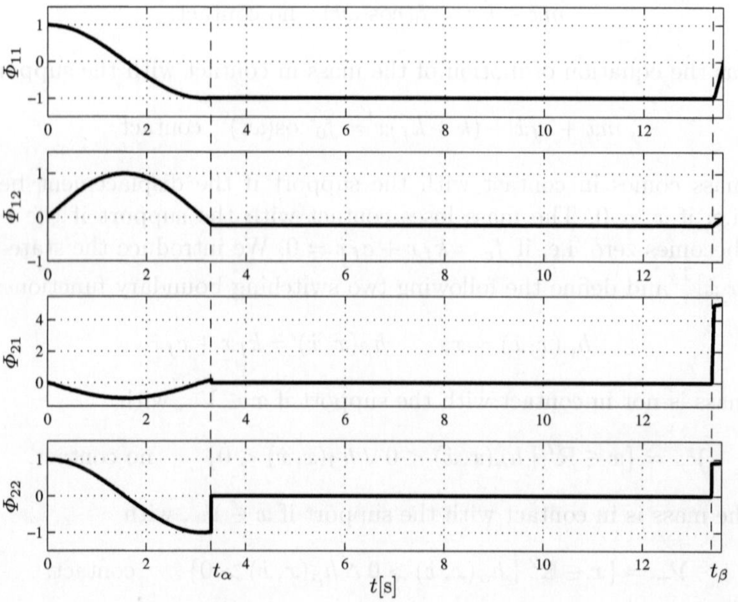

Fig. 7.9. Fundamental solution matrix along the limit cycle in Figure 7.8.

7.4 Example II: The Discontinuous Support

As a second example we will consider a mass-spring system with a discontinuous support (Figure 7.10a). The support is massless, has a spring stiffness k_f and damping coefficient c_f, which makes the support a first-order system. The displacement of the mass relative to the equilibrium position is denoted by x and of the support by y. The system has two possible modes: the mass is in contact with the support or the mass is not in contact with the support. Let f_c denote the contact force between mass and support. If the mass is not in contact the following holds:

$$x < y \quad \text{and} \quad f_c = 0,$$

and if the mass is in contact, we have:

$$x = y \quad \text{and} \quad f_c = k_f y + c_f \dot{y} = k_f x + c_f \dot{x} \geq 0.$$

The support, being a first-order system, relaxes to the equilibrium state, $y = 0$, if the mass is not in contact with the support. If we assume that the relaxation time of the support is much smaller than the time between two contact events, then we can neglect the free motion of the support. It is therefore assumed that the support is at rest at the moment that contact is made. This assumption reduces the system to a second-order differential equation. The free motion of the mass satisfies the ordinary differential equation

$$m\ddot{x} + kx = f_0 \cos(\omega t) \quad \text{no contact,} \tag{7.73}$$

whereas the equation of motion of the mass in contact with the support is

$$m\ddot{x} + c_f\dot{x} + (k + k_f)x = f_0 \cos(\omega t) \quad \text{contact.} \tag{7.74}$$

The mass comes in contact with the support if the displacement becomes zero, i.e. if $x = 0$. The mass loses contact with the support if the contact force becomes zero, i.e. if $f_c = k_f x + c_f \dot{x} = 0$. We introduce the state-vector $\boldsymbol{x} = \begin{bmatrix} x \ \dot{x} \end{bmatrix}^{\mathrm{T}}$ and define the following two switching boundary functions:

$$h_\alpha(x, \dot{x}) = x, \qquad h_\beta(x, \dot{x}) = k_f x + c_f \dot{x}. \tag{7.75}$$

The mass is not in contact with the support if $\boldsymbol{x} \in \mathcal{V}_-$, with

$$\mathcal{V}_- = \{\boldsymbol{x} \in \mathbb{R}^2 \mid h_\alpha(x, \dot{x}) < 0 \vee h_\beta(x, \dot{x}) < 0\} \qquad \text{no contact,}$$

and the mass is in contact with the support if $\boldsymbol{x} \in \mathcal{V}_+$, with

$$\mathcal{V}_+ = \{\boldsymbol{x} \in \mathbb{R}^2 \mid h_\alpha(x, \dot{x}) > 0 \wedge h_\beta(x, \dot{x}) > 0\} \qquad \text{contact.}$$

The switching boundary Σ, which divides the state-space $\mathbb{R}^2$ in the subspaces $\mathcal{V}_-$ and $\mathcal{V}_+$, consists of the union of two surfaces Σ_α and Σ_β. The switching boundary Σ_α is defined by

$$\Sigma_\alpha = \{\boldsymbol{x} \in \mathbb{R}^2 \mid h_\alpha(x, \dot{x}) = 0, h_\beta(x, \dot{x}) \geq 0\}, \tag{7.76}$$

and has the normal $\boldsymbol{n}_\alpha = \begin{bmatrix} 1 \ 0 \end{bmatrix}^{\mathrm{T}}$. The switching boundary Σ_β is defined by

$$\Sigma_\beta = \{\boldsymbol{x} \in \mathbb{R}^2 \mid h_\alpha(x, \dot{x}) \geq 0, h_\beta(x, \dot{x}) = 0\}, \tag{7.77}$$

and has the normal $\boldsymbol{n}_\beta = \begin{bmatrix} k_f \ c_f \end{bmatrix}^{\mathrm{T}}$.

The switching boundary Σ, consisting of Σ_α and Σ_β, and the subspaces $\mathcal{V}_-$ and $\mathcal{V}_+$ are depicted in Figure 7.10b. Remark that the switching boundary Σ is non-smooth at the origin.

The state equation of this non-autonomous discontinuous system reads as

$$\dot{\boldsymbol{x}}(t) = \boldsymbol{f}(t, \boldsymbol{x}(t)) = \begin{cases} \boldsymbol{f}_-(t, \boldsymbol{x}(t)), & \boldsymbol{x} \in \mathcal{V}_-, \\ \boldsymbol{f}_+(t, \boldsymbol{x}(t)), & \boldsymbol{x} \in \mathcal{V}_+, \end{cases} \tag{7.78}$$

with

$$\boldsymbol{f}_-(t, \boldsymbol{x}) = \begin{bmatrix} \dot{x} \\ -\dfrac{k}{m}x + \dfrac{f_0}{m}\cos(\omega t) \end{bmatrix}, \tag{7.79}$$

and

$$\boldsymbol{f}_+(t, \boldsymbol{x}) = \begin{bmatrix} \dot{x} \\ -\dfrac{k + k_f}{m}x - \dfrac{c_f}{m}\dot{x} + \dfrac{f_0}{m}\cos(\omega t) \end{bmatrix}. \tag{7.80}$$

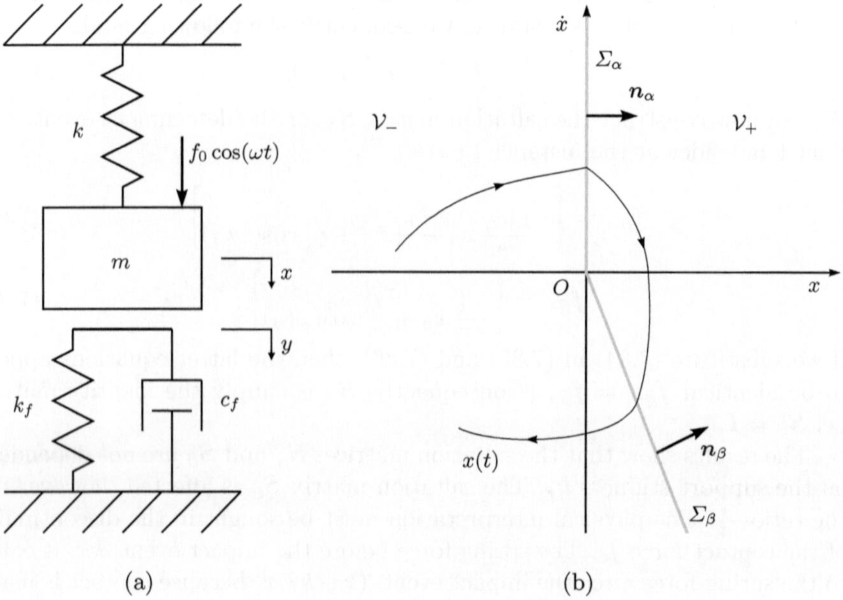

Fig. 7.10. Mass with discontinuous support.

System (7.78), which is discontinuous for $\boldsymbol{x} \in \Sigma$, can be extended to a differential inclusion with Filippov's convex method as described in Chapter 3.

We first consider the impact event, which is the transition from the mode without contact to the mode with contact. We assume that trajectory $\boldsymbol{x}(t)$ crosses Σ, leaving $\mathcal{V}_-$ and entering $\mathcal{V}_+$, at $t = t_\alpha$. The trajectory crosses therefore the Σ_α part of Σ at this instance. We can now construct the saltation matrix $\boldsymbol{S}_\alpha$ of the impact event. The right-hand sides at the instance t_α are

$$\boldsymbol{f}_{\alpha-} = \begin{bmatrix} \dot{x}_\alpha \\ -\frac{k}{m} x_\alpha + \frac{f_0}{m} \cos(\omega t_\alpha) \end{bmatrix}, \tag{7.81}$$

$$\boldsymbol{f}_{\alpha+} = \begin{bmatrix} \dot{x}_\alpha \\ -\frac{k+k_f}{m} x_\alpha - \frac{c_f}{m} \dot{x}_\alpha + \frac{f_0}{m} \cos(\omega t_\alpha) \end{bmatrix}. \tag{7.82}$$

The saltation matrix $\boldsymbol{S}_\alpha$ becomes

$$\begin{aligned} \boldsymbol{S}_\alpha &= \boldsymbol{I} + \frac{(\boldsymbol{f}_{\alpha+} - \boldsymbol{f}_{\alpha-})\boldsymbol{n}_\alpha^{\mathrm{T}}}{\boldsymbol{n}_\alpha^{\mathrm{T}} \boldsymbol{f}_{\alpha-}} \\ &= \boldsymbol{I} + \begin{bmatrix} 0 & 0 \\ -\frac{c_f}{m} & 0 \end{bmatrix} = \begin{bmatrix} 1 & 0 \\ -\frac{c_f}{m} & 1 \end{bmatrix}. \end{aligned} \tag{7.83}$$

We now consider the transition from the mode with contact to the mode without contact, being the detachment event. We assume that the state vector crosses Σ, leaving $\mathcal{V}_+$ and entering $\mathcal{V}_-$, at $t = t_\beta$. The state crosses therefore the Σ_β part of Σ at this instance. Consequently, the following holds

$$f_c = k_f x_\beta + c_f \dot{x}_\beta = 0. \tag{7.84}$$

We can now construct the saltation matrix $\boldsymbol{S}_\beta$ for the detachment event. The right-hand sides at the instance t_β are

$$\boldsymbol{f}_{\beta-} = \begin{bmatrix} \dot{x}_\beta \\ -\frac{k+k_f}{m} x_\beta - \frac{c_f}{m}\dot{x}_\beta + \frac{f_0}{m}\cos(\omega t_\beta) \end{bmatrix}, \tag{7.85}$$

$$\boldsymbol{f}_{\beta+} = \begin{bmatrix} \dot{x}_\beta \\ -\frac{k}{m} x_\beta + \frac{f_0}{m}\cos(\omega t_\beta) \end{bmatrix}. \tag{7.86}$$

If we substitute (7.84) in (7.85) and (7.86), then the latter equations appear to be identical $\boldsymbol{f}_{\beta-} = \boldsymbol{f}_{\beta+}$. Consequently, $\boldsymbol{S}_\beta$ is simply the identity matrix, i.e. $\boldsymbol{S}_\beta = \boldsymbol{I}$.

The results show that the saltation matrices $\boldsymbol{S}_\alpha$ and $\boldsymbol{S}_\beta$ are not dependent on the support stiffness k_f. The saltation matrix $\boldsymbol{S}_\alpha$ is affected, however, by the ratio $\frac{c_f}{m}$. The physical interpretation must be sought in the discontinuity of the contact force f_c. The spring force before the impact event, kx, is equal to the spring force after the impact event, $(k + k_f)x$, because contact is made when $x = 0$. But the damping force before the impact event, being zero, is not equal to the damping force after the impact event, $c_f \dot{x}$. The contact force f_c will be continuous for the transition from contact to no-contact (detachment), which is the reason that $\boldsymbol{S}_\beta$ is equal to the identity matrix.

If the damping coefficient c_f is set to zero, then the system reduces to a second-order system with discontinuous stiffness, which is a non-smooth continuous system. In this case, the switching boundaries Σ_α and Σ_β do not form an angle and Σ is a smooth switching boundary. The saltation matrices $\boldsymbol{S}_\alpha$ and $\boldsymbol{S}_\beta$ are both equal to the identity matrix in this case. It can be concluded that the jumps in the fundamental solution matrix are not caused by the discontinuous stiffness but by the discontinuous damping term.

7.5 Summary

In this chapter, it has been shown that the local stability of periodic solutions depends (for the hyperbolic case) on the eigenvalues of the fundamental solution matrix (called Floquet multipliers). A review of fundamental solution matrices of smooth systems has been given. Subsequently, the concept of the fundamental solution matrix has been extended to Filippov systems and it has been shown that the time evolution of the fundamental solution matrix undergoes a jump on switching boundaries. The jump of the fundamental solution matrix can be expressed by means of the saltation matrix. In the next two chapters we focus on bifurcations of equilibria and periodic solutions. Saltation matrices turn out to be very important to understand the bifurcation behaviour of periodic solutions of Filippov systems.

8

Bifurcations of Equilibria in Non-smooth Continuous Systems

Dynamical systems can possess stationary states or equilibria which can be stable or unstable. It is often desirable to know how the equilibria of a system change when a parameter of the system is changed. The number and stability of equilibria can change at a certain critical parameter value. Loosely speaking, this qualitative change in the structural behaviour of the system is called *bifurcation.*

The theory of bifurcations of equilibria in smooth vector fields is well understood [52, 54, 75, 133]. However, little is known about bifurcations of equilibria in non-smooth continuous vector fields. In this chapter, we will study bifurcations of equilibria occurring in *non-smooth continuous systems.* It will be shown that a bifurcation in a non-smooth continuous system can be accompanied by a jump of an eigenvalue over the imaginary axis under the variation of a parameter. We will try to compare the bifurcations found in non-smooth systems with bifurcations of smooth systems. Bifurcations of equilibria will be used as a stepping stone to bifurcations of periodic solutions in discontinuous systems of Filippov-type in the next chapter.

The theory of bifurcations of equilibria in smooth systems will be briefly discussed in Section 8.1. Much attention will be paid to the different definitions that exist for the term bifurcation. Section 8.2 introduces the type of non-smooth systems which will be considered. Some results on piecewise linear systems are presented in Section 8.3. Subsequently, the basic idea of a discontinuous bifurcation is presented in Section 8.4. Bifurcation theorems for planar piecewise linear systems with a single switching boundary are derived in Section 8.5. The general case is far more complex and no rigourous results on bifurcations will be given for general non-smooth continuous systems with more than one switching boundary. The complications of bifurcations in non-smooth continuous systems will be demonstrated by a number of examples in Sections 8.6 and 8.7. The bifurcations arising in non-smooth continuous systems are compared with classical bifurcations of smooth systems. Conclusions are drawn in the discussion of Section 8.8.

8.1 Bifurcations in Smooth Systems

In this chapter, we consider bifurcations of equilibria of autonomous systems which depend on one single parameter μ:

$$\dot{\boldsymbol{x}} = \boldsymbol{f}(\boldsymbol{x}, \mu). \tag{8.1}$$

Let n denote the dimension of the system, i.e. $\boldsymbol{x} \in \mathbb{R}^n$. The system (8.1) is called *smooth* if $\boldsymbol{f}(\boldsymbol{x}, \mu)$ is differentiable up to any order in both $\boldsymbol{x}$ and μ. Equilibria of (8.1) are solutions of the algebraic equations

$$\boldsymbol{0} = \boldsymbol{f}(\boldsymbol{x}, \mu). \tag{8.2}$$

In order to illustrate graphically the dependence of an equilibrium $\boldsymbol{x}$ on μ, we require a scalar measure of the n-vector $\boldsymbol{x}$. We shall use the notation $[\boldsymbol{x}]$ for such a measure of $\boldsymbol{x}$. Examples are $[\boldsymbol{x}] = x_1$ and $[\boldsymbol{x}] = \|\boldsymbol{x}\|$. A diagram depicting $[\boldsymbol{x}]$ versus μ, where $(\boldsymbol{x}, \mu)$ solves equation (8.2), will be called a *bifurcation diagram.* The continuous curves of solutions of (8.2) under variation of μ are called *branches.* The branches of smooth systems are continuous and smooth but can split into more branches. On a regular point of a branch, that is on a point where the branch does not split or turn around, we can define the *slope* of the branch. We will use the following abbreviations

$$\boldsymbol{J}(\boldsymbol{x}, \mu) := \frac{\partial \boldsymbol{f}(\boldsymbol{x}, \mu)}{\partial \boldsymbol{x}} \text{ and } \boldsymbol{f}_{,\mu}(\boldsymbol{x}, \mu) := \frac{\partial \boldsymbol{f}(\boldsymbol{x}, \mu)}{\partial \mu}. \tag{8.3}$$

Both derivatives exist for a smooth system. Using the implicit function theorem [75, 133] it follows that, provided that the Jacobian matrix $\boldsymbol{J}(\boldsymbol{x}, \mu)$ is non-singular, locally (8.2) is equivalent to writing $\boldsymbol{x}$ as a function of μ, i.e. $\boldsymbol{0} = \boldsymbol{f}(\boldsymbol{x}(\mu), \mu)$. Then it follows from differentiating (8.2) with respect to μ that

$$\boldsymbol{J}(\boldsymbol{x}, \mu)\frac{d\boldsymbol{x}}{d\mu} + \boldsymbol{f}_{,\mu}(\boldsymbol{x}, \mu) = \boldsymbol{0}. \tag{8.4}$$

As $\boldsymbol{J}(\boldsymbol{x}, \mu)$ is non-singular we can solve for $d\boldsymbol{x}/d\mu$. A point $(\boldsymbol{x}, \mu)$ is called regular if $\det(\boldsymbol{J}(\boldsymbol{x}, \mu)) \neq 0$. The slope s of a branch on a regular point can be found from

$$s = \frac{d[\boldsymbol{x}]}{d\mu} = \frac{\partial[\boldsymbol{x}]}{\partial \boldsymbol{x}}\frac{d\boldsymbol{x}}{d\mu} = -\frac{\partial[\boldsymbol{x}]}{\partial \boldsymbol{x}}\boldsymbol{J}^{-1}(\boldsymbol{x}, \mu)\boldsymbol{f}_{,\mu}(\boldsymbol{x}, \mu). \tag{8.5}$$

The scalar measure $[\boldsymbol{x}]$ should of course be such that it is indeed differentiable with respect to $\boldsymbol{x}$ at all equilibrium points on the branch. The choice $[\boldsymbol{x}] = \|\boldsymbol{x}\|$ can cause problems as it is not everywhere differentiable.

Some definitions from bifurcation theory will be repeated. Seydel [133] defines a *bifurcation point* in the following way:

Definition 8.1 (Geometric definition of a bifurcation [133]). *A bifurcation point (with respect to μ) is a solution $(\boldsymbol{x}^*, \mu^*)$, where the number of equilibria or (quasi-) periodic solutions changes when μ passes μ^*.*

Remark: A quasi-periodic solution is a solution that can be expressed as a countable sum of time-periodic functions with an incommensurate set of base frequencies [52, 116, 133].

Definition 8.1 is to be understood that also the number of equilibria and (quasi-) periodic solutions at the point under consideration have to be taken into account. Consider for instance the bifurcation diagram depicted in Figure 8.1b. In this case there are two equilibria for $\mu < 0$, one equilibrium for $\mu = 0$ (which is the point under consideration) and two equilibria for $\mu > 0$. The point $(x, \mu) = (0, 0)$ is therefore a bifurcation point because the number of equilibria changes at this point for varying μ (the change is: 2–1–2). We conclude that if branches intersect, then their intersection point must be a bifurcation point.

Likewise, the system $\dot{x} = \mu x$ has one equilibrium for $\mu < 0$, an infinite number of equilibria for $\mu = 0$ (which is the point under consideration) and one equilibrium for $\mu > 0$ (the change is: 1–∞–1). The point $(x, \mu) = (0, 0)$ is therefore a bifurcation point. Definition 8.1 is a purely geometric definition of a bifurcation, which does not use any knowledge about the stability of the limit sets. Stability might be exchanged at a bifurcation point but this is not necessary. Consider for instance the two-dimensional system

$$\begin{aligned} \dot{x}_1 &= \mu x_1, \\ \dot{x}_2 &= x_2. \end{aligned}$$

Clearly, there is a bifurcation at $(x, \mu) = (0, 0)$ but stability is not exchanged (the equilibrium changes from a saddle to an unstable node). For one-dimensional smooth systems however, a bifurcation is accompanied by an exchange of stability.

Another example is the saddle–node (or turning point) bifurcation depicted in Figure 8.1a: there are zero equilibria for $\mu < 0$, one equilibrium for $\mu = 0$ and two equilibria for $\mu > 0$. One could be tempted to think that a bifurcation occurs if the slope of the branch becomes vertical. A counter example is the system

$$\dot{x} = f(x, \mu) = \mu - x^3.$$

The Jacobian is $\boldsymbol{J}(x, \mu) = -3x^2$ and therefore $\boldsymbol{J}(0, 0) = 0$. Furthermore, $f_{,\mu} \neq 0$ so the slope at $(0, 0)$ is vertical. But, if we study the bifurcation diagram in Figure 8.2, then we see that $(0, 0)$ is not a bifurcation point according to Definition 8.1. Such a point is called a hysteresis point [133]. Remark that the hysteresis effect is caused by the fact that

$$\left.\frac{\partial^2 f(x, \mu)}{\partial x^2}\right|_{\substack{x=0 \\ \mu=0}} = 0.$$

A bifurcation diagram can be misleading. Two branches can cross each other in the two-dimensional bifurcation diagram without intersecting in the

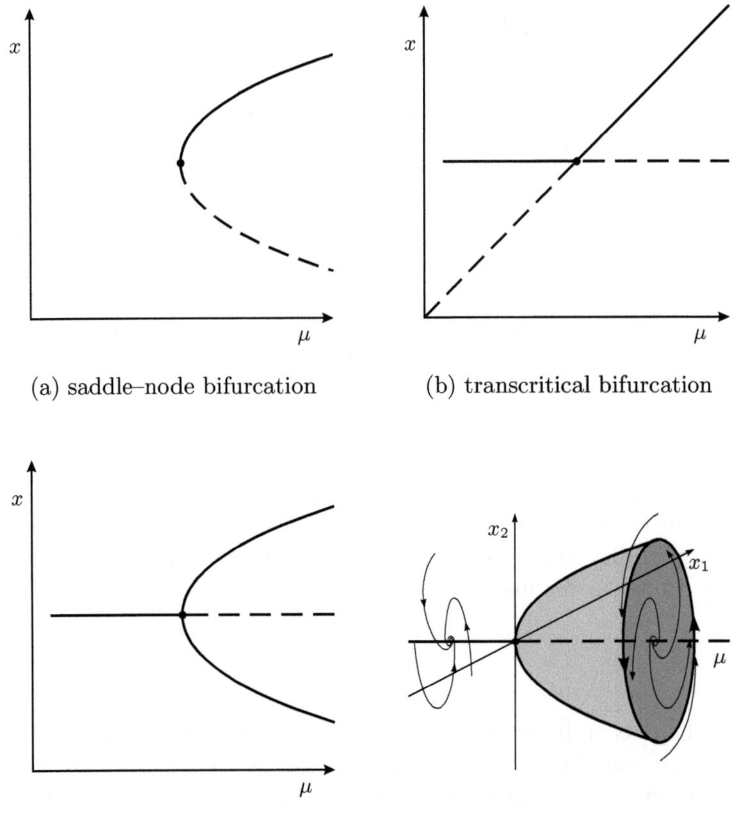

(a) saddle–node bifurcation (b) transcritical bifurcation

(c) pitchfork bifurcation (d) Hopf bifurcation

Fig. 8.1. Bifurcations of equilibria in smooth systems.

multi-dimensional space. Such a point is not a bifurcation point. The problem is caused by the projection of the multi-dimensional state-parameter space on the two-dimensional bifurcation diagram and can be circumvented by another choice for $[\boldsymbol{x}]$. We will make the assumption that a crossing of two branches in the bifurcation diagram represents an intersection of the branches in the multi-dimensional space. Note that this is never a problem if $n = 1$.

Bifurcations in the sense of Definition 8.1 expose a topological change of the phase portrait of the system as its parameter passes the bifurcation point. We will adopt the following definition for topological equivalence taken from [75]:

Definition 8.2 (Topological equivalence [75]). *A dynamical system* $\dot{\boldsymbol{x}} = \boldsymbol{f}(\boldsymbol{x})$, $\boldsymbol{x} \in \mathbb{R}^n$, *is topologically equivalent in a region* $U \subset \mathbb{R}^n$ *to a dynamical system* $\dot{\boldsymbol{y}} = \boldsymbol{g}(\boldsymbol{y})$, $\boldsymbol{y} \in \mathbb{R}^n$, *in a region* $V \subset \mathbb{R}^n$ *if there is a homeomorphism*

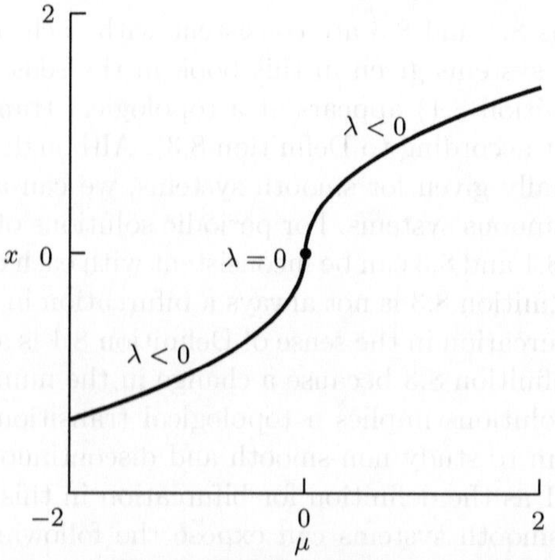

Fig. 8.2. Hysteresis point.

$h : U \to V$, $h(U) = V$, mapping trajectories of the first system in U onto trajectories of the second system in V, preserving the direction of time.

Remark: A homeomorphism is an invertible map such that both the map and its inverse are continuous.

In [75] a definition of a bifurcation is given based on topological non-equivalence of the phase portrait (Definition 8.2):

Definition 8.3 (Definition of bifurcation through topological equivalence [75]). *The appearance of topologically nonequivalent phase portraits under variation of the bifurcation parameter μ is called a bifurcation.*

A third definition of a bifurcation is based on (the loss of) structural stability [52]. An equilibrium is structurally stable if the Jacobian matrix (or the linearized system around the equilibrium) does not have eigenvalues on the imaginary axis.

Definition 8.4 (Definition of bifurcation through structural stability [52]). *A value μ^* of equation (8.1) for which the solution of (8.1) is not structurally stable is a bifurcation value of μ.*

Definition 8.4 is in some sense unsatisfactory. The system $\dot{x} = f(x, \mu) = 0$ is structurally unstable for all values of μ. Definition 8.4 suggests that all values of μ are bifurcation points. No bifurcation exists according to Definitions 8.1 and 8.3. Also a hysteresis point would be a bifurcation point according to Definition 8.4. Definition 8.4 will therefore not be used. It is noted that, for smooth systems, if a point is a bifurcation point according to Definitions 8.1 and 8.3, then it is also a bifurcation according to Definition 8.4. The implication does not hold in the other direction.

The Definitions 8.1 and 8.3 are consistent with each other for all the examples of smooth systems given in this book in the sense that a bifurcation of branches (Definition 8.1) appears at a topological transition point (being a bifurcation point according to Definition 8.3). Although Definitions 8.1, 8.2 and 8.3 are originally given for smooth systems, we can apply them to non-smooth or discontinuous systems. For periodic solutions of discontinuous systems, Definitions 8.1 and 8.3 can be inconsistent with each other. A bifurcation in the sense of Definition 8.3 is not always a bifurcation in the sense of Definition 8.1. But a bifurcation in the sense of Definition 8.1 is always a bifurcation in the sense of Definition 8.3 because a change in the number of equilibria or (quasi)-periodic solutions implies a topological transition of the phase portrait. With the aim to study non-smooth and discontinuous systems, we will take Definition 8.1 as the definition for bifurcation in this book.

Equilibria of smooth systems can expose the following bifurcations (see Figure 8.1):

a) *Saddle–node bifurcation* or *Turning point bifurcation*: a branch of equilibria turns around under variation of a parameter,
b) *Transcritical bifurcation*: two branches of equilibria intersect under variation of a parameter, such that both branches exist on either side of the bifurcation point in the bifurcation diagram,
c) *Pitchfork bifurcation*: two branches of equilibria intersect under variation of a parameter, such that one branch exists on either side of the bifurcation point in the bifurcation diagram and the other branch turns around at the intersection point,
d) *Hopf bifurcation*: a branch of periodic solutions emanates from an equilibrium branch.

Bifurcations (a)-(c) are static bifurcations, at which only branches of equilibria meet, and (d) is a dynamic bifurcation of an equilibrium point where a branch of periodic solutions is created at the bifurcation point. Each of the bifurcations (a)-(c) leads, for one-dimensional systems, to an exchange of stability of an equilibrium. A Hopf bifurcation can only be present in a two- or higher-dimensional system and leads, for a two-dimensional smooth system to an exchange of stability. A bifurcation of an equilibrium branch in a higher-dimensional system causes the equilibrium to gain or lose stability within an eigenspace (not necessarily implying an exchange of stability of the equilibrium). Bifurcations of equilibria are therefore directly associated with an eigenvalue (or pair of complex conjugated eigenvalues) that moves from the left-half complex plane to the right-half complex plane (or vice versa) under variation of a parameter. The Jacobian matrices of smooth systems are smooth functions of the state vector and parameter. The eigenvalues of the Jacobian matrix will therefore also depend continuously (but not necessarily smooth) on the parameter. A bifurcation of an equilibrium point of a smooth system occurs when one eigenvalue (or a pair of them) passes the imaginary axis when a parameter is varied. The scenario is depicted in Figure 8.3a where

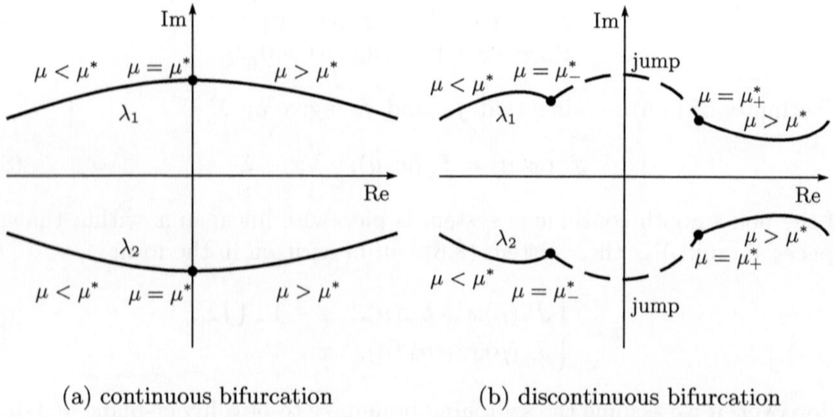

Fig. 8.3. Eigenvalue paths at a continuous bifurcation (left) and a discontinuous single crossing bifurcation (right).

a pair of complex conjugated eigenvalues passes the imaginary axis when a parameter μ is varied and a Hopf bifurcation occurs at some critical value $\mu = \mu^*$. The bifurcations occurring in smooth systems are called *continuous bifurcations* in this book because the eigenvalues behave continuously.

8.2 Non-smooth Continuous Systems

An autonomous dynamical system of the form (8.1)

$$\dot{\boldsymbol{x}} = \boldsymbol{f}(\boldsymbol{x}, \mu),$$

dependent on a parameter μ, is called a *non-smooth continuous system* if $\boldsymbol{f}(\boldsymbol{x}, \mu)$ is continuous in $\boldsymbol{x}$ but non-smooth on one or more switching boundaries Σ. The system

$$\dot{x} = |x + \mu| + 3x^2$$

is for instance a non-smooth continuous system due to the absolute value operator. Non-smooth continuous systems are therefore a subclass of Filippov systems (see Chapter 3). Non-smooth continuous system with a single switching boundary can generally be put in the form

$$\dot{\boldsymbol{x}} = \begin{cases} \boldsymbol{f}_-(\boldsymbol{x}, \mu), & \boldsymbol{x} \in \mathcal{V}_- \bigcup \Sigma, \\ \boldsymbol{f}_+(\boldsymbol{x}, \mu), & \boldsymbol{x} \in \mathcal{V}_+ \end{cases} \tag{8.6}$$

depending on a parameter μ and being continuous on the switching boundary Σ. The switching boundary function $h(\boldsymbol{x}, \mu)$ defines the subspaces

$$\begin{aligned} \mathcal{V}_- &= \{\boldsymbol{x} \in \mathbb{R}^n \mid h(\boldsymbol{x}, \mu) < 0\}, \\ \Sigma &= \{\boldsymbol{x} \in \mathbb{R}^n \mid h(\boldsymbol{x}, \mu) = 0\}, \\ \mathcal{V}_+ &= \{\boldsymbol{x} \in \mathbb{R}^n \mid h(\boldsymbol{x}, \mu) > 0\}. \end{aligned} \tag{8.7}$$

Continuity of (8.6) requires that $\boldsymbol{f}_+$ and $\boldsymbol{f}_-$ agree on Σ

$$\boldsymbol{f}_-(\boldsymbol{x}, \mu) = \boldsymbol{f}_+(\boldsymbol{x}, \mu), \quad \forall \boldsymbol{x} \in \Sigma. \tag{8.8}$$

If the non-smooth continuous system is piecewise linear in $\boldsymbol{x}$ within the subspaces $\mathcal{V}_-$ and $\mathcal{V}_+$, then system (8.6) can be written in the form

$$\dot{\boldsymbol{x}} = \begin{cases} \boldsymbol{J}_-(\mu)\boldsymbol{x} + \boldsymbol{b}_-(\mu), & \boldsymbol{x} \in \mathcal{V}_- \bigcup \Sigma, \\ \boldsymbol{J}_+(\mu)\boldsymbol{x} + \boldsymbol{b}_+(\mu), & \boldsymbol{x} \in \mathcal{V}_+. \end{cases} \tag{8.9}$$

Moreover, if we assume the switching boundary to be a hyper-plane $h(\boldsymbol{x}, \mu) = \boldsymbol{w}^{\mathrm{T}}(\mu)\boldsymbol{x} + w(\mu)$, then the continuity condition (8.8) yields

$$-\left(\boldsymbol{J}_+(\mu) - \boldsymbol{J}_-(\mu)\right)\boldsymbol{x} = \boldsymbol{b}_+(\mu) - \boldsymbol{b}_-(\mu), \qquad \forall \boldsymbol{x} \in \Sigma, \tag{8.10}$$

or, for non-singular $(\boldsymbol{J}_+(\mu) - \boldsymbol{J}_-(\mu))$,

$$-\boldsymbol{w}^{\mathrm{T}}(\mu)\left(\boldsymbol{J}_+(\mu) - \boldsymbol{J}_-(\mu)\right)^{-1}\left(\boldsymbol{b}_+(\mu) - \boldsymbol{b}_-(\mu)\right) + w(\mu) = \boldsymbol{0}. \tag{8.11}$$

Non-smooth continuous systems are nonlinear and generally not piecewise linear. If the equilibrium is located on a switching boundary, then the nonlinear system can locally be approximated by a piecewise linear system of the form (8.9).

8.3 Co-existing Equilibria for a Single Switching Boundary

The co-existence of equilibria *for a subclass* of piecewise linear systems with only one switching boundary can conveniently be analyzed with a method developed by Feigin [30, 32, 33]. The method was developed to analyze co-existence of fixed points in piecewise linear mappings used to study periodic solutions [24]. The method can, with little adjustments, be applied to the analysis of equilibria of a special class of continuous piecewise linear differential equations with a single switching boundary, as is shown in [25] and in the sequel of this section.

Consider a piecewise linear system of the following special form

$$\dot{\boldsymbol{x}} = \begin{cases} \boldsymbol{J}_-\boldsymbol{x} + \boldsymbol{c}\mu, & \boldsymbol{x} \in \mathcal{V}_- \bigcup \Sigma, \\ \boldsymbol{J}_+\boldsymbol{x} + \boldsymbol{c}\mu, & \boldsymbol{x} \in \mathcal{V}_+, \end{cases} \tag{8.12}$$

depending on a parameter μ and being continuous on the switching boundary Σ. Let the switching function be given by $h(\boldsymbol{x}) = x_n$, not being dependent on μ, which defines the subspaces $\mathcal{V}_-$, Σ and $\mathcal{V}_+$ (8.7). Note that the class of piecewise linear systems of the form (8.12) is very special. The Jacobian matrices do not depend on μ and it holds that $\boldsymbol{b}_-(\mu) = \boldsymbol{b}_+(\mu) = \boldsymbol{c}\mu$, which is linear in μ. The system has only a single switching boundary which is the hyper-plane $x_n = 0$. A system of the form (8.12) with an arbitrary switching boundary can always be transformed to have $x_n = 0$ as switching boundary. We explicitly state that the non-smooth continuous systems considered in the next sections are generally not of the form (8.12) but belong to the more general class of piecewise linear systems (8.9) with one or more switching boundaries.

Continuity at the switching boundary of system (8.12) implies that the first $n-1$ columns of the Jacobians agree, i.e. $\boldsymbol{J}_{-_{ik}} = \boldsymbol{J}_{+_{ik}}, \forall i = 1 \ldots n, k = 1 \ldots n-1$. Assuming that $\boldsymbol{J}_-$ and $\boldsymbol{J}_+$ are non-singular, at most one equilibrium can exist in each of the subspaces $\mathcal{V}_-$ and $\mathcal{V}_+$ because the system is linear within these subspaces. Let $\boldsymbol{x}_- \in \mathcal{V}_- \bigcup \Sigma$ and $\boldsymbol{x}_+ \in \mathcal{V}_+$ be equilibria of system (8.12), i.e.

$$\begin{aligned} \boldsymbol{0} &= \boldsymbol{J}_-\boldsymbol{x}_- + \boldsymbol{c}\mu, \\ \boldsymbol{0} &= \boldsymbol{J}_+\boldsymbol{x}_+ + \boldsymbol{c}\mu. \end{aligned} \tag{8.13}$$

If $\boldsymbol{J}_-$ and $\boldsymbol{J}_+$ are non-singular, then we can solve for $\boldsymbol{x}_-$ and $\boldsymbol{x}_+$

$$\boldsymbol{x}_- = -\boldsymbol{J}_-^{-1}\boldsymbol{c}\mu = -\frac{\operatorname{adj}(\boldsymbol{J}_-)}{\det(\boldsymbol{J}_-)}\boldsymbol{c}\mu, \qquad \boldsymbol{x}_+ = -\boldsymbol{J}_+^{-1}\boldsymbol{c}\mu = -\frac{\operatorname{adj}(\boldsymbol{J}_+)}{\det(\boldsymbol{J}_+)}\boldsymbol{c}\mu, \tag{8.14}$$

where $\operatorname{adj}(\boldsymbol{A})$ denotes the adjoint matrix of $\boldsymbol{A}$. It follows from (8.14) that the elements of $\boldsymbol{x}_-$ and $\boldsymbol{x}_+$ can be expressed as

$$x_{-_k} = \frac{b_{-_k}}{\det(\boldsymbol{J}_-)}\mu, \qquad x_{+_k} = \frac{b_{+_k}}{\det(\boldsymbol{J}_+)}\mu, \tag{8.15}$$

with $b_{-_k} = [-\operatorname{adj}(\boldsymbol{J}_-)\boldsymbol{c}]_k$ and $b_{+_k} = [-\operatorname{adj}(\boldsymbol{J}_+)\boldsymbol{c}]_k$. It holds that $b_{-_n} = b_{+_n} =: b_n$ because the matrices $\boldsymbol{J}_-$ and $\boldsymbol{J}_+$ differ only in the n-th column (see also [24]). The n-th elements of the equilibria $\boldsymbol{x}_-$ and $\boldsymbol{x}_+$

$$x_{-_n} = \frac{b_n}{\det(\boldsymbol{J}_-)}\mu, \qquad x_{+_n} = \frac{b_n}{\det(\boldsymbol{J}_+)}\mu, \tag{8.16}$$

are therefore only a function of b_n, the determinant of the Jacobian and the parameter μ. We have to require that $x_{-_n} \leq 0$ and $x_{+_n} > 0$ in order to let the equilibria be admissible $\boldsymbol{x}_- \in \mathcal{V}_- \bigcup \Sigma$ and $\boldsymbol{x}_+ \in \mathcal{V}_+$. Only one of the equilibria exists for $\mu = 0$, i.e. $\boldsymbol{x}_{-,\mu=0} = \boldsymbol{0}$, being located at the switching boundary. The two equilibria co-exist for $\mu < 0$ *or* $\mu > 0$ if the elements x_{-_n} and x_{+_n} have opposite signs. Co-existence of the equilibria for $\mu < 0$ ($\mu > 0$) implies non-existence of equilibria for $\mu > 0$ ($\mu < 0$). A necessary and sufficient condition for co-existence of equilibria of system (8.12) is

$$\det(\boldsymbol{J}_-)\det(\boldsymbol{J}_+) < 0. \tag{8.17}$$

The equilibria exist for opposite signs of μ if

$$\det(\boldsymbol{J}_-)\det(\boldsymbol{J}_+) > 0. \tag{8.18}$$

The co-existence of equilibria of system (8.12) is therefore determined by the signs of the determinants of the Jacobian matrices. Moreover, the sign of the determinant of the Jacobian depends solely on the number of real valued eigenvalues, because $\det(\boldsymbol{J}) = \lambda_1\lambda_2\ldots\lambda_n$ and $\lambda\bar{\lambda} \geq 0$ (in which $\bar{\lambda}$ is the complex conjugate of λ). If the number of negative real valued eigenvalues is odd, then the determinant of the non-singular Jacobian is negative. If the number of negative real valued eigenvalues is even, then the determinant is positive. A non-singular Jacobian with only complex conjugated eigenvalues has therefore a positive determinant.

The co-existence conditions of Feigin (i.e. (8.17) and (8.18)) have direct consequences for the bifurcation behaviour of system (8.12). If condition (8.17) is satisfied, then a bifurcation in the sense of definition 8.1 must exist for $\mu = 0$ because the number of equilibria for $\mu < 0$ is different from the number of equilibria for $\mu > 0$. The branch of equilibria turns around at $\mu = 0$ and the bifurcation point can appropriately be named as turning point bifurcation. If condition (8.18) is satisfied, then the branch of equilibria for $\mu < 0$ continues for $\mu > 0$. Bifurcations of fixed points of piecewise linear mappings were (partly) classified by means of conditions (8.17) and (8.18) in the work of Feigin and di Bernardo [24, 25, 30, 32, 33].

8.4 Discontinuous Bifurcation: The Basic Idea

Some aspects and definitions of bifurcations of equilibria in smooth systems have been briefly recalled in Section 8.1. Bifurcations of equilibria in smooth systems are associated with an eigenvalue (or pair of complex conjugated eigenvalues) that passes the imaginary axis under variation of a parameter. The bifurcation analysis of dynamical systems therefore hinges on the dependency of the Jacobian matrix (and its eigenvalues) on system parameters. Non-smooth continuous systems possess switching boundaries on which the vector field is non-smooth and for which the classical Jacobian matrix can not be obtained. In this section we will try to enlarge the concept of bifurcation to the class of non-smooth continuous systems.

Consider an autonomous non-smooth continuous system of the form (8.6), having a single switching boundary Σ. Let $\boldsymbol{x}_\mu$ be an equilibrium point of (8.6) for some value of μ, i.e.

$$\boldsymbol{f}_-(\boldsymbol{x}_\mu, \mu) = \boldsymbol{0} \text{ for } \boldsymbol{x}_\mu \in \mathcal{V}_- \cup \Sigma \quad \text{or} \quad \boldsymbol{f}_+(\boldsymbol{x}_\mu, \mu) = \boldsymbol{0} \text{ for } \boldsymbol{x}_\mu \in \mathcal{V}_+.$$

If $\boldsymbol{x}_\mu$ is not on Σ, then we can find a single-valued Jacobian matrix $\boldsymbol{J}(\boldsymbol{x}_\mu, \mu)$

$$\boldsymbol{J}(\boldsymbol{x}_\mu, \mu) = \begin{cases} \boldsymbol{J}_-(\boldsymbol{x}_\mu, \mu) = \frac{\partial \boldsymbol{f}_-(\boldsymbol{x},\mu)}{\partial \boldsymbol{x}}\Big|_{\boldsymbol{x}=\boldsymbol{x}_\mu}, & \boldsymbol{x}_\mu \in \mathcal{V}_-, \\ \boldsymbol{J}_+(\boldsymbol{x}_\mu, \mu) = \frac{\partial \boldsymbol{f}_+(\boldsymbol{x},\mu)}{\partial \boldsymbol{x}}\Big|_{\boldsymbol{x}=\boldsymbol{x}_\mu}, & \boldsymbol{x}_\mu \in \mathcal{V}_+, \end{cases}$$

that locally defines the vector field around the equilibrium point $\boldsymbol{x}_\mu$ if the matrix $\boldsymbol{J}(\boldsymbol{x}_\mu, \mu)$ does not have eigenvalues on the imaginary axis. The matrices $\boldsymbol{J}_-(\boldsymbol{x}, \mu)$ and $\boldsymbol{J}_+(\boldsymbol{x}, \mu)$ are the Jacobian matrices on either side of Σ associated with the vector field in $\mathcal{V}_-$ and $\mathcal{V}_+$. If $\boldsymbol{x}_\mu \in \Sigma$, then the local vector field is determined by *two* Jacobian matrices, i.e. $\boldsymbol{J}_-(\boldsymbol{x}_\mu, \mu)$ and $\boldsymbol{J}_+(\boldsymbol{x}_\mu, \mu)$. Assume that we vary μ such that the equilibrium point $\boldsymbol{x}_\mu$ moves from $\mathcal{V}_-$ to $\mathcal{V}_+$ via Σ. Let $\boldsymbol{x}_\Sigma$ denote the unique equilibrium on Σ for $\mu = \mu_\Sigma$, i.e.

$$\begin{aligned} &\boldsymbol{x}_\mu \in \mathcal{V}_-, && \mu < \mu_\Sigma, \\ &\boldsymbol{x}_\mu = \boldsymbol{x}_\Sigma \in \Sigma, && \mu = \mu_\Sigma, \\ &\boldsymbol{x}_\mu \in \mathcal{V}_+, && \mu > \mu_\Sigma. \end{aligned}$$

The Jacobian matrix $\boldsymbol{J}(\boldsymbol{x}_\mu, \mu)$ varies as μ is varied and is discontinuous at $\mu = \mu_\Sigma$ for which $\boldsymbol{x}_\mu = \boldsymbol{x}_\Sigma$. Loosely speaking, we say that $\boldsymbol{J}(\boldsymbol{x}_\mu, \mu)$ 'jumps' at $\mu = \mu_\Sigma$ from $\boldsymbol{J}_-(\boldsymbol{x}_\Sigma, \mu_\Sigma)$ to $\boldsymbol{J}_+(\boldsymbol{x}_\Sigma, \mu_\Sigma)$. A jump of the Jacobian matrix under the influence of a parameter implies a jump of the eigenvalues. In Section 2.3 we elaborated how we can define a generalized differential $\partial \boldsymbol{f}$. Similarly, a *generalized Jacobian* $\boldsymbol{J}(\boldsymbol{x}, \mu)$ was defined in (2.12) as the transpose of the generalized differential of $\boldsymbol{f}$ with respect to $\boldsymbol{x}$

$$\boldsymbol{J}(\boldsymbol{x}, \mu) = \left(\partial_{\boldsymbol{x}} \boldsymbol{f}(\boldsymbol{x}, \mu)\right)^{\mathrm{T}}, \tag{8.19}$$

which is set-valued at $(\boldsymbol{x}_\Sigma, \mu_\Sigma)$. The generalized Jacobian at $(\boldsymbol{x}_\Sigma, \mu_\Sigma)$ is therefore the closed convex hull of $\boldsymbol{J}_-(\boldsymbol{x}_\Sigma, \mu_\Sigma)$ and $\boldsymbol{J}_+(\boldsymbol{x}_\Sigma, \mu_\Sigma)$

$$\begin{aligned} \boldsymbol{J}(\boldsymbol{x}_\Sigma, \mu_\Sigma) &= \overline{\mathrm{co}}\{\boldsymbol{J}_-(\boldsymbol{x}_\Sigma, \mu_\Sigma), \boldsymbol{J}_+(\boldsymbol{x}_\Sigma, \mu_\Sigma)\} \\ &= \{(1-q)\boldsymbol{J}_-(\boldsymbol{x}_\Sigma, \mu_\Sigma) + q\boldsymbol{J}_+(\boldsymbol{x}_\Sigma, \mu_\Sigma), \forall q \in [0,1]\}. \end{aligned} \tag{8.20}$$

In fact, (8.20) defines *how* the Jacobian 'jumps' at Σ. The generalized Jacobian is, for a system with a single switching boundary, a convex combination of two matrices $\boldsymbol{J}_-(\boldsymbol{x}, \mu)$ and $\boldsymbol{J}_+(\boldsymbol{x}, \mu)$ if $\boldsymbol{x} \in \Sigma$. To be more precise, (8.20) gives the set of values which the generalized Jacobian can attain on Σ. From the set-valued generalized Jacobian we can obtain the set-valued eigenvalues. We can look upon $\mathrm{eig}(\boldsymbol{J}(\boldsymbol{x}_\Sigma, \mu_\Sigma))$ together with (8.20) as if it gives a *unique* path of eigenvalues 'during' the jump as q is varied from 0 to 1.

Systems with multiple switching boundaries can possess equilibria located on the crossing of two or more switching boundaries. Two switching boundaries Σ_1 and Σ_2 divide the state space in four subspaces $\mathcal{V}_{++}$, $\mathcal{V}_{+-}$, $\mathcal{V}_{-+}$ and $\mathcal{V}_{--}$ (Figure 8.4). The generalized Jacobian of an equilibrium point $\boldsymbol{x}_\Sigma \in \Sigma_1 \bigcap \Sigma_2$ located on the crossing of Σ_1 and Σ_2, is the convex hull of four Jacobian matrices

$$\begin{aligned} \boldsymbol{J}(\boldsymbol{x}_\Sigma, \mu_\Sigma) &= \overline{\mathrm{co}}\{\boldsymbol{J}_{++}(\boldsymbol{x}_\Sigma, \mu_\Sigma), \boldsymbol{J}_{+-}(\boldsymbol{x}_\Sigma, \mu_\Sigma), \boldsymbol{J}_{-+}(\boldsymbol{x}_\Sigma, \mu_\Sigma), \boldsymbol{J}_{++}(\boldsymbol{x}_\Sigma, \mu_\Sigma)\} \\ &= \{q_1(\boldsymbol{J}_{-+} - \boldsymbol{J}_{--}) + q_2(\boldsymbol{J}_{+-} - \boldsymbol{J}_{--}) + \boldsymbol{J}_{--}, \forall q_1, q_2 \in [0,1]\}, \end{aligned} \tag{8.21}$$

where use has been made of the continuity condition $\boldsymbol{J}_{+-} - \boldsymbol{J}_{--} = \boldsymbol{J}_{++} - \boldsymbol{J}_{-+}$ and in which the dependency of $(\boldsymbol{x}_\Sigma, \mu_\Sigma)$ has been omitted for brevity.

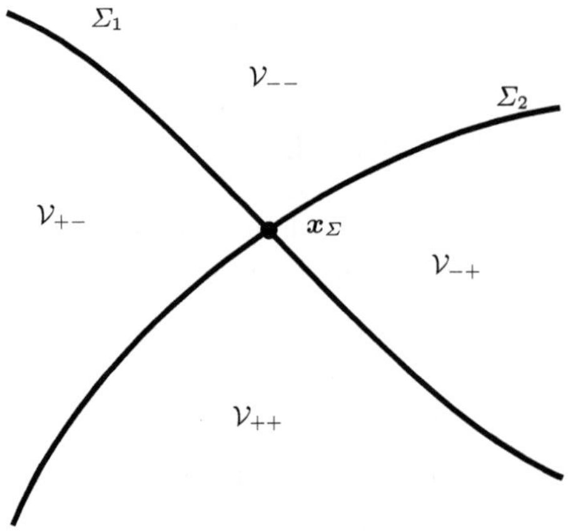

Fig. 8.4. State space with two switching boundaries.

Consequently, a generalized Jacobian of a point on k switching boundaries is dependent on k auxiliary variables q_i, $i = 1 \dots k$. If the equilibrium point $\boldsymbol{x}_\mu$ of a system with two switching boundaries moves under the variation of μ from one subspace to another, e.g.

$$\begin{aligned} &\boldsymbol{x}_\mu \in \mathcal{V}_{--}, && \mu < \mu_\Sigma, \\ &\boldsymbol{x}_\mu = \boldsymbol{x}_\Sigma \in \Sigma_1 \bigcap \Sigma_2, && \mu = \mu_\Sigma, \\ &\boldsymbol{x}_\mu \in \mathcal{V}_{++}, && \mu > \mu_\Sigma, \end{aligned}$$

then the generalized Jacobian (8.21) does not define a *unique path* of eigenvalues 'during' the jump as q_2 can be varied independently from q_1. The values of q_1 and q_2 both vary from 0 to 1 as x_Σ moves from $\mathcal{V}_{--}$ to $\mathcal{V}_{++}$ but (q_1, q_2) is unknown. Still, (8.21) defines a jump of the eigenvalues of which only the begin and end points are known.

Example 8.1
Consider the equilibrium point $(x_1, x_2) = (0, 0)$ for $\mu = 0$ of the system

$$\begin{aligned} \dot{x}_1 &= x_2, \\ \dot{x}_2 &= -|x_1 + 4\mu| - |x_2 - \mu| - \tfrac{1}{2}x_2 + 5\mu, \end{aligned} \tag{8.22}$$

which is located on the crossing of two switching boundaries $\Sigma_1 = \{x_1 = -4\mu\}$ and $\Sigma_2 = \{x_1 = \mu\}$. The Jacobian matrices in the four subspaces surrounding the equilibrium point are

$$\boldsymbol{J}_{++} = \begin{bmatrix} 0 & 1 \\ -1 & -\frac{3}{2} \end{bmatrix}, \quad \boldsymbol{J}_{+-} = \begin{bmatrix} 0 & 1 \\ -1 & \frac{1}{2} \end{bmatrix}, \quad \boldsymbol{J}_{-+} = \begin{bmatrix} 0 & 1 \\ 1 & -\frac{3}{2} \end{bmatrix}, \quad \boldsymbol{J}_{--} = \begin{bmatrix} 0 & 1 \\ 1 & \frac{1}{2} \end{bmatrix}. \tag{8.23}$$

The generalized Jacobian of the equilibrium point $\boldsymbol{x}_\Sigma = \boldsymbol{0}$ is set-valued and is the convex hull of four Jacobian matrices

$$\begin{aligned} \boldsymbol{J}(\boldsymbol{x}_\Sigma, 0) &= \overline{\text{co}}\{\boldsymbol{J}_{++}, \boldsymbol{J}_{+-}, \boldsymbol{J}_{-+}, \boldsymbol{J}_{++}\} \\ &= \{q_1(\boldsymbol{J}_{-+} - \boldsymbol{J}_{--}) + q_2(\boldsymbol{J}_{+-} - \boldsymbol{J}_{--}) + \boldsymbol{J}_{--}, \forall q_1, q_2 \in [0, 1]\}. \end{aligned} \tag{8.24}$$

The set-valued generalized Jacobian has set-valued eigenvalues. The set is two-dimensional in the sense that it is dependent on two auxiliary variables q_1 and q_2. The set of eigenvalues is therefore not a one-dimensional path but a two-dimensional subspace in the complex plane (see Figure 8.5). The bifurcation point of system (8.22) will be analyzed in Example 8.5.

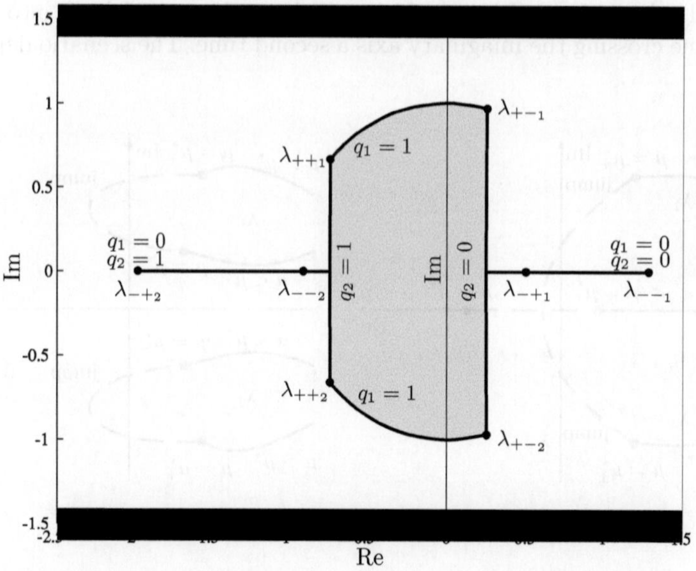

Fig. 8.5. Set of eigenvalues of the generalized Jacobian (8.24).

It is important to realize that for smooth systems the eigenvalues are single-valued functions of the parameter μ and that the eigenvalues are set-valued functions in μ for non-smooth continuous systems. An eigenvalue can pass the imaginary axis while varying μ, leading to a classical bifurcation, but it can also cross the imaginary axis during its jump defined by the generalized Jacobian. Examples will be given in the next sections where jumps of eigenvalues over the imaginary axis lead to non-classical bifurcations.

We will name a bifurcation associated by a jump of an eigenvalue (or a pair of them) over the imaginary axis a *discontinuous bifurcation.* If the system possesses (locally) only one switching boundary, then one can speak of a *path* of the eigenvalues. A typical scenario of a discontinuous bifurcation is depicted in Figure 8.3b where the unique path of a pair of complex conjugated eigenvalues on the jump is indicated by the dashed lines. The eigenvalue-path 'during' the jump is determined by the eigenvalues of the convex hull of the Jacobian matrices $\boldsymbol{J}_-(\boldsymbol{x}_\Sigma, \mu_\Sigma)$ and $\boldsymbol{J}_+(\boldsymbol{x}_\Sigma, \mu_\Sigma)$.

The possibility of the eigenvalues to become set-valued greatly complicates the bifurcation behaviour as the eigenvalue(s) could also cross the imaginary axis multiple times during its jump. The bifurcations associated with eigenvalues that cross the imaginary axis multiple times will be called multiple crossing bifurcations. Two examples (with a one-dimensional path of eigenvalues) are depicted in Figure 8.6a and b. Figure 8.6a shows a pair of complex conjugated eigenvalues that crosses the imaginary axis during the jump, splits into two real valued eigenvalues of which one eigenvalue moves back into the left-half plane crossing the imaginary axis a second time. The scenario depicted in

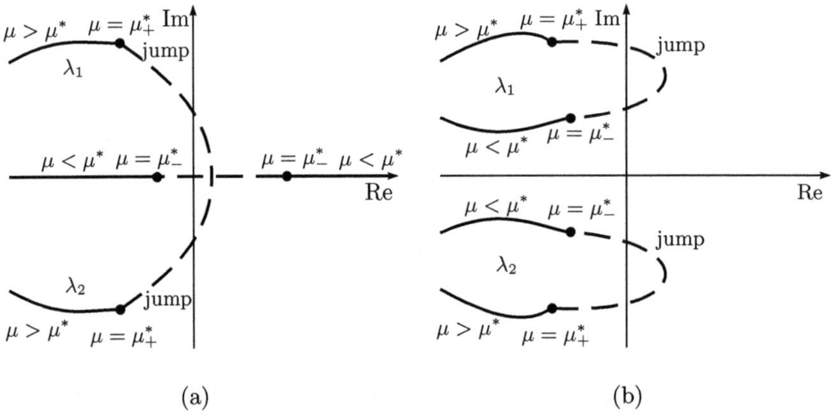

Fig. 8.6. Eigenvalue paths with multiple crossings at a bifurcation.

Figure 8.6a corresponds to a discontinuous bifurcation which is a combination of a classical Hopf bifurcation and a classical saddle–node bifurcation, as we will see in Section 8.7. A pair of complex conjugated eigenvalues crosses the imaginary axis twice during the jump in Figure 8.6b. The scenarios depicted in Figure 8.6a and b are multiple crossing bifurcations of set-valued eigenvalues which form a one-dimensional path in the complex plane. The bifurcation point of system (8.22) in Example 8.1 has set-valued eigenvalues that form an area in the complex plane (see Figure 8.5). The area of eigenvalues contains an interval of the imaginary axis. The set of eigenvalues of Example 8.1 has therefore more than one intersection point with the imaginary axis and the associated bifurcation will also be denoted by the term multiple crossing bifurcation. A discontinuous bifurcation can therefore be a single crossing bifurcation, which behaves very much like a classical bifurcation, or it can be a multiple crossing bifurcation being far more complex.

We call the type of bifurcation, at which set-valued eigenvalues cross the imaginary axis, a discontinuous bifurcation because the eigenvalues behave discontinuous at the bifurcation point. A bifurcation point, as defined by Definition 8.1, is called a *discontinuous bifurcation point* if the eigenvalues at the bifurcation point are set-valued and contain a value on the imaginary axis.

Definition 8.5 (Discontinuous Bifurcation). *Let $\boldsymbol{x}_\mu$ be an equilibrium of a non-smooth continuous system $\dot{\boldsymbol{x}} = \boldsymbol{f}(\boldsymbol{x}, \mu)$ with an arbitrary number of switching boundaries Σ_j, $j = 1, \ldots, k$. Let $\boldsymbol{x}_\mu = \boldsymbol{x}_\Sigma$ for $\mu = \mu_\Sigma$ be an equilibrium located on one or more switching boundaries, i.e. $\boldsymbol{x}_\Sigma \in \Sigma_1 \bigcap \Sigma_2 \cdots \bigcap \Sigma_l$, $1 \leq l \leq k$. Let $n_E(\mu)$ denote the number of co-existing equilibria, $n_P(\mu)$ denote the number of co-existing periodic solutions and $n_Q(\mu)$ denote the number of co-existing quasi-periodic solutions of $\dot{\boldsymbol{x}} = \boldsymbol{f}(\boldsymbol{x}, \mu)$. A bifurcation occurs at $\mu = \mu_\Sigma$ if*

$$(\lim_{\mu \uparrow \mu_\Sigma} n_E(\mu) \neq n_E(\mu)) \bigvee (\lim_{\mu \downarrow \mu_\Sigma} n_E(\mu) \neq n_E(\mu)),$$

or

$$(\lim_{\mu \uparrow \mu_\Sigma} n_P(\mu) \neq n_P(\mu)) \bigvee (\lim_{\mu \downarrow \mu_\Sigma} n_P(\mu) \neq n_P(\mu)),$$

or

$$(\lim_{\mu \uparrow \mu_\Sigma} n_Q(\mu) \neq n_Q(\mu)) \bigvee (\lim_{\mu \downarrow \mu_\Sigma} n_Q(\mu) \neq n_Q(\mu)).$$

The bifurcation point $(\boldsymbol{x}_\Sigma, \mu_\Sigma)$ is a discontinuous bifurcation point if the generalized Jacobian $\boldsymbol{J}(\boldsymbol{x}_\Sigma, \mu_\Sigma)$ is set-valued and if there exists an i such that

$$\operatorname{Re}(\lambda_i) \ni 0, \qquad \boldsymbol{\lambda} = \operatorname{eig}(\boldsymbol{J}(\boldsymbol{x}_\Sigma, \mu_\Sigma)).$$

A (discontinuous) bifurcation can be a single or a multiple crossing bifurcation. Let $\mathcal{S}_\lambda = \operatorname{eig}(\boldsymbol{J}(\boldsymbol{x}_\Sigma, \mu_\Sigma))$ denote the subspace in the complex plane of the set-valued eigenvalues and let Im^+ be the subspace of purely imaginary numbers with positive imaginary part containing the origin.

Definition 8.6 (Single Crossing Bifurcation). *If $\mathcal{S}_\lambda \bigcap \mathrm{Im}^+$ comprises only one element, then the bifurcation is a single crossing bifurcation.*

Definition 8.7 (Multiple Crossing Bifurcation). *If $\mathcal{S}_\lambda \bigcap \mathrm{Im}^+$ comprises more than one element, then the bifurcation is a multiple crossing bifurcation.*

Some important fundamental questions arise at this point:

1. Does a jump of an eigenvalue (or a pair of them) over the imaginary axis under the influence of a parameter *imply* a bifurcation in the sense of Definition 8.1?
2. Does a continuous bifurcation for a smooth approximating system exist if the non-smooth system exposes a discontinuous bifurcation?

3. Is the discontinuous bifurcation of the non-smooth system related to the continuous bifurcation of a smooth approximating system (assuming that it exists)?
4. Do discontinuous bifurcations exist that do not have a continuous counterpart? In other words: do discontinuous bifurcations exist that behave qualitatively different from any continuous bifurcation?
5. Can we classify a discontinuous bifurcation by inspecting the point(s) where the set of eigenvalues of the generalized Jacobian intersects the imaginary axis?

The answers to these questions are not straightforward. Some examples which illustrate some problems related to these questions will be studied in the next sections. Single crossing bifurcations will be discussed in Section 8.6. Continuous bifurcations belong to the class of single crossing bifurcations. A number of multiple crossing bifurcations will be studied in Section 8.7. We will return to the fundamental questions in the last section.

8.5 Planar Systems with a Single Switching Boundary

In this section we will rigourously prove some results on bifurcations in planar continuous piecewise linear systems with a single switching boundary. Consider a planar piecewise linear system of the special form (8.12)

$$\dot{\boldsymbol{x}} = \begin{cases} \boldsymbol{J}_- \boldsymbol{x} + \boldsymbol{c}\mu, & \boldsymbol{x} \in \mathcal{V}_- \bigcup \Sigma, \\ \boldsymbol{J}_+ \boldsymbol{x} + \boldsymbol{c}\mu, & \boldsymbol{x} \in \mathcal{V}_+, \end{cases} \tag{8.25}$$

with $\boldsymbol{x} = \begin{bmatrix} x_1 \ x_2 \end{bmatrix}^{\mathrm{T}}$ being continuous on the switching boundary $\Sigma = \{\boldsymbol{x} \in \mathbb{R}^2 \mid x_2 = 0\}$. The continuity conditions on the switching boundary require that $J_{-11} = J_{+11}$ and $J_{-21} = J_{+21}$. Assuming that the Jacobian matrices $\boldsymbol{J}_-$ and $\boldsymbol{J}_+$ are non-singular, at most one equilibrium can exist in each of the subspaces $\mathcal{V}_-$ and $\mathcal{V}_+$ because the system is linear and hyperbolic within these subspaces. In Section 8.3 it has been proven that the equilibria co-exist for $\mu < 0$ or for $\mu > 0$ if

$$\det(\boldsymbol{J}_-)\det(\boldsymbol{J}_+) < 0, \tag{8.26}$$

and one equilibrium exists for all μ if

$$\det(\boldsymbol{J}_-)\det(\boldsymbol{J}_+) > 0. \tag{8.27}$$

The determinant/trace of a convex combination of $\boldsymbol{J}_-$ and $\boldsymbol{J}_+$ is a convex combination of the determinants/traces of $\boldsymbol{J}_-$ and $\boldsymbol{J}_+$ as follows from the following proposition.

Proposition 8.8. *For a planar continuous piecewise linear system of the form (8.25) it holds that*

$$\det(q\boldsymbol{J}_+ + (1-q)\boldsymbol{J}_-) = q\det(\boldsymbol{J}_+) + (1-q)\det(\boldsymbol{J}_-), \tag{8.28}$$

and

$$\operatorname{trace}(q\boldsymbol{J}_+ + (1-q)\boldsymbol{J}_-) = q\operatorname{trace}(\boldsymbol{J}_+) + (1-q)\operatorname{trace}(\boldsymbol{J}_-). \tag{8.29}$$

Proof: It follows from the continuity conditions that

$$q\boldsymbol{J}_+ + (1-q)\boldsymbol{J}_- = \begin{bmatrix} J_{-11} & qJ_{+12} + (1-q)J_{-12} \\ J_{-21} & qJ_{+22} + (1-q)J_{-22} \end{bmatrix}. \tag{8.30}$$

Evaluation of the determinant and the trace completes the proof. □

The co-existence criterion (8.26) suggests that a turning point bifurcation can occur in the system (8.25).

Theorem 8.9 (Turning Point Bifurcation Theorem). *If a planar continuous piecewise linear system of the form (8.25) with a single switching boundary undergoes a turning point bifurcation, then the path of the set-valued eigenvalues crosses the imaginary axis through the origin.*

Proof: If there exists a turning point bifurcation, then it follows from (8.26) that $\det(\boldsymbol{J}_-)\det(\boldsymbol{J}_+) < 0$. The turning point $\boldsymbol{x}^*$ must be located on the switching boundary Σ. The generalized Jacobian $\boldsymbol{J}$ of the equilibrium on Σ is the set $\boldsymbol{J} = \overline{\text{co}}(\boldsymbol{J}_-, \boldsymbol{J}_+) = \{\boldsymbol{J}_q \mid \boldsymbol{J}_q = q\boldsymbol{J}_+ + (1-q)\boldsymbol{J}_-,\ q \in [0,1]\}$. It therefore must hold, using Proposition 8.8, that there exists a $q \in [0,1]$ for which $\det(\boldsymbol{J}_q) = 0$. Consequently, there exists a set-valued eigenvalue λ_i of the generalized Jacobian $\boldsymbol{J}$ containing the origin, i.e.

$$0 \in \lambda_i, \qquad \boldsymbol{\lambda} = \operatorname{eig}(\boldsymbol{J}).$$

The path of the set-valued eigenvalues of the generalized Jacobian at the discontinuous turning point bifurcation, therefore passes the origin. □

A pitchfork bifurcation can not occur in a system of the form (8.25) because at most two equilibria can co-exist for one value of μ. Similarly, a transcritical bifurcation of (8.25) is impossible because the equilibria $\boldsymbol{x}_- = -\boldsymbol{J}_-^{-1}\boldsymbol{c}\mu$, $\boldsymbol{x}_+ = -\boldsymbol{J}_+^{-1}\boldsymbol{c}\mu$ can not exist for $\mu < 0$ *and* $\mu > 0$. Another possibility for bifurcation in (8.25) is a Hopf bifurcation, at which a limit cycle is created/destroyed.

Proposition 8.10. *If a planar continuous piecewise linear system of the form (8.25) has a limit cycle then it must hold that*

$$\operatorname{trace}(\boldsymbol{J}_+)\operatorname{trace}(\boldsymbol{J}_-) < 0. \tag{8.31}$$

Proof: The proof follows from Bendixson's criterion [52, 71], which states if the trace of the Jacobian matrix does not change sign and is not identical to zero in a simply connected region D, then there does not exist a periodic solution which lies entirely in D. Consequently, if a continuous piecewise linear system with a single switching boundary has a periodic solution, then the trace of the Jacobian matrix must change or be equal to zero. Bendixson's criterion can be derived from Green's theorem and has originally been stated for smooth dynamical systems. Green's theorem relates a line integral to a surface integral. A system of the form (8.25) is non-smooth on a switching boundary Σ. A line integral along a periodic solution Γ is not affected by the non-smoothness on Σ because the set of points of Γ which are on Σ is of measure zero. Similarly, the surface integral over the interior of Γ is not affected by the non-smoothness on Σ because the area of Σ is zero. Bendixson's criterion can therefore be used for continuous non-smooth systems. The trace of the Jacobian in a piecewise linear system of the form (8.25) is constant in each of the subspaces $\mathcal{V}_-$ and $\mathcal{V}_+$. For a periodic solution it must therefore hold that

$$\operatorname{trace}(\boldsymbol{J}_+)\operatorname{trace}(\boldsymbol{J}_-) \leq 0. \tag{8.32}$$

A limit cycle, if it exists, must be partly in $\mathcal{V}_-$ and be partly in $\mathcal{V}_+$ and the Jacobian matrices $\boldsymbol{J}_-$ and $\boldsymbol{J}_+$ must have eigenvalues with non-zero real part, i.e. $\operatorname{trace}(\boldsymbol{J}_-) \neq 0$ and $\operatorname{trace}(\boldsymbol{J}_+) \neq 0$. □

Theorem 8.11 (Hopf Bifurcation Theorem). *If a planar continuous piecewise linear system with a single switching boundary of the form (8.25) undergoes a Hopf bifurcation at $\mu = 0$, such that a path of equilibria exists for $\mu < 0$ and $\mu \geq 0$ and a limit cycle exists for $\mu < 0$ or for $\mu > 0$, then the path of the set-valued generalized Jacobian at the bifurcation point must pass the imaginary axis with a complex conjugated pair of eigenvalues.*

Proof: If an equilibrium exists for $\mu < 0$ as well as for $\mu > 0$ then it must hold that $\det(\boldsymbol{J}_+)\det(\boldsymbol{J}_-) > 0$. If a limit cycle exists in a planar system, then there must at least be one equilibrium located in the interior of this limit cycle. As there is only one equilibrium, it follows from index theory [52] that the equilibrium can not be a saddle point. Consequently, it must hold that if the equilibrium is located in $\mathcal{V}_+$ (or $\mathcal{V}_-$) then it holds that $\det(\boldsymbol{J}_+) > 0$ (or $\det(\boldsymbol{J}_-) > 0$). It follows from $\det(\boldsymbol{J}_+)\det(\boldsymbol{J}_-) > 0$ that $\det(\boldsymbol{J}_+) > 0$ and $\det(\boldsymbol{J}_-) > 0$. The determinant of every convex combination $\boldsymbol{J}_q = q\boldsymbol{J}_+ + (1-q)\boldsymbol{J}_-$ is due to Proposition 8.8 also positive $\det(\boldsymbol{J}_q) > 0$ for $q \in [0,1]$. Furthermore, the existence of a limit cycle implies that $\operatorname{trace}(\boldsymbol{J}_+)\operatorname{trace}(\boldsymbol{J}_-) < 0$ (see Proposition 8.10). It therefore must hold, using Proposition 8.8, that there exists a $q \in [0,1]$ for which $\operatorname{trace}(\boldsymbol{J}_q) = 0$. The eigenvalues of $\boldsymbol{J}_q$ are zeros of the characteristic equation

$$\lambda^2 - \operatorname{trace}(\boldsymbol{J}_q)\lambda + \det(\boldsymbol{J}_q) = 0, \tag{8.33}$$

or

$$\lambda_{1,2} = \frac{1}{2}\operatorname{trace}(\boldsymbol{J}_q) \pm \frac{1}{2}\sqrt{\operatorname{trace}(\boldsymbol{J}_q)^2 - 4\det(\boldsymbol{J}_q)}.$$

Consequently, there exists a $q \in [0,1]$ for which there is a complex conjugated pair of eigenvalues $\lambda_{1,2} = \pm i\sqrt{\det(\boldsymbol{J}_q)}$ on the imaginary axis which belongs to the matrix $\boldsymbol{J}_q$. The path of the set-valued generalized Jacobian must therefore pass the imaginary axis with a complex conjugated pair of eigenvalues. □

Planar piecewise linear non-smooth continuous systems with a single switching boundary can therefore show two types of single crossing bifurcations: a discontinuous turning point bifurcation and a discontinuous Hopf bifurcation. A discontinuous multiple crossing bifurcation of a system of type (8.25) will be studied in Example 8.7 of Section 8.7.

8.6 Single Crossing Bifurcations

A treatise of some discontinuous bifurcations of equilibria will be given in the next subsections. For each of the continuous bifurcations (a)-(d) (see Figure 8.1) we try to find a similar discontinuous (single crossing) bifurcation occurring in a non-smooth continuous system. The non-smooth system should be as simple as possible and will therefore be chosen as a piecewise linear continuous function. First the continuous bifurcation is briefly treated, and then its discontinuous counterpart is discussed. The insight in discontinuous bifurcations of equilibria of non-smooth continuous systems is important in its own right but will also be of value for the understanding of bifurcations of periodic solutions of discontinuous systems of Filippov-type in the next chapter.

8.6.1 Saddle–node Bifurcation

The smooth scalar system

$$\dot{x} = f(x, \mu) = \mu - x^2 \tag{8.34}$$

has two equilibria for $\mu > 0$

$$x = \sqrt{\mu}, \qquad x = -\sqrt{\mu}.$$

The Jacobian $\boldsymbol{J}(x) = -2x$ becomes singular at $x = 0$. There exists a bifurcation at $(x, \mu) = (0, 0)$ in the $x - \mu$ space (Figure 8.7a), which is known as a saddle–node bifurcation point (or turning point bifurcation). The upper branch in Figure 8.7a is stable (solid line) and the lower one is unstable (dashed line). At a continuous saddle–node bifurcation, $f_{,\mu}$ (cf. (8.3)) does not belong to the range of the matrix $\boldsymbol{J}$ (see Theorem 3.1 in [75]). Hence, the matrix $[\boldsymbol{J}\ f_{,\mu}]$ has rank $n = 1$. This can be geometrically interpreted as stating that the *continuation problem* is unique, i.e. we can follow the branch up to the bifurcation point and uniquely continue on the other part of the branch. However, the fact that the continuation problem is unique does not necessarily imply that $[\boldsymbol{J}\ f_{,\mu}]$ has full rank.

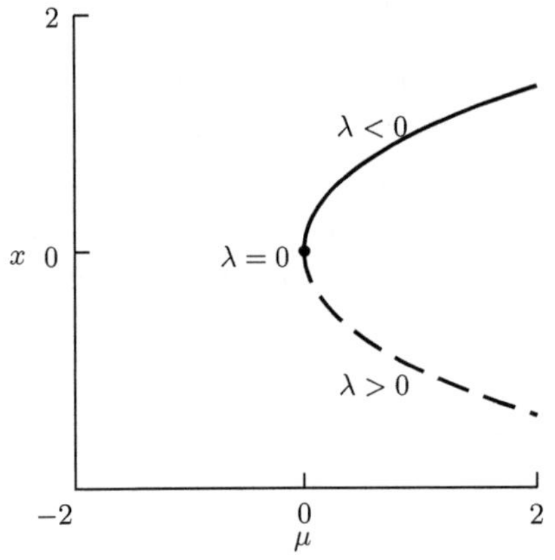

(a) smooth

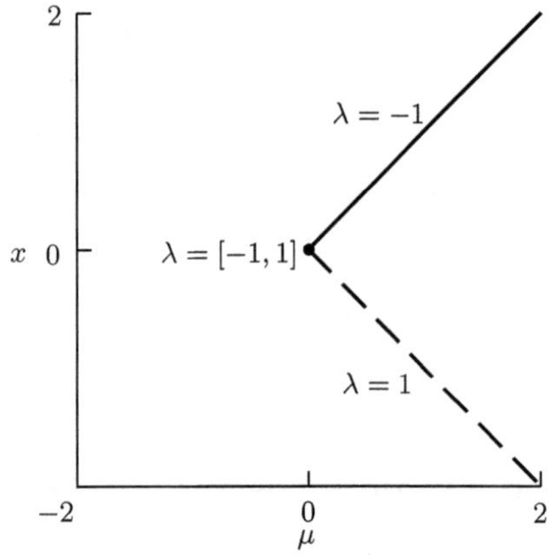

(b) discontinuous

Fig. 8.7. Saddle–node bifurcation.

We now replace the term x^2 by $|x|$ which yields a non-smooth system

$$\dot{x} = f(x, \mu) = \mu - |x|, \tag{8.35}$$

which has again two equilibria for $\mu > 0$

$$x = \mu, \qquad x = -\mu.$$

with the generalized set-valued Jacobian $\boldsymbol{J}(x, \mu) = -\operatorname{Sign} x$ and $f_{,\mu}(x, \mu) = 1$. The set-valued generalized Jacobian at $(x, \mu) = (0, 0)$ is the set

$$\boldsymbol{J}(0, 0) = \{\boldsymbol{J}_q = -2q + 1, \forall\, 0 \leq q \leq 1\}. \tag{8.36}$$

The matrix $\boldsymbol{J}_q = -2q + 1$ becomes singular at $q = \frac{1}{2}$ and it therefore holds that $0 \in \boldsymbol{J}(0, 0)$. The bifurcation diagram is depicted in Figure 8.7b and looks similar to the one for the continuous version. Again, there is a stable branch and an unstable branch but they now meet at an acute angle. From inspection of the bifurcation diagram we see that a static bifurcation (in the sense of Definition 8.1) exists at $(x, \mu) = (0, 0)$. We also conclude that the single eigenvalue on the bifurcation point is set-valued, i.e. $\lambda = [-1, 1]$. Where for the smooth case the eigenvalue passed the origin, the set-valued eigenvalue of the non-smooth system 'jumps' over the imaginary axis through the origin. For this reason, we will call the point $(x, \mu) = (0, 0)$ a *discontinuous* bifurcation point. The matrix $\left[\boldsymbol{J}_q \; f_{,\mu}(0, 0)\right]$ for $q = \frac{1}{2}$ has rank $n = 1$ similar to the smooth case. However, it seems not justified to conclude from this that the continuation problem (i.e. the possibility to follow the branch after the bifurcation point) is unique because the slope of the branch is not properly defined at the discontinuous bifurcation point. The jump of the eigenvalue and the acute conjunction of branches are properties of discontinuous bifurcations which we will also encounter for bifurcations of periodic solutions. The non-smooth continuous system (8.35) is of the type (8.25). The discontinuous bifurcation is therefore a discontinuous turning point bifurcation for which Theorem 8.9 holds.

It should be noted that, if we select to smoothen the non-smooth system with the following arctangent function

$$\dot{x} \approx \mu - \frac{2}{\pi}\arctan(\varepsilon x)x \approx \mu - \frac{2}{\pi}\varepsilon x^2 + O(x^4),$$

then the resulting bifurcation will be a continuous saddle–node bifurcation for all ε as can be seen from the expansion around the bifurcation point ($x = 0$). The choice of an arctangent function to smoothen the non-smooth term $|x|$ is arbitrary. A Taylor series approximation of an arbitrary smooth function $g(x)$, that approximates the term $|x|$, will generally be of the form

$$|x| \approx g(0) + g'x + \frac{1}{2}g''x^2 + O(x^3).$$

which yields as approximation for $f(x, \mu)$ for small values of $|x|$

$$f(x, \mu) \approx \mu - g(0) - g'x - \frac{1}{2}g''x^2 + O(x^3).$$

We now transform the latter equation with $y = x + \frac{g'}{g''}$ and $\mu^* = \mu - g(0) + \frac{1}{2}\frac{g'^2}{g''}$ to

$$f(y, \mu^*) \approx \mu^* - \frac{1}{2}g''y^2 + O\left((y - \frac{g'}{g''})^3\right),$$

which is the normal form of the saddle–node bifurcation. A regularization of (8.35) with an arbitrary smooth function will therefore reveal a saddle–node bifurcation. The discontinuous bifurcation in Figure 8.7b resembles the continuous saddle–node bifurcation in Figure 8.7a and we will call it therefore a discontinuous saddle–node bifurcation (or turning point bifurcation).

8.6.2 Transcritical Bifurcation

Consider the scalar smooth system

$$\dot{x} = f(x, \mu) = \mu x - x^2, \tag{8.37}$$

with the two equilibria $x = 0$ and $x = \mu$. The Jacobian of (8.37)

$$\boldsymbol{J}(x, \mu) = \mu - 2x$$

has the single eigenvalues

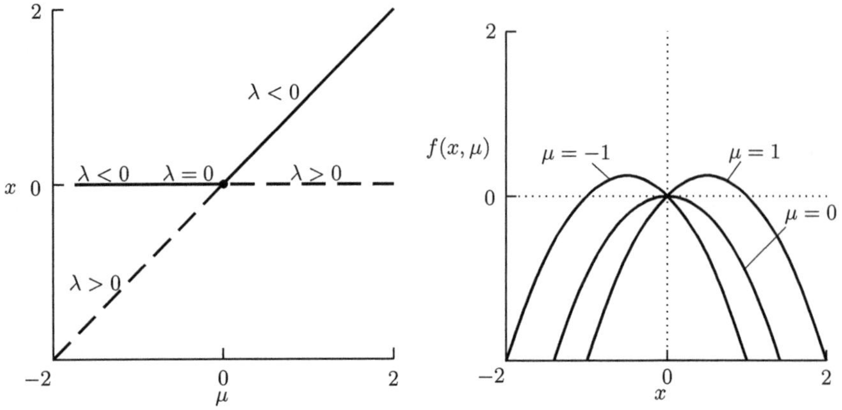

(a) bifurcation diagram (b) $f(x, \mu)$

Fig. 8.8. Transcritical bifurcation, continuous.

$$\begin{aligned} \lambda &= \mu, \quad \text{at } x = 0, \\ \lambda &= -\mu, \text{ at } x = \mu. \end{aligned}$$

The static bifurcation, shown in Figure 8.8a, is a transcritical bifurcation point at which two branches of equilibria exchange stability. The function $f(x, \mu)$ is depicted in Figure 8.8b for $\mu = -1$, $\mu = 0$ and $\mu = 1$. The system (8.37) has two distinct equilibria for $\mu \neq 0$, where one is always in the origin. At the bifurcation point ($\mu = 0$), the two equilibria coincide. The two equilibria exchange stability when the bifurcation point is passed. At a continuous transcritical bifurcation point, $f_{,\mu}$ does belong to the range of the matrix $\boldsymbol{J}$. The matrix $\left[\boldsymbol{J}\ f_{,\mu}\right]$ has rank $n - 1 = 0$ at $(x, \mu) = (0, 0)$. A second branch therefore crosses the bifurcation point, which makes the continuation problem non-unique.

We now study the following non-smooth system

$$\dot{x} = f(x, \mu) = |\frac{1}{2}\mu| - |x - \frac{1}{2}\mu|. \tag{8.38}$$

This non-smooth system approximates the parabola in Figure 8.8b by a piecewise linear curve (a tent) as is depicted in Figure 8.9b. The lines are bold where the curves overlap each other. The non-smooth system (8.38) has the same equilibria ($x = 0$ and $x = \mu$) as the smooth system (8.37). From inspection of the bifurcation diagram depicted in Figure 8.9a, we see that a static bifurcation (in the sense of Definition 8.1) exists at $(x, \mu) = (0, 0)$ as the number of equilibria changes (the change is 2–1–2). The bifurcation of the non-smooth system depicted in Figure 8.9a is similar to the transcritical bifurcation in Figure 8.8a. The generalized Jacobian of (8.38) is

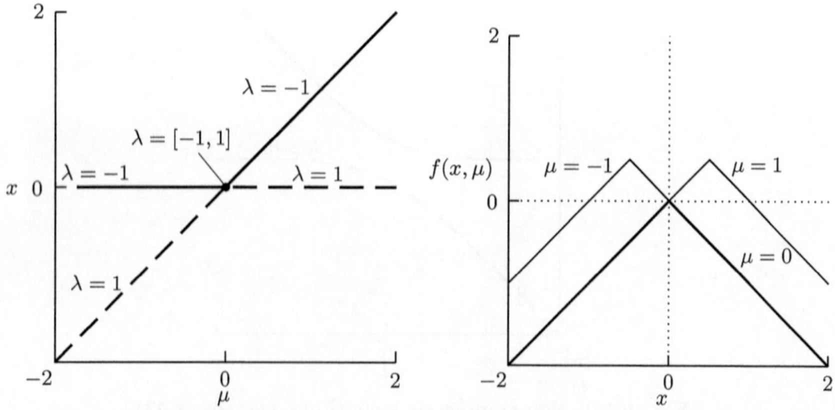

(a) bifurcation diagram (b) $f(x,\mu)$

Fig. 8.9. Transcritical bifurcation, discontinuous.

$$J(x,\mu) = -\operatorname{Sign}(x - \frac{1}{2}\mu),$$

which is set-valued at $(x,\mu) = (0,0)$. The generalized Jacobian has the eigenvalues

$$\begin{aligned} \lambda &= -1, && \text{at } x = 0 \text{ if } \mu < 0,\\ \lambda &= 1, && \text{at } x = 0 \text{ if } \mu > 0,\\ \lambda &= 1, && \text{at } x = \mu \text{ if } \mu < 0,\\ \lambda &= -1, && \text{at } x = \mu \text{ if } \mu > 0,\\ \lambda &= [-1,1], && \text{at } (x,\mu) = (0,0). \end{aligned}$$

Where for the continuous transcritical bifurcation the eigenvalue passed the origin, the set-valued eigenvalue of the non-smooth system 'jumps' over the imaginary axis through the origin. For this reason, we will call the point $(x,\mu) = (0,0)$ a discontinuous bifurcation point (Figure 8.9a). Because the structure of the branches around the discontinuous bifurcation point resembles the structure of the transcritical bifurcation, we will call this bifurcation a discontinuous *transcritical* bifurcation. Note that the non-smooth continuous system (8.38) is not of the type (8.25).

The equilibrium point $(x,\mu) = (0,0)$ is located on the intersection of two switching boundaries $\mu = 0$ and $x - \frac{1}{2}\mu = 0$ in the (x,μ) space. Two auxiliary variables, q_1 and q_2, are needed to describe the set-valuedness of $J(0,0)$ and $f_{,\mu}(0,0)$.

The set-valued Jacobian at $(x,\mu) = (0,0)$ is described by the set

$$J(0,0) = \{J_q = -2q_1 + 1, \forall 0 \le q_1 \le 1\}, \tag{8.39}$$

of which $J_q = -2q_1 + 1$ becomes singular at $q_1 = \frac{1}{2}$. Furthermore,

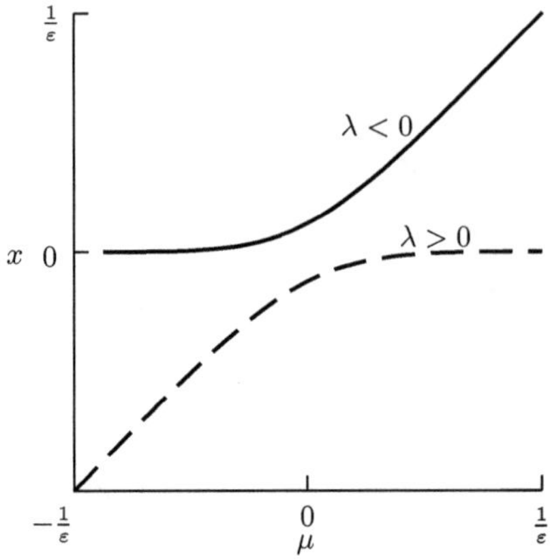

Fig. 8.10. Non-symmetric smoothing, system (8.43).

$$f_{,\mu}(x,\mu) = \frac{1}{2}\,\mathrm{Sign}(\mu) + \frac{1}{2}\,\mathrm{Sign}(x - \frac{1}{2}\mu), \tag{8.40}$$

which is set-valued at the bifurcation point. The set $f_{,\mu}(0,0)$ can be described by the auxiliary variables q_1 and q_2

$$\begin{aligned} f_{,\mu}(0,0) &= \{f_{,\mu q} = \frac{1}{2}(2q_2 - 1) + \frac{1}{2}(2q_1 - 1), \forall 0 \le q_1 \le 1, \forall 0 \le q_2 \le 1\} \\ &= \{f_{,\mu q} = q_2 + q_1 - 1, \forall 0 \le q_1 \le 1, \forall 0 \le q_2 \le 1\}. \end{aligned} \tag{8.41}$$

The first auxiliary variable, q_1, will be varied to satisfy the condition $\det(\boldsymbol{J}_q) = 0$ and the second auxiliary variable, q_2, will be varied to ensure that $\left[\boldsymbol{J}_q\ f_{,\mu q}\right]$ has rank $n-1=0$. The matrix $\left[\boldsymbol{J}_q\ f_{,\mu q}\right]$ has rank $n-1=0$ at $q_1 = \frac{1}{2}, q_2 = \frac{1}{2}$. Because of the non-smoothness of the problem it is not justified to conclude that the continuation problem is non-unique as for the smooth case but a resemblance is present.

If we smoothen the non-smooth system (8.38) with the following particular arctangent function

$$\dot{x} \approx \frac{1}{\pi}\arctan(\frac{1}{2}\varepsilon\mu)\mu - \frac{2}{\pi}\arctan(\varepsilon(x - \frac{1}{2}\mu))(x - \frac{1}{2}\mu) \approx \frac{2}{\pi}\varepsilon(\mu x - x^2),$$

then the resulting bifurcation will be a continuous transcritical bifurcation for all ε as can be seen from the expansion around the bifurcation point ($x = 0$, $\mu = 0$). The smoothened system can be transformed to the normal form with the time transformation $\tau = \varepsilon t$.

However, not every smoothing function gives a transcritical bifurcation. Consider for instance the following non-symmetric smoothing:

$$\begin{aligned} |\tfrac{1}{2}\mu| &\approx \tfrac{2}{\pi}\arctan(\tfrac{1}{2}\varepsilon\mu)\tfrac{1}{2}\mu + \tfrac{1}{\varepsilon}, \\ |x - \tfrac{1}{2}\mu| &\approx \tfrac{2}{\pi}\arctan(\varepsilon(x - \tfrac{1}{2}\mu))(x - \tfrac{1}{2}\mu), \end{aligned} \tag{8.42}$$

which gives

$$\dot{x} \approx \frac{2}{\pi}\varepsilon(\mu x - x^2) + \frac{1}{\varepsilon}, \tag{8.43}$$

for $|x| < \varepsilon$ and $|\mu| < \varepsilon$. Equation (8.43) has two branches in the bifurcation diagram for varying μ but the branches do not intersect (Figure 8.10). No bifurcation exists for (8.43).

8.6.3 Pitchfork Bifurcation

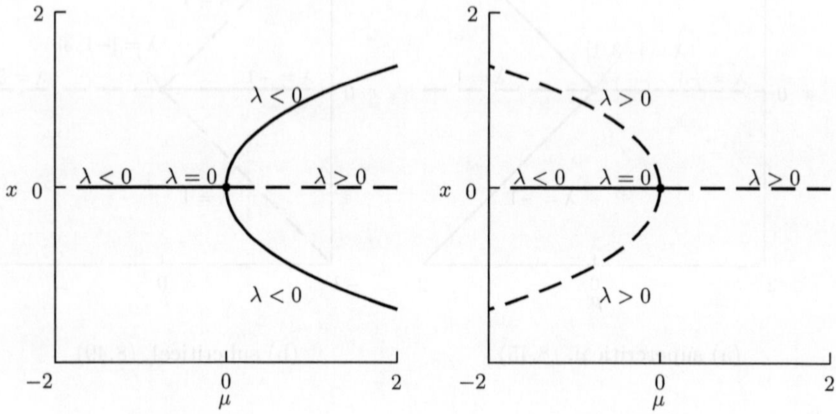

(a) supercritical, $\alpha = -1$ (b) subcritical, $\alpha = 1$

Fig. 8.11. Pitchfork bifurcation, continuous.

We consider the smooth system

$$\dot{x} = f(x, \mu) = \mu x + \alpha x^3, \tag{8.44}$$

where the constant α will be taken as $\alpha = \pm 1$. There is one equilibrium for $\frac{\mu}{\alpha} \geq 0$ and there are three equilibria for $\frac{\mu}{\alpha} < 0$, i.e.

$$\begin{aligned} x &= 0 && \text{trivial equilibrium,} \\ x &= \pm\sqrt{-\tfrac{\mu}{\alpha}} && \text{for } \tfrac{\mu}{\alpha} < 0. \end{aligned}$$

The Jacobian

$$\boldsymbol{J} = \mu + 3\alpha x^2$$

has the single eigenvalues

$$\begin{aligned} \lambda &= \mu, && \text{at } x = 0, \\ \lambda &= -2\mu, && \text{at } x = \pm\sqrt{-\tfrac{\mu}{\alpha}} \text{ for } \tfrac{\mu}{\alpha} < 0. \end{aligned}$$

For $\alpha < 0$ there is a supercritical pitchfork bifurcation (Figure 8.11a) and for $\alpha > 0$ a subcritical pitchfork bifurcation (Figure 8.11b). At a continuous pitchfork bifurcation point, $f_{,\mu}$ does belong to the range of the matrix $\boldsymbol{J}$ (see Theorem 3.1 in [75]). The matrix $[\boldsymbol{J}\ f_{,\mu}]$ has rank $n-1=0$ at $(x,\mu)=(0,0)$, which is consistent with the fact that two branches intersect at the bifurcation point.

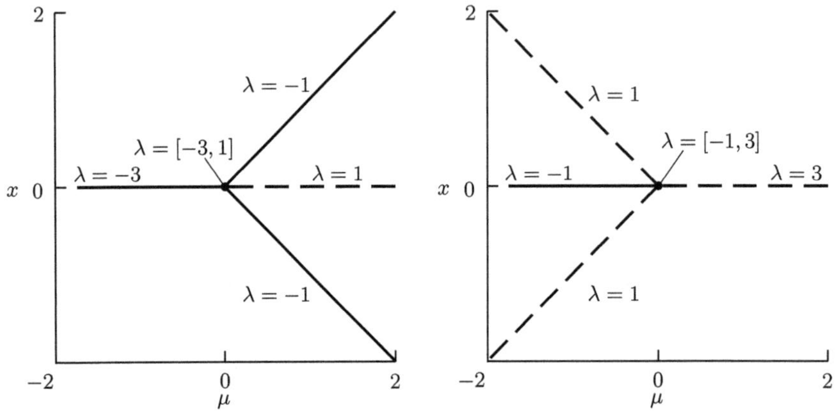

(a) supercritical, (8.45)

(b) subcritical, (8.49)

Fig. 8.12. Pitchfork bifurcation, discontinuous.

We now study the following non-smooth system

$$\dot{x} = f(x,\mu) = -x + |x + \frac{1}{2}\mu| - |x - \frac{1}{2}\mu|, \tag{8.45}$$

which has one equilibrium for $\mu \leq 0$ and three equilibria for $\mu > 0$

$$\begin{aligned} x &= 0 \quad \text{trivial equilibrium,} \\ x &= \pm\mu \ \text{ for } \mu > 0. \end{aligned}$$

From inspection of the bifurcation diagram, depicted in Figure 8.12a, we observe that a static bifurcation (in the sense of Definition 8.1) exists at $(x,\mu)=(0,0)$ as the number of equilibria changes (the change is 1–1–3). The generalized Jacobian of (8.45)

$$\boldsymbol{J}(x,\mu) = -1 + \mathrm{Sign}(x + \frac{1}{2}\mu) - \mathrm{Sign}(x - \frac{1}{2}\mu)$$

has the single eigenvalues

$$\begin{aligned} \lambda &= -3, \text{ at } x = 0, \mu < 0, \\ \lambda &= 1, \quad \text{at } x = 0, \mu > 0, \\ \lambda &= -1, \text{ at } x = \pm\mu, \mu > 0, \end{aligned}$$

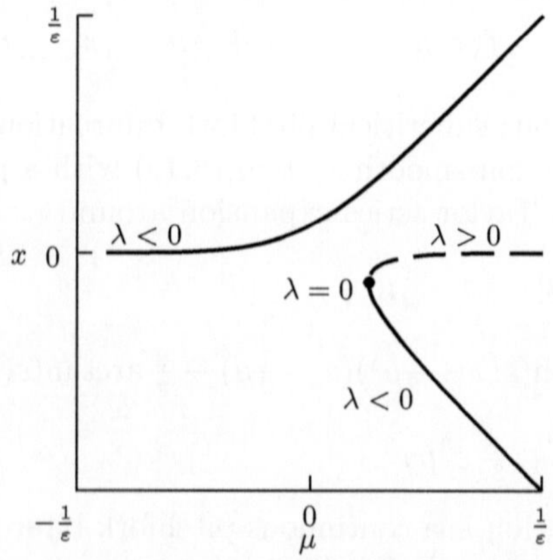

Fig. 8.13. Non-symmetric smoothing, (8.51).

and is set-valued at $(x, \mu) = (0, 0)$. As there are two switching boundaries where the vector field is non-smooth we need two auxiliary variables to describe the set-valued Jacobian at $(x, \mu) = (0, 0)$

$$\begin{aligned} \boldsymbol{J}(0,0) &= \{\boldsymbol{J}_q = -1 + (-2q_1 + 1) - (-2q_2 + 1), \forall 0 \le q_1 \le 1, \forall 0 \le q_2 \le 1\} \\ &= \{\boldsymbol{J}_q = 2(q_2 - q_1) - 1, \forall 0 \le q_1 \le 1, \forall 0 \le q_2 \le 1\}, \end{aligned} \tag{8.46}$$

of which $\boldsymbol{J}_q = 2(q_2 - q_1) - 1$ becomes singular at $q_2 - q_1 = \frac{1}{2}$. Furthermore,

$$f_{,\mu}(x, \mu) = \frac{1}{2}\operatorname{Sign}(x + \frac{1}{2}\mu) + \frac{1}{2}\operatorname{Sign}(x - \frac{1}{2}\mu), \tag{8.47}$$

which is set-valued at the bifurcation point. The set $f_{,\mu}(0,0)$ can be expressed as

$$\begin{aligned} f_{,\mu}(0,0) &= \{f_{,\mu q} = \frac{1}{2}(2q_1 - 1) + \frac{1}{2}(2q_2 - 1), \forall 0 \le q_1 \le 1, \forall 0 \le q_2 \le 1\} \\ &= \{f_{,\mu q} = q_1 + q_2 - 1, \forall 0 \le q_1 \le 1, \forall 0 \le q_2 \le 1\}. \end{aligned} \tag{8.48}$$

The matrix $\left[\boldsymbol{J}_q \; f_{,\mu q}\right]$ has rank $n - 1 = 0$ at $q_1 = \frac{3}{4}, q_2 = \frac{1}{4}$. The continuation problem is clearly non-unique as can be seen from the bifurcation diagram (Figure 8.12a). However, to conclude the non-uniqueness of the continuation problem from the rank condition seems not justified because of the non-smoothness of the problem.

The bifurcation diagram of system (8.45) is shown in Figure 8.12a. The bifurcation point is a discontinuous bifurcation point because the eigenvalue 'jumps' over the imaginary axis and will be named as a discontinuous *pitchfork* bifurcation because it resembles the continuous pitchfork bifurcation. Similarly, the system

$$\dot{x} = f(x, \mu) = x + |x + \frac{1}{2}\mu| - |x - \frac{1}{2}\mu| \tag{8.49}$$

exhibits a discontinuous subcritical pitchfork bifurcation (Figure 8.12b).

We smoothen the non-smooth system (8.45) with a particular arctangent function and apply a Taylor series expansion around $(x = 0,\ \mu = 0)$

$$\begin{aligned}\dot{x} &= -x + |x + \tfrac{1}{2}\mu| - |x - \tfrac{1}{2}\mu| \\ &\approx -x + \tfrac{2}{\pi}\arctan(\varepsilon(x + \tfrac{1}{2}\mu))(x + \tfrac{1}{2}\mu) - \tfrac{2}{\pi}\arctan(\varepsilon(x - \tfrac{1}{2}\mu))(x - \tfrac{1}{2}\mu) \\ &\approx (-1 + \tfrac{4}{\pi}\varepsilon\mu)x - \tfrac{8}{3\pi}\varepsilon^3\mu x^3.\end{aligned}$$

The resulting bifurcation is a continuous pitchfork bifurcation with the bifurcation point at $(x = 0,\ \mu = \frac{\pi}{4\varepsilon})$. The bifurcation point of the smoothened system therefore approaches the origin as ε is increased.

However, not every smoothing function gives a pitchfork bifurcation. Consider for instance the following non-symmetric smoothing:

$$\begin{aligned}|x + \tfrac{1}{2}\mu| &\approx \tfrac{2}{\pi}\arctan(\varepsilon(x + \tfrac{1}{2}\mu))(x + \tfrac{1}{2}\mu) + \tfrac{1}{\varepsilon}, \\ |x - \tfrac{1}{2}\mu| &\approx \tfrac{2}{\pi}\arctan(\varepsilon(x - \tfrac{1}{2}\mu))(x - \tfrac{1}{2}\mu),\end{aligned} \tag{8.50}$$

which gives

$$\dot{x} \approx (-1 + \frac{4}{\pi}\varepsilon\mu)x - \frac{8}{3\pi}\varepsilon^3\mu x^3 + \frac{1}{\varepsilon}, \tag{8.51}$$

for $|x| < \varepsilon$ and $|\mu| < \varepsilon$. Equation (8.51) has two branches close to the origin in the bifurcation diagram for varying μ, but the branches do not intersect (Figure 8.13). Only a saddle–node bifurcation exists for (8.51).

8.6.4 Hopf Bifurcation

At a Hopf bifurcation point the equilibrium loses its stability and a periodic solution is born (or destroyed). First, we consider the smooth planar system

$$\begin{aligned}\dot{x}_1 &= \mu x_1 - \omega x_2 + (\alpha x_1 - \beta x_2)(x_1^2 + x_2^2), \\ \dot{x}_2 &= \omega x_1 + \mu x_2 + (\beta x_1 + \alpha x_2)(x_1^2 + x_2^2),\end{aligned} \tag{8.52}$$

where μ, ω, α and β are constants. We will study the equilibria and periodic solutions of system (8.52) for different values of μ. This system has an equilibrium $\boldsymbol{x} = [x_1, x_2]^{\mathrm{T}} = [0, 0]^{\mathrm{T}}$ for all values of μ and the Jacobian matrix of the linearized system around the equilibrium is

$$\boldsymbol{J} = \begin{bmatrix} \mu & -\omega \\ \omega & \mu \end{bmatrix},$$

with the eigenvalues $\lambda_{1,2} = \mu \pm i\omega$. For $\mu < 0$, the equilibrium is asymptotically stable. When μ is increased to $\mu = 0$ the equilibrium becomes non-hyperbolic, and for $\mu > 0$ the equilibrium becomes unstable. By using the transformation

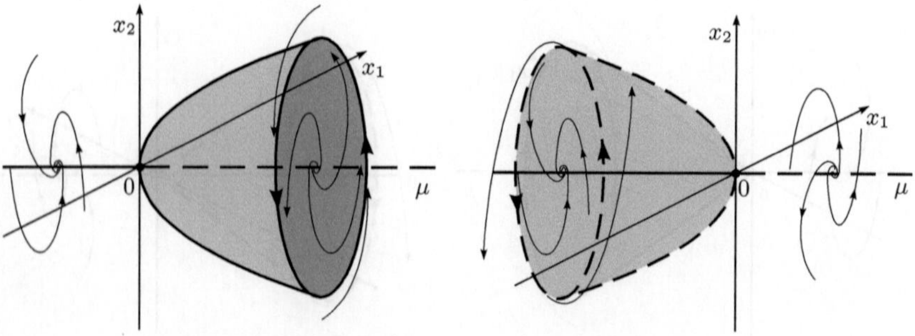

a s ercritical $\alpha < 0$ *b s bcritical* $\alpha > 0$

Fig. 8.14. Hopf bifurcation, continuous.

$$x_1 = r\cos\theta \text{ and } x_2 = r\sin\theta, \tag{8.53}$$

we transform (8.52) into

$$\dot{r} = \mu r + \alpha r^3, \tag{8.54}$$

$$\dot{\theta} = \omega + \beta r^2. \tag{8.55}$$

The trivial equilibrium of (8.54) corresponds to the equilibrium of (8.52), and the nontrivial equilibrium ($r = \pm\sqrt{|\mu/\alpha|}$) of (8.54) corresponds to a periodic solution of (8.52). In the latter case, r is the amplitude and $\dot{\theta}$ is the frequency of the periodic solution that is created by the Hopf bifurcation. The transformation (8.53) therefore transforms the Hopf bifurcation into a pitchfork bifurcation. The bifurcation diagram for the Hopf bifurcation is depicted in Figure 8.14 and the bifurcation diagram for the transformed system (8.54) is identical to Figure 8.11 where x should be replaced by r.

We now study the following non-smooth continuous system

$$\begin{aligned}\dot{x}_1 &= -x_1 - \omega x_2 + \tfrac{x_1}{\sqrt{x_1^2+x_2^2}}\left(|\sqrt{x_1^2+x_2^2}+\tfrac{1}{2}\mu| - |\sqrt{x_1^2+x_2^2}-\tfrac{1}{2}\mu|\right),\\ \dot{x}_2 &= \omega x_1 - x_2 + \tfrac{x_2}{\sqrt{x_1^2+x_2^2}}\left(|\sqrt{x_1^2+x_2^2}+\tfrac{1}{2}\mu| - |\sqrt{x_1^2+x_2^2}-\tfrac{1}{2}\mu|\right),\end{aligned} \tag{8.56}$$

which is dependent on the parameters μ and ω. We will study the equilibria and periodic solutions of system (8.56) for different values of μ. The non-smooth system (8.56) has the same equilibrium as the smooth system with the same stability. We transform the system (8.56) with the transformation (8.53) into

$$\dot{r} = -r + |r + \frac{1}{2}\mu| - |r - \frac{1}{2}\mu|, \tag{8.57}$$

$$\dot{\theta} = \omega. \tag{8.58}$$

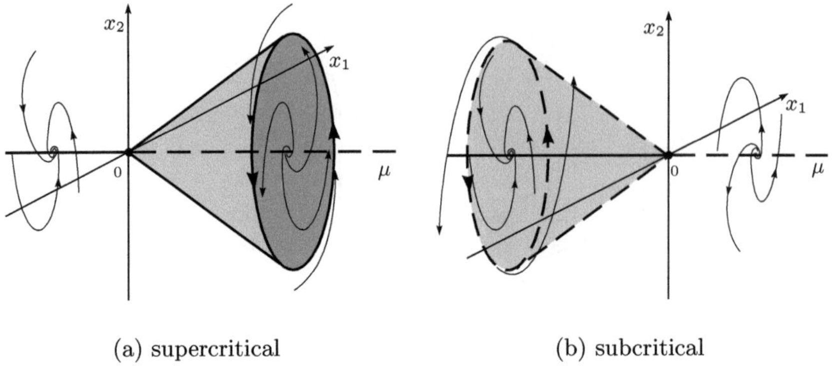

Fig. 8.15. Hopf bifurcation, discontinuous.

Equation (8.57) is identical to the non-smooth system (8.45) exposing a discontinuous pitchfork bifurcation. The scenario for the discontinuous Hopf bifurcation is depicted in Figure 8.15 and the scenario for (8.57) is identical to Figure 8.12.

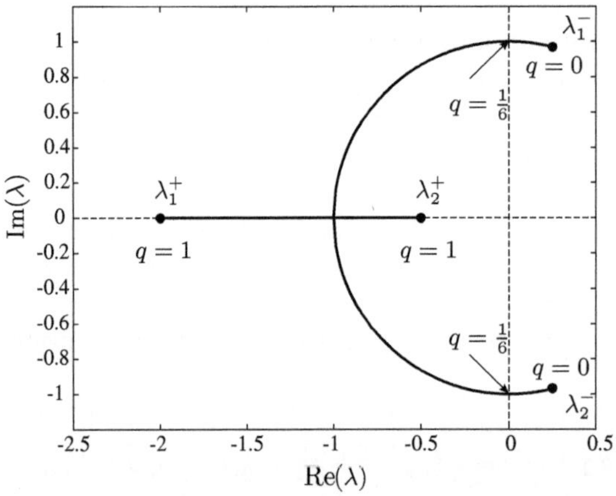

Fig. 8.16. Eigenvalue-path of the discontinuous Hopf bifurcation of system (8.59).

The discontinuous Hopf bifurcation of the preceding example occurred in a system with two switching boundaries. A discontinuous Hopf bifurcation can, however, also occur in a system with a single switching boundary. Consider the second order system

$$\begin{aligned} \dot{x}_1 &= x_2, \\ \dot{x}_2 &= -x_2 - \tfrac{3}{2}|x_2 - \mu| - x_1, \end{aligned} \tag{8.59}$$

which has only one switching boundary $\Sigma = \{(x_1, x_2) \in \mathbb{R}^2 \mid x_2 - \mu = 0\}$. Note that system (8.59) can easily be transformed to be of the type (8.25). The system has only one equilibrium

$$x_1 = -\frac{3}{2}|\mu|, \quad x_2 = 0, \tag{8.60}$$

which exists for all $\mu \in \mathbb{R}$. The generalized Jacobian of the system is

$$\boldsymbol{J}(\boldsymbol{x}, \mu) = \begin{bmatrix} 0 & 1 \\ -1 & -1 - \frac{3}{2}\,\mathrm{Sign}(x_2 - \mu) \end{bmatrix}. \tag{8.61}$$

The generalized Jacobian at the equilibrium jumps from $\boldsymbol{J}_-$ to $\boldsymbol{J}_+$ when μ is increased from $\mu < 0$ to $\mu > 0$ with

$$\begin{aligned} \boldsymbol{J}_- &= \begin{bmatrix} 0 & 1 \\ -1 & \frac{1}{2} \end{bmatrix}, \quad \lambda^-_{1,2} = \tfrac{1}{4} \pm \tfrac{1}{4}\sqrt{15}, \\ \boldsymbol{J}_+ &= \begin{bmatrix} 0 & 1 \\ -1 & -\frac{5}{2} \end{bmatrix}, \quad \lambda^+_1 = -2,\ \lambda^+_2 = -\tfrac{1}{2}. \end{aligned} \tag{8.62}$$

The equilibrium is therefore a stable node for $\mu < 0$ and an unstable focus for $\mu > 0$ and therefore loses stability at $\mu = 0$. The generalized Jacobian at the equilibrium point $x_1 = x_2 = 0$ for $\mu = 0$ can be expressed as the set-valued matrix $\boldsymbol{J}(\boldsymbol{0}) = \{\boldsymbol{J}_q, q \in [0, 1]\}$ with $\boldsymbol{J}_q = q\boldsymbol{J}_+ + (1 - q)\boldsymbol{J}_-$. The two eigenvalues are therefore set-valued at the equilibrium point $x_1 = x_2 = 0$ for $\mu = 0$ and form a one-dimensional path in the complex plane parameterized by the auxiliary variable q and given by

$$\lambda_1 = \{\lambda_{q1},\ q \in [0, 1]\}, \qquad \lambda_2 = \{\lambda_{q2},\ q \in [0, 1]\}, \tag{8.63}$$

with $\lambda_{q1,2} = \mathrm{eig}\,\boldsymbol{J}_q$:

$$\lambda_{q1,2} = \begin{cases} \frac{1}{4} - \frac{3}{2}q \pm \frac{1}{4}i\sqrt{-(1 - 6q)^2 + 16}, & q \le \frac{5}{6}, \\ \frac{1}{4} - \frac{3}{2}q \pm \frac{1}{4}\sqrt{(1 - 6q)^2 - 16}, & q > \frac{5}{6}. \end{cases} \tag{8.64}$$

The path of the eigenvalues is depicted in Figure 8.16. The eigenvalues cross the imaginary axis for $q = \frac{1}{6}$, which causes a discontinuous Hopf bifurcation. The phase plane of system (8.59) is shown in Figure 8.17 for $\mu = -1$ and for $\mu = 1$. The phase plane for $\mu = -1$ shows the equilibrium as a stable node, while the phase plane for $\mu = 1$ shows an unstable focus as well as a stable limit cycle, created by the discontinuous Hopf bifurcation.

In Section 8.3, a criterion is given for the co-existence of equilibria for a special class of piecewise linear systems to which system (8.59) belongs. Equilibria do not co-exist for system (8.59) according to condition (8.18), using the fact that

$$\det(\boldsymbol{J}_+)\det(\boldsymbol{J}_-) > 0, \tag{8.65}$$

which indeed agrees with the bifurcation scenario of system (8.59). Furthermore, Proposition 8.10 and the Hopf Bifurcation Theorem 8.11, presented in Section 8.5, apply for system (8.59). Indeed, it holds that

$$\mathrm{trace}(\boldsymbol{J}_+)\,\mathrm{trace}(\boldsymbol{J}_-) < 0, \tag{8.66}$$

and the set-valued eigenvalues therefore contain a crossing with the imaginary axis as a complex conjugated pair.

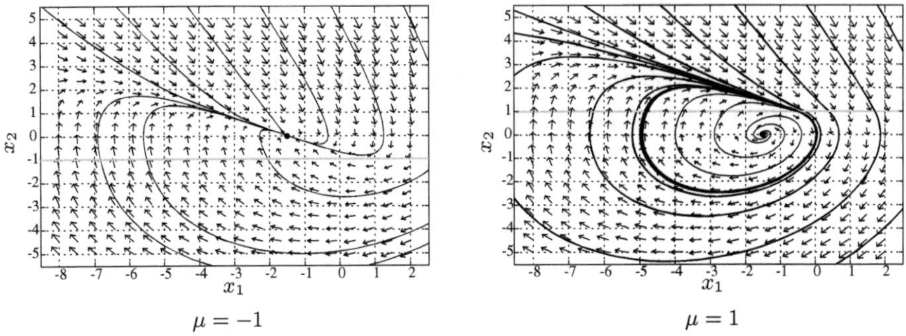

Fig. 8.17. Discontinuous Hopf bifurcation of system (8.59).

8.7 Multiple Crossing Bifurcations

The bifurcations in the previous section are all characterized by a single crossing of the eigenvalue(s) through the imaginary axis. If the eigenvalues are set-valued, which is the case for the discontinuous bifurcations, then the set of eigenvalues forms in the preceding examples a one-dimensional path in the complex plane. The eigenvalue(s) either moved continuously through the imaginary axis under the variation of a parameter (being a continuous bifurcation) or a one-dimensional path of eigenvalues crossed the imaginary axis during a jump (leading to a discontinuous bifurcation). Non-smooth continuous systems can also exhibit bifurcations of equilibria for which a one-dimensional path of eigenvalue(s) crosses multiple times the imaginary axis, as was already pointed out in Section 8.4. Equilibria of non-smooth continuous systems with multiple switching boundaries can have set-valued eigenvalues which do not form a one-dimensional path but a two-dimensional area in the complex plane. Such a set of eigenvalues, which forms an area in the complex plane, can contain part of the imaginary axis leading to a multiple crossing bifurcation. Multiple crossing bifurcations are much more complex than single crossing bifurcations and do not have a smooth counterpart. In the sequel, we will discuss a number of two-dimensional systems showing multiple crossing bifurcations.

Example 8.2 (Combined Hopf and pitchfork behaviour)
Consider the two-dimensional non-smooth continuous system

$$\begin{aligned} \dot{x}_1 &= x_2, \\ \dot{x}_2 &= -x_1 + |x_1 + \mu| - |x_1 - \mu| - x_2 - |x_2 + \mu| + |x_2 - \mu|. \end{aligned} \tag{8.67}$$

The system has three equilibria for $\mu > 0$

$$\begin{array}{ll} \text{trivial equilibrium banch} & x_1 = 0, \quad x_2 = 0, \\ \text{non-trivial equilibrium banch} & x_1 = \pm 2\mu, \ x_2 = 0. \end{array} \tag{8.68}$$

For $\mu \leq 0$, the only equilibrium is the trivial equilibrium point $x_1 = 0, x_2 = 0$. The point $(\boldsymbol{x}, \mu) = (\boldsymbol{0}, 0)$ is a bifurcation point according to Definition 8.1. The generalized Jacobian matrix of the system

$$\boldsymbol{J}(\boldsymbol{x}, \mu) = \begin{bmatrix} 0 & 1 \\ J_{21} & J_{22} \end{bmatrix}, \tag{8.69}$$

with

$$\begin{aligned} J_{21} &= -1 + \mathrm{Sign}(x_1 + \mu) - \mathrm{Sign}(x_1 - \mu), \\ J_{22} &= -1 - \mathrm{Sign}(x_2 + \mu) + \mathrm{Sign}(x_2 - \mu), \end{aligned} \tag{8.70}$$

is set-valued at four different switching boundaries. The bifurcation point is located at the crossing of the four switching boundaries, which causes the generalized Jacobian at the bifurcation point to be dependent on four auxiliary variables q_i

$$\boldsymbol{J}(\boldsymbol{0}, 0) = \{\boldsymbol{J}_q, \ q_i \in [0, 1], \ i = 1, \ldots, 4\}, \tag{8.71}$$

with

$$\boldsymbol{J}_q = \begin{bmatrix} 0 & 1 \\ -1 - 2q_1 + 2q_2 & -1 - 2q_3 + 2q_4 \end{bmatrix}. \tag{8.72}$$

The set-valued generalized Jacobian at the bifurcation point defines a set of eigenvalues in the complex plane. This set of eigenvalues is spanned in the complex plane by four auxiliary variables. It is therefore not possible to speak of a 'path' of eigenvalues. The set of eigenvalues is only a path if the bifurcation point is located on only one switching boundary or if the system is one-dimensional, which forces the set of eigenvalues to be on the real axis. The fact that the set of eigenvalues at the bifurcation point is higher-dimensional tremendously complicates the analysis of the system.

The Jacobian matrix at the trivial branch ($x_1 = x_2 = 0$) is $\boldsymbol{J}_-^{\mathrm{tr}}$ for $\mu < 0$ with

$$\boldsymbol{J}_-^{\mathrm{tr}} = \boldsymbol{J}(\boldsymbol{0}, \mu < 0) = \begin{bmatrix} 0 & 1 \\ -3 & 1 \end{bmatrix}, \quad \lambda_{1,2} = \frac{1}{2} \pm i\frac{1}{2}\sqrt{11}, \tag{8.73}$$

and $\boldsymbol{J}_+^{\mathrm{tr}}$ for $\mu > 0$ with

$$\boldsymbol{J}_+^{\mathrm{tr}} = \boldsymbol{J}(\boldsymbol{0}, \mu > 0) = \begin{bmatrix} 0 & 1 \\ 1 & -3 \end{bmatrix}, \quad \lambda_{1,2} = -\frac{3}{2} \pm \frac{1}{2}\sqrt{13} \approx \{0.30, -3.30\}. \tag{8.74}$$

The trivial equilibrium is therefore an unstable focus for $\mu < 0$ and a saddle point for $\mu > 0$. The Jacobian matrix on the non-trivial branches is

$$\begin{aligned}\boldsymbol{J}^{\text{non}} &= \boldsymbol{J}([\pm 2\mu, 0]^{\mathrm{T}}, \mu > 0) = \begin{bmatrix} 0 & 1 \\ -1 & -3 \end{bmatrix}, \\ \lambda_{1,2} &= -\tfrac{3}{2} \pm \tfrac{1}{2}\sqrt{5} \approx \{-0.38, -2.62\}.\end{aligned} \tag{8.75}$$

Equilibria on the non-trivial branches are therefore stable nodes.

Using the transformation

$$y_1 = \frac{x_1}{\mu}, \quad y_2 = \frac{x_2}{\mu}, \tag{8.76}$$

we can transform system (8.67) for $\mu < 0$ into

$$\begin{aligned}\dot{y}_1 &= y_2, \\ \dot{y}_2 &= -y_1 - |y_1 + 1| + |y_1 - 1| - y_2 + |y_2 + 1| - |y_2 - 1|,\end{aligned} \tag{8.77}$$

and for $\mu > 0$ into

$$\begin{aligned}\dot{y}_1 &= y_2, \\ \dot{y}_2 &= -y_1 + |y_1 + 1| - |y_1 - 1| - y_2 - |y_2 + 1| + |y_2 - 1|.\end{aligned} \tag{8.78}$$

The transformed systems are independent of μ for $\mu \neq 0$. Equilibria and

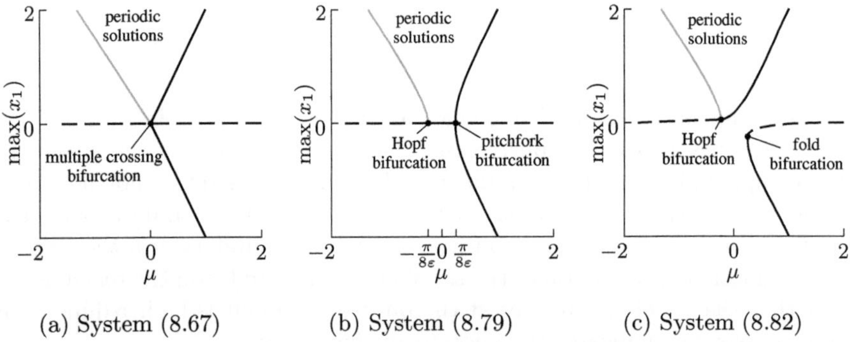

Fig. 8.18. Bifurcation diagrams of the non-smooth system (8.67) (left) and the smooth approximating systems (8.79) (middle) and (8.82) (right).

periodic solutions of (8.77) and (8.78) are after an inverse transformation with (8.76) also equilibria and periodic solutions of system (8.67). The location of the equilibria of system (8.67) scale therefore with μ. But also all periodic solutions of system (8.67) scale with μ. This means that the shape of a periodic solution of (8.67) does not change for varying μ, but the size of the periodic solution scales with μ. The period time is independent of μ. The bifurcation diagram of system (8.67) is depicted in Figure 8.18a. Branches of equilibria are indicated by black lines and periodic branches by grey lines. Stable branches are indicated by solid lines and unstable branches by dashed lines. The periodic solution and the period time have been found by numerical simulation. The point $(x_1, x_2, \mu) = (0, 0, 0)$ is a bifurcation point where two branches of equilibria bifurcate from the trivial branch, similar to a pitchfork bifurcation, and also a periodic solution is created at the bifurcation point. The magnitude $\max(x_1)$ varies linearly in μ for all branches as was expected from the transformation. The period time of the periodic solution is $T = 4.03$ s and is independent of μ. The phase plane of system (8.67) is shown in Figure 8.19 for three different values of μ. The system has for $\mu < 0$ an unstable focus and a stable periodic solution. The periodic solution disappears for $\mu = 0$ and the trivial equilibrium point turns into a stable focus. The trivial equilibrium point becomes a saddle point for $\mu > 0$ and two additional stable nodes appear at $\boldsymbol{x} = \begin{bmatrix} \pm 2\mu & 0 \end{bmatrix}^{\mathrm{T}}$. The stable and unstable invariant manifolds of the saddle point $\boldsymbol{x} = \boldsymbol{0}$ are depicted by thick lines in Figure 8.19 and the unstable invariant manifolds form heteroclinic connections between the saddle point and the stable nodes. The multiple crossing bifurcation has the following bifurcation structure:

$$\left.\begin{array}{r}\text{unstable focus}\\ \text{periodic solution}\end{array}\right\} \xrightarrow{\text{multiple crossing bifurcation of (8.67)}} \left\{\begin{array}{l}\text{stable node}\\ \text{saddle}\\ \text{stable node}\end{array}\right.$$

The bifurcation structure is different from any known bifurcation structure of (co-dimension 1) bifurcations in smooth systems.

Example 8.3 (Symmetric smooth approximation of (8.67))
We now study a smooth system

$$\begin{aligned}\dot{x}_1 =& \; x_2,\\ \dot{x}_2 =& -x_1 + \frac{2}{\pi}\arctan(\varepsilon(x_1+\mu))(x_1+\mu) - \frac{2}{\pi}\arctan(\varepsilon(x_1-\mu))(x_1-\mu)\\ & - x_2 - \frac{2}{\pi}\arctan(\varepsilon(x_2+\mu))(x_2+\mu) + \frac{2}{\pi}\arctan(\varepsilon(x_2-\mu))(x_2-\mu),\end{aligned} \tag{8.79}$$

which is a smooth approximation of system (8.67). The smooth system (8.79) can be expanded in a Taylor series around $x_1 = x_2 = 0$ and $\mu \ll 1$

$$\begin{aligned}\dot{x}_1 &= x_2,\\ \dot{x}_2 &\approx (-1 + \tfrac{8}{\pi}\varepsilon\mu)x_1 - \tfrac{64}{3\pi}\varepsilon^3\mu x_1^3 + (-1 - \tfrac{8}{\pi}\varepsilon\mu)x_2 + \tfrac{64}{3\pi}\varepsilon^3\mu x_2^3.\end{aligned} \tag{8.80}$$

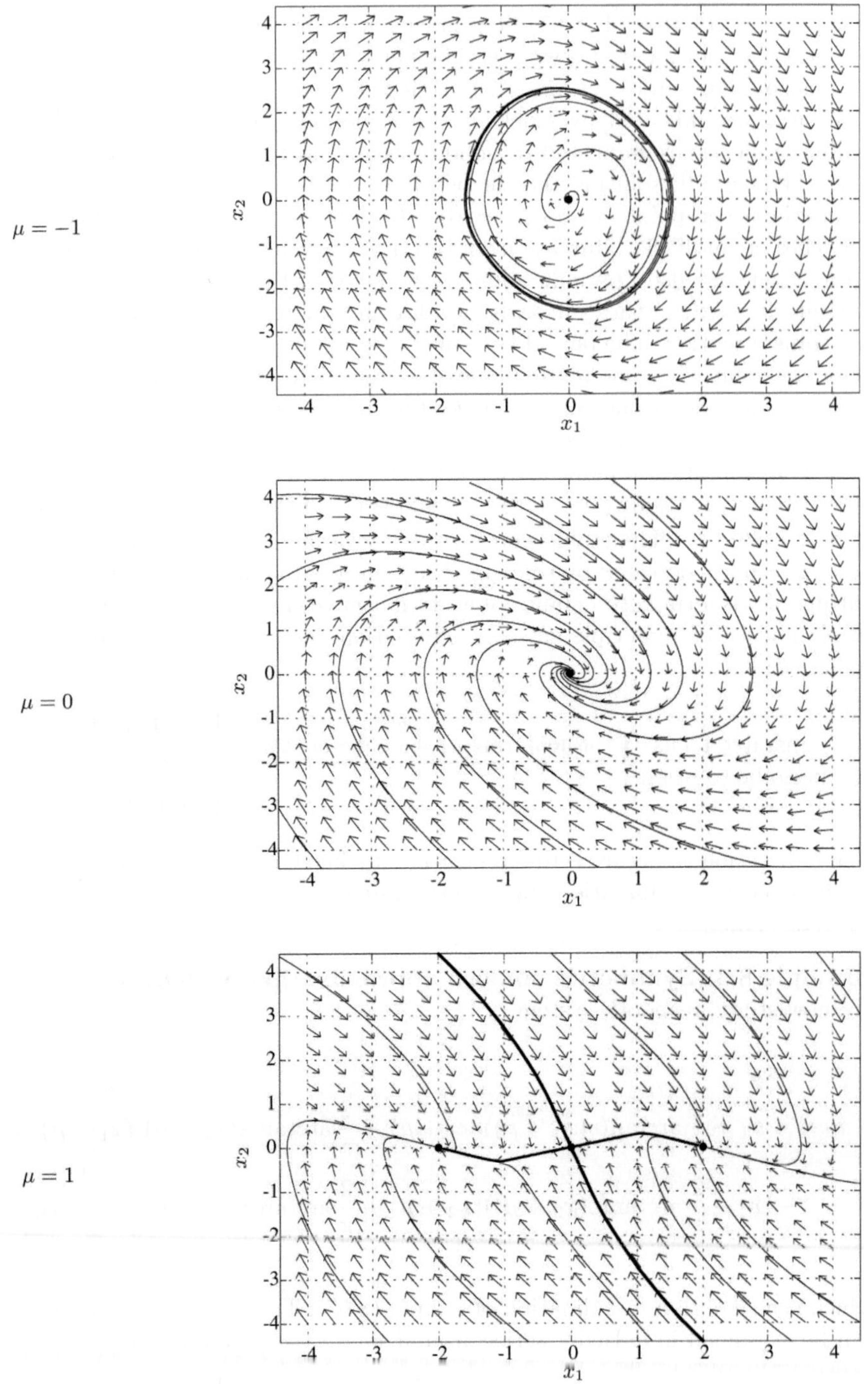

Fig. 8.19. Multiple crossing bifurcation of system (8.67).

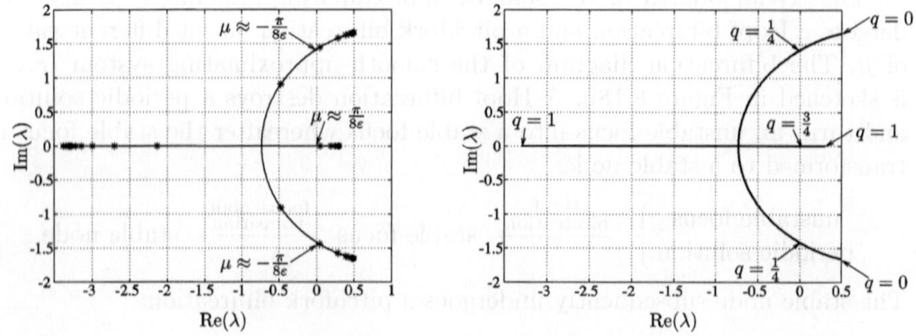

(a) Eigenvalues of the approximating system.

(b) Path of the set-valued eigenvalues of $\boldsymbol{J}(0,0)^{\mathrm{tr}}$.

Fig. 8.20. Eigenvalue path of system (8.67).

The chosen regularization of the non-smooth terms in (8.79) is symmetric in the sense that it preserves the symmetry of the system. The smooth approximating system (8.79), therefore, also has the trivial branch of equilibria $(x_1, x_2) = (0, 0)$. The location of the eigenvalues on the trivial branch have been computed numerically (with $\varepsilon = 10$) for varying μ and are plotted (indicated by '$*$') in the complex plane in Figure 8.20a. The eigenvalue-path of the convex combination of the Jacobian matrices (8.74) of the non-smooth system (8.67)

$$\boldsymbol{J}^{\mathrm{tr}} = \{(\boldsymbol{J}_+^{\mathrm{tr}} - \boldsymbol{J}_-^{\mathrm{tr}})q + \boldsymbol{J}_-^{\mathrm{tr}},\ q \in [0, 1]\}, \tag{8.81}$$

which uses only a *single* auxiliary variable q, is indicated by a solid line in Figures 8.20a and b. The eigenvalues of the smooth approximating system seem to be almost located on the eigenvalue-path of the above convex combination of $\boldsymbol{J}_+^{\mathrm{tr}}$ and $\boldsymbol{J}_-^{\mathrm{tr}}$ of the non-smooth system (8.67) (the $*$-signs are located on the solid line in Figure 8.20a). We observe that the eigenvalues of the convex combination of $\boldsymbol{J}_+^{\mathrm{tr}}$ and $\boldsymbol{J}_-^{\mathrm{tr}}$ cross the imaginary axis twice. At $q = \frac{1}{4}$ a pair of complex conjugated eigenvalues passes the imaginary axis and at $q = \frac{3}{4}$ a single eigenvalue passes the origin. The coincidence of the eigenvalue-path of the smooth approximating system with the eigenvalue-path of the convex combination (8.81) suggests that the bifurcation behaviour of this particular system might be studied by the eigenvalue-path of $\boldsymbol{J}^{\mathrm{tr}}$, which uses only a *single* auxiliary variable. The set $\boldsymbol{J}^{\mathrm{tr}}$ is a subset of $\boldsymbol{J}(\boldsymbol{0}, 0)$, being dependent on *four* auxiliary variables. The reason why this bifurcation can possibly be analyzed with a single auxiliary variable might be sought in the symmetries of the system. Bifurcation points on multiple switching boundaries can generally not be studied by a convex combination of two Jacobian matrices as will be demonstrated in Example 8.5.

The trivial branch of the smooth approximating system (8.79) also undergoes a Hopf bifurcation and a pitchfork bifurcation but at different values of μ. The bifurcation diagram of the smooth approximating system (8.79) is sketched in Figure 8.18b. A Hopf bifurcation destroys a periodic solution and turns an unstable focus into a stable focus whereafter the stable focus is transformed to a stable node:

$$\left.\begin{array}{r}\text{unstable focus}\\ \text{periodic solution}\end{array}\right\} \xrightarrow[]{\substack{\text{Hopf}\\ \text{bifurcation}}} \text{stable focus} \xrightarrow[]{\substack{\text{focus–node}\\ \text{transition}}} \text{stable node}$$

The stable node subsequently undergoes a pitchfork bifurcation:

$$\text{stable node} \xrightarrow[]{\substack{\text{pitchfork}\\ \text{bifurcation}}} \begin{cases}\text{stable node}\\ \text{saddle}\\ \text{stable node}\end{cases}$$

The Hopf bifurcation is approximately located at $\mu = -\frac{\pi}{8\varepsilon}$ and the pitchfork bifurcation approximately at $\mu = \frac{\pi}{8\varepsilon}$. The two bifurcations approach each other for increasing ε. The two bifurcations seem to occur simultaneously in the non-smooth system (8.67).

Example 8.4 (Non-symmetric smooth approximation of (8.67))
We now study another smooth approximating system of system (8.67) using a non-symmetric regularization:

$$\begin{aligned}\dot{x}_1 &= x_2,\\ \dot{x}_2 &= -x_1 + \frac{2}{\pi}\arctan(\varepsilon(x_1+\mu))(x_1+\mu) - \frac{2}{\pi}\arctan(\varepsilon(x_1-\mu))(x_1-\mu)\\ &\quad - x_2 - \frac{2}{\pi}\arctan(\varepsilon(x_2+\mu))(x_2+\mu) + \frac{2}{\pi}\arctan(\varepsilon(x_2-\mu))(x_2-\mu) + \frac{1}{\varepsilon}.\end{aligned} \tag{8.82}$$

The non-symmetric regularization does not preserve the symmetry of the system. The bifurcation diagram of the smooth approximating system (8.79) is sketched in Figure 8.18c. A Hopf bifurcation destroys a periodic solution and creates a stable focus:

$$\left.\begin{array}{r}\text{unstable focus}\\ \text{periodic solution}\end{array}\right\} \xrightarrow[]{\substack{\text{Hopf}\\ \text{bifurcation}}} \text{stable focus}$$

The stable focus is transformed into a stable node. Two other branches are created by a saddle–node bifurcation:

$$\begin{array}{lll}\text{stable focus} \xrightarrow[]{\substack{\text{focus–node}\\ \text{transition}}} \text{stable node} & & \text{stable node}\\ \qquad\qquad\qquad\quad\ \emptyset & \xrightarrow[]{\substack{\text{saddle–node}\\ \text{bifurcation}}} & \begin{cases}\text{saddle}\\ \text{stable node}\end{cases}\end{array}$$

The multiple crossing bifurcation of Example 8.2 shows a similarity with a Hopf and a pitchfork bifurcation. Different smooth approximating systems (Examples 8.3 and 8.4), however, give a different sequence of continuous bifurcations. Consequently, we can not name the multiple crossing bifurcation in the non-smooth system (8.67) by the term 'Hopf–pitchfork bifurcation'. The terminology for multiple crossing bifurcation points will become even more problematic in the following examples.

Example 8.5 (Combined Hopf and turning point behaviour)
Consider the two-dimensional non-smooth continuous system

$$\begin{aligned} \dot{x}_1 &= x_2, \\ \dot{x}_2 &= -|x_1 + 4\mu| - |x_2 - \mu| - \tfrac{1}{2}x_2 + 5\mu, \end{aligned} \tag{8.83}$$

which is piecewise linear with two switching boundaries $\Sigma_1 = \{x_1 + 4\mu = 0\}$ and $\Sigma_2 = \{x_2 - \mu = 0\}$. The system (8.83) has for $\mu > 0$ two distinct equilibria

$$\begin{aligned} &\text{equilibrium 1: } x_1 = -\mu, \;\; x_2 = 0, \text{ unstable focus,} \\ &\text{equilibrium 2: } x_1 = -9\mu, x_2 = 0, \text{ saddle point.} \end{aligned} \tag{8.84}$$

The two equilibria agree at $\mu = 0$ and the system has no equilibria for $\mu < 0$. The phase plane of the system is shown in Figure 8.21 for $\mu = -1$, $\mu = 0$ and $\mu = 1$. The stable and unstable invariant manifolds of the saddle point are depicted with thick solid lines. The phase plane for $\mu = 1$ also shows a stable periodic solution. All trajectories, invariant manifolds and periodic solutions in Figure 8.21 are obtained by numerical integration. The corresponding bifurcation diagram (Figure 8.22) reveals that two unstable equilibrium branches and one stable periodic branch meet each other at $\mu = 0$. Apparently, a bifurcation (in the sense of Definition 8.1) occurs at $\mu = 0$. The bifurcation point exhibits the behaviour of a turning point, as an equilibrium branch turns around, as well as the behaviour of a Hopf bifurcation, because a branch of periodic solutions is created at the bifurcation point. The discontinuous bifurcation has the structure:

$$\emptyset \xrightarrow{\text{multiple crossing bifurcation}} \begin{cases} \text{saddle} \\ \text{unstable focus} \\ \text{stable periodic solution} \end{cases}$$

The system is piecewise linear and has the generalized Jacobian matrix

$$\boldsymbol{J}(\boldsymbol{x}, \mu) = \begin{bmatrix} 0 & 1 \\ -\operatorname{Sign}(x_1 + 4\mu) & -\operatorname{Sign}(x_2 - \mu) - \frac{1}{2} \end{bmatrix}, \tag{8.85}$$

which gives the set-valued Jacobian matrix at the bifurcation point

$$\boldsymbol{J}(\boldsymbol{0}, 0) = \{\boldsymbol{J}_q,\ q_1, q_2 \in [0,1]\}, \text{ where } \boldsymbol{J}_q = \begin{bmatrix} 0 & 1 \\ -2q_1 + 1 & -2q_2 + \frac{1}{2} \end{bmatrix}. \tag{8.86}$$

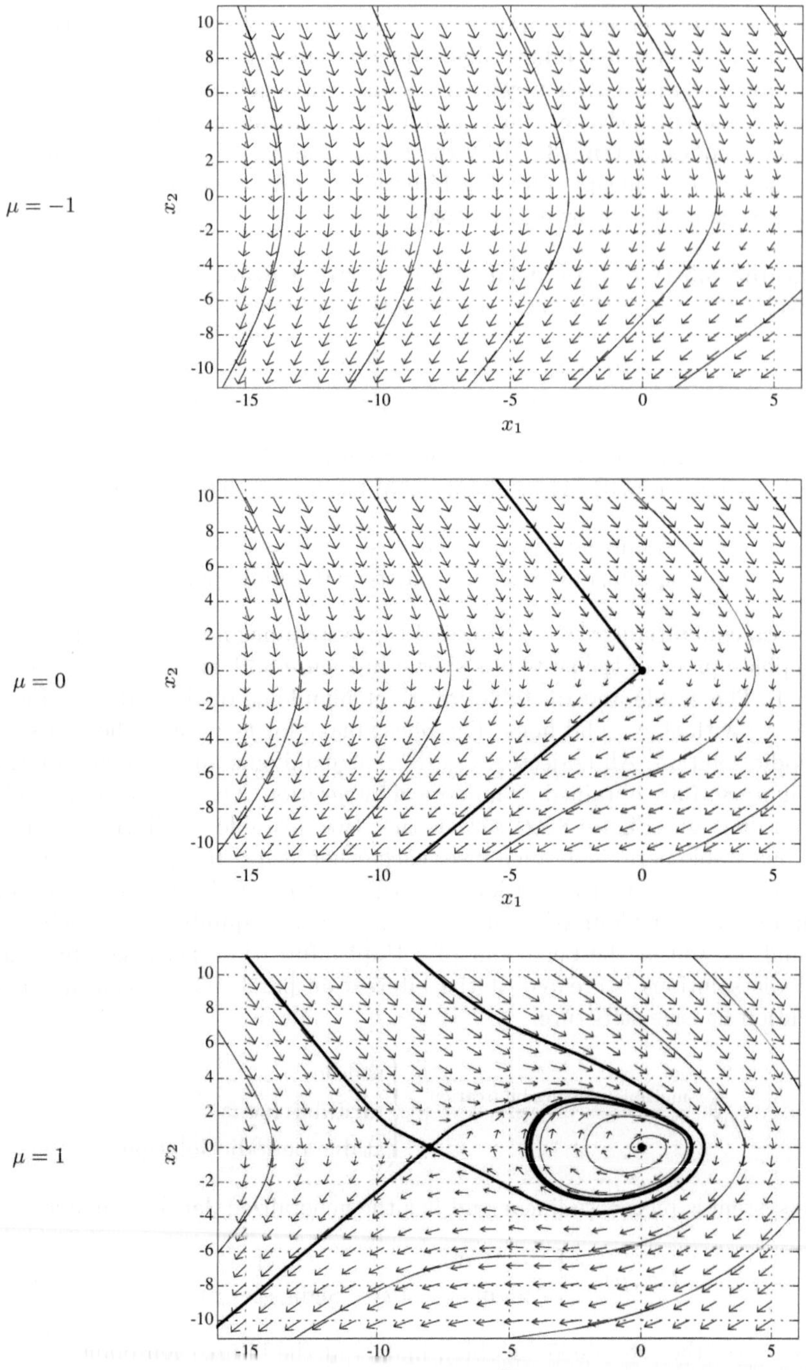

Fig. 8.21. Multiple crossing bifurcation of system (8.83).

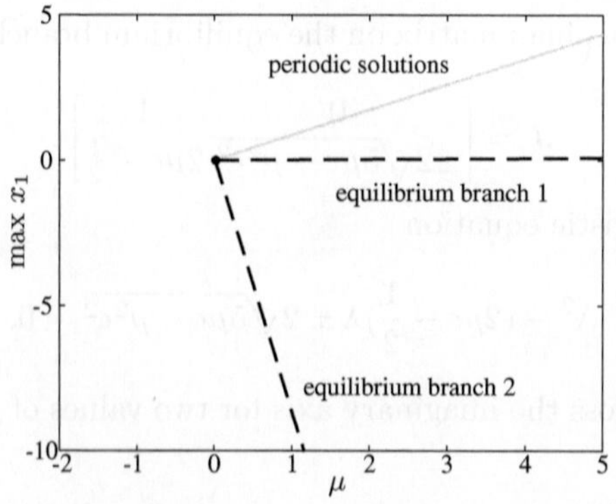

Fig. 8.22. Bifurcation diagram of system (8.83).

The fact that the bifurcation point is located on *two* switching boundaries makes the eigenvalues dependent on two auxiliary variables (q_1 and q_2). The set-valued eigenvalues of the set-valued Jacobian $\boldsymbol{J}(\boldsymbol{0}, 0)$ have been studied in Example 8.1. The set of eigenvalues of $\boldsymbol{J}(\boldsymbol{0}, 0)$ forms an area in the complex plane which intersects the imaginary axis as is depicted in Figure 8.5. The discontinuous bifurcation of system (8.83) is therefore a multiple crossing bifurcation, showing the behaviour of both a Hopf bifurcation and a turning point bifurcation.

Example 8.6 (Smooth approximation of system (8.83))
Consider the following smooth approximation of system (8.83):

$$\begin{aligned}\dot{x}_1 &= x_2,\\ \dot{x}_2 &= -\frac{2}{\pi}\arctan\left(\varepsilon(x_1+4\mu)\right)(x_1+4\mu)\\ &\quad -\frac{2}{\pi}\arctan\left(\varepsilon(x_2-\mu)\right)(x_2-\mu)-\frac{1}{2}x_2+5\mu.\end{aligned} \tag{8.87}$$

The smooth system (8.87) will be studied in the neighbourhood of the discontinuous bifurcation point ($x_1 = 0, x_2 = 0$) at $\mu = 0$ of the non-smooth system (8.83) using Taylor series

$$\begin{aligned}\dot{x}_1 &= x_2,\\ \dot{x}_2 &= -\epsilon(x_1+4\mu)^2-\epsilon(x_2-\mu)^2-\tfrac{1}{2}x_2+5\mu,\end{aligned} \tag{8.88}$$

for $\epsilon|x_1+4\mu| \ll 1$, $\epsilon|x_2-\mu| \ll 1$ with the abbreviation $\epsilon = \frac{2}{\pi}\varepsilon$. System (8.88) has for $0 < \mu < \frac{5}{\epsilon}$ two equilibria

$$x_1 = -4\mu \pm \frac{1}{\epsilon}\sqrt{5\mu\epsilon-\mu^2\epsilon^2}, \qquad x_2 = 0. \tag{8.89}$$

Apparently, the smoothened system exhibits (for this particular smoothing) a turning point in the form of a saddle–node bifurcation at ($x_1 = 0, x_2 = 0$)

with $\mu = 0$. The Jacobian matrix on the equilibrium branches of system (8.88) is

$$\boldsymbol{J} = \begin{bmatrix} 0 & 1 \\ \pm 2\sqrt{5\mu\epsilon - \mu^2\epsilon^2} & 2\mu\epsilon - \frac{1}{2} \end{bmatrix}, \tag{8.90}$$

with the characteristic equation

$$\lambda^2 - (2\mu\epsilon - \frac{1}{2})\lambda \pm 2\sqrt{5\mu\epsilon - \mu^2\epsilon^2} = 0. \tag{8.91}$$

The eigenvalues cross the imaginary axis for two values of μ (only considering $\mu < \frac{5}{\epsilon}$)

$$\begin{aligned} &\mu = 0: \ \lambda_1 = 0, \ \lambda_2 = -\tfrac{1}{2}, \text{ saddle–node bifurcation,} \\ &\mu = \tfrac{1}{4\epsilon}: \lambda_{1,2} = \pm i\sqrt[4]{\tfrac{19}{4}}, \quad \text{Hopf bifurcation on equilibrium branch 1.} \end{aligned} \tag{8.92}$$

The path of the eigenvalues gives rise to a saddle–node bifurcation at $\mu = 0$, causing the equilibrium branch to turn around, and a Hopf bifurcation at $\mu = \frac{1}{4\epsilon}$, creating a branch of periodic solutions. The Hopf bifurcation point approaches the saddle–node bifurcation point for increasing values of ϵ. The logical structure of the bifurcation diagram of the smooth approximating system shows two continuous single crossing bifurcations. A saddle–node bifurcation creates a saddle and a stable node, which is transformed into a stable focus:

$$\emptyset \xrightarrow{\text{saddle–node bifurcation}} \begin{cases} \text{saddle} \\ \text{stable node} \xrightarrow[\text{transition}]{\text{node–focus}} \text{stable focus} \end{cases}$$

Subsequently, the stable focus undergoes a Hopf bifurcation:

$$\text{stable focus} \xrightarrow{\text{Hopf bifurcation}} \begin{cases} \text{unstable focus} \\ \text{stable periodic solution} \end{cases}$$

The discontinuous bifurcation point of the non-smooth system (8.83) exhibits the bifurcation behaviour of both a turning point bifurcation and a Hopf bifurcation. The branches of a periodic solution, a saddle point and a focus meet each other at the discontinuous bifurcation point. The particular smooth approximating system studied in this example shows a saddle–node bifurcation (turning point bifurcation) and a Hopf bifurcation. Other smooth approximating systems of (8.83) may have a different sequence of continuous bifurcations.

The systems considered in Examples 8.2 and 8.5 show multiple crossing bifurcations. The bifurcation points of systems (8.67) and (8.83) are located on more than one switching boundary. As a result, the generalized Jacobian matrix at the bifurcation point is dependent on multiple auxiliary variables and a (unique) 'path' of the eigenvalues can not be obtained. A multiple crossing bifurcation can however also occur in a system with a single switching boundary as will be shown in the following example.

Example 8.7 (Multiple crossing bifurcation with a turning point) Consider the two-dimensional non-smooth continuous system

$$\begin{aligned} \dot{x}_1 &= x_1 + 2|x_1| + x_2, \\ \dot{x}_2 &= x_1 + 2|x_1| + \tfrac{1}{2}x_2 + \mu, \end{aligned} \tag{8.93}$$

which is piecewise linear and has a single switching boundary $\Sigma = \{\boldsymbol{x} \in \mathbb{R}^2 \mid x_1 = 0\}$. Note that system (8.93) is of the type (8.25) (after a simple coordinate transformation). The system (8.93) has for $\mu < 0$ two distinct equilibria

$$\begin{aligned} &\text{equilibrium 1: } x_1 = -\tfrac{2}{3}\mu,\ x_2 = 2\mu, \\ &\text{equilibrium 2: } x_1 = 2\mu, \quad x_2 = 2\mu, \end{aligned} \tag{8.94}$$

and has no equilibria for $\mu > 0$. The generalized Jacobian matrix of the system is

$$\boldsymbol{J}(x_1) = \begin{bmatrix} 1 + 2\,\mathrm{Sign}(x_1) & 1 \\ 1 + 2\,\mathrm{Sign}(x_1) & \frac{1}{2} \end{bmatrix}, \tag{8.95}$$

which takes a constant value at each side of the switching boundary

$$\boldsymbol{J}_- = \begin{bmatrix} -1 & 1 \\ -1 & \frac{1}{2} \end{bmatrix} \quad \text{for } x_1 < 0, \quad \lambda_{1,2} = -\frac{1}{4} \pm i\sqrt{\frac{7}{16}}, \tag{8.96}$$

$$\boldsymbol{J}_+ = \begin{bmatrix} 3 & 1 \\ 3 & \frac{1}{2} \end{bmatrix} \quad \text{for } x_1 > 0, \quad \lambda_{1,2} = \frac{7}{4} \pm \sqrt{\frac{73}{16}} \approx \{-0.386, 3.886\}. \tag{8.97}$$

The generalized Jacobian at the bifurcation point $\boldsymbol{x} = \boldsymbol{0}$ is the closed convex hull of the Jacobians on each side of the switching boundary

$$\boldsymbol{J}(\boldsymbol{0}) = \overline{\mathrm{co}}(\boldsymbol{J}_-, \boldsymbol{J}_+) = \{(1-q)\boldsymbol{J}_- + q\boldsymbol{J}_+,\ \forall q \in [0,1]\}. \tag{8.98}$$

The eigenvalues $\lambda_{1,2}$ of $\boldsymbol{J}(\boldsymbol{0})$ are set-valued and form a *path* in the complex plane with q as path parameter. The path of eigenvalues of the generalized Jacobian at $\boldsymbol{x} = \boldsymbol{0}$ is depicted in Figure 8.23. The eigenvalues of $\boldsymbol{J}_q$ are purely complex for $q = \frac{1}{8}$ and one eigenvalue crosses the origin for $q = \frac{1}{4}$. The path of the eigenvalues of $\boldsymbol{J}(\boldsymbol{0})$ shows that the discontinuous bifurcation point is a multiple crossing bifurcation. With the previous examples in mind, one might suggests that the behaviour of this multiple crossing bifurcation is the combination of two single crossing bifurcations, a Hopf bifurcation and a turning point bifurcation. Figure 8.24 depicts the phase plane of (8.93) for $\mu = -1$, $\mu = 0$ and $\mu = 1$. The two equilibria (8.94) are present for $\mu = -1$ of which one equilibrium is a stable focus and the other equilibrium is a saddle point. The invariant manifolds of the saddle point show an interesting behaviour. An unstable invariant half-manifold of the saddle point is spiralling

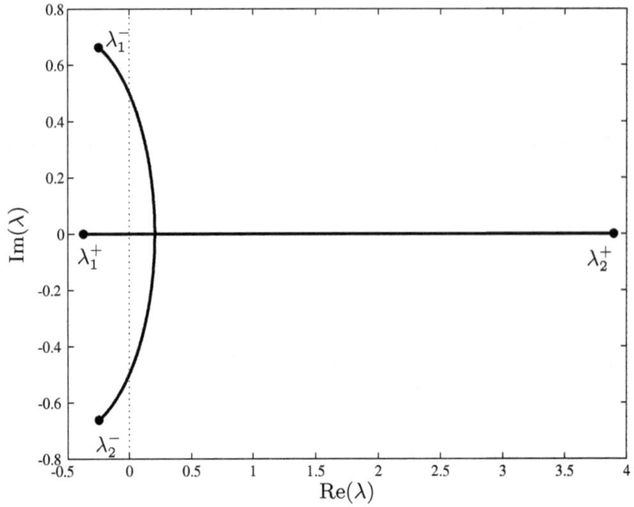

Fig. 8.23. Path of the eigenvalues of $\boldsymbol{J}(\mathbf{0})$ (8.98) of system (8.93).

towards the stable focus while one of the stable invariant half-manifolds is folded to the other stable invariant half-manifold. The two equilibria collide to one equilibrium for $\mu = 0$ and only two invariant half-manifolds (a stable and an unstable) remain. An equilibrium having only two invariant half-manifolds is a peculiarity of non-smooth continuous systems. No equilibrium or periodic solution exists for $\mu = 1$. The multiple crossing bifurcation has the structure:

$$\left.\begin{array}{l}\text{stable focus}\\ \text{saddle}\end{array}\right\} \xrightarrow{\text{multiple crossing bifurcation of (8.93)}} \emptyset$$

Clearly, the behaviour of a turning point bifurcation is present in the bifurcation scenario observed in Figure 8.24 and Theorem 8.9 applies. The bifurcation scenario does not show a creation (or destruction) of a periodic solution under variation of μ, i.e. no behaviour of a Hopf bifurcation. The multiple crossing bifurcation is therefore not simply the combination of two single crossing bifurcations.

Example 8.8 (Smooth approximation of system (8.93))
Some insight in the bifurcation behaviour depicted in Figure 8.24 can be obtained by considering a smooth approximation of system (8.93)

$$\begin{aligned}\dot{x}_1 &= x_1 + \tfrac{4}{\pi}\arctan(\varepsilon x_1)x_1 + x_2,\\ \dot{x}_2 &= x_1 + \tfrac{4}{\pi}\arctan(\varepsilon x_1)x_1 + \tfrac{1}{2}x_2 + \mu.\end{aligned} \tag{8.99}$$

Of course, we have to keep in mind that (8.99) is just one particular smooth approximation of (8.93). Figure 8.25 and 8.26 show the phase planes of (8.99) for six different values of μ in the neighbourhood of $\mu = 0$ (using $\varepsilon = 20$).

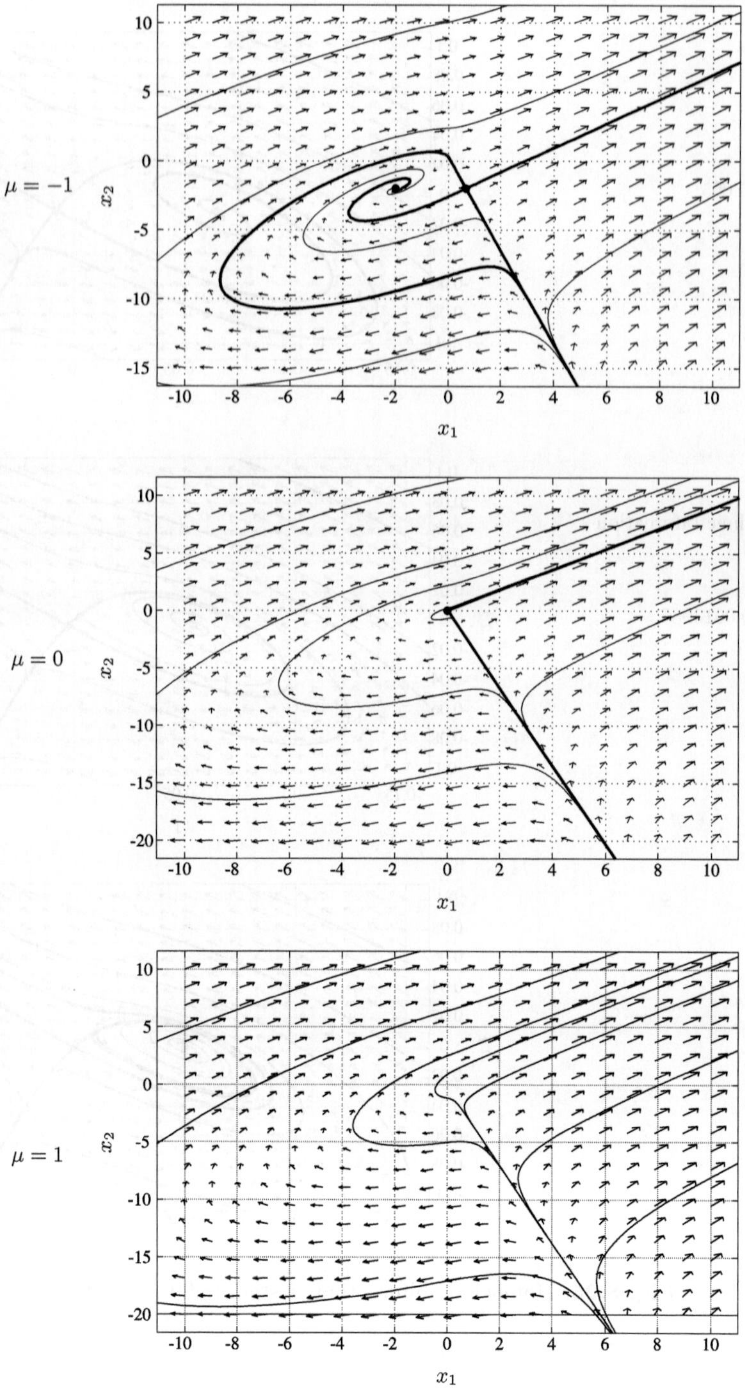

Fig. 8.24. Multiple crossing bifurcation of system (8.93).

$\mu = -0.005$

Homoclinic bifurcation

$\mu = -0.0013987$

$\mu = 0$

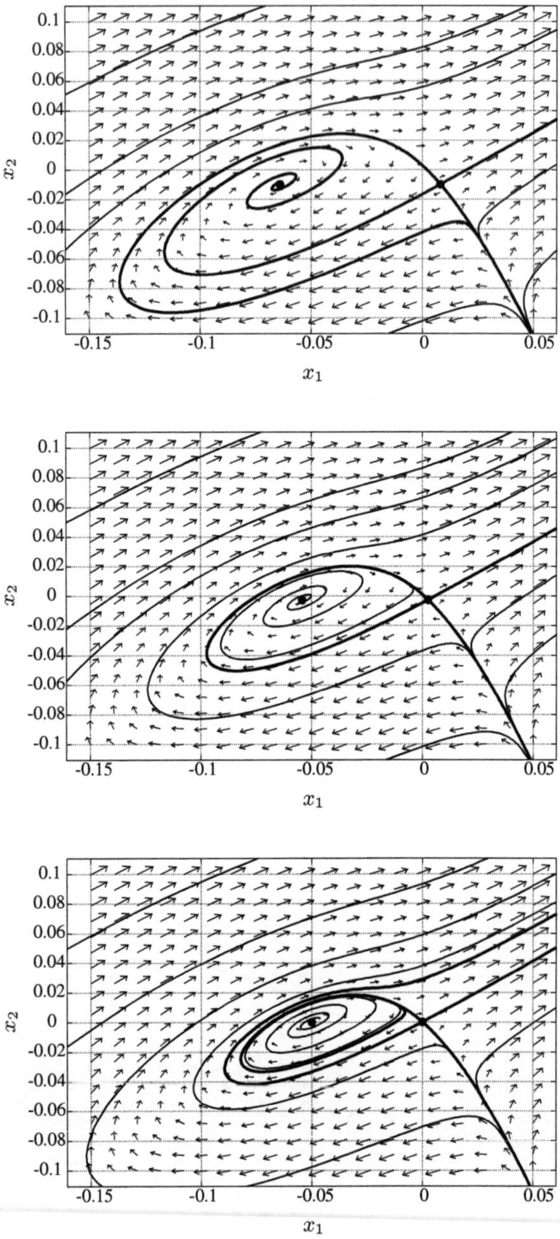

Fig. 8.25. Homoclinic bifurcation, system (8.99), $\varepsilon = 20$.

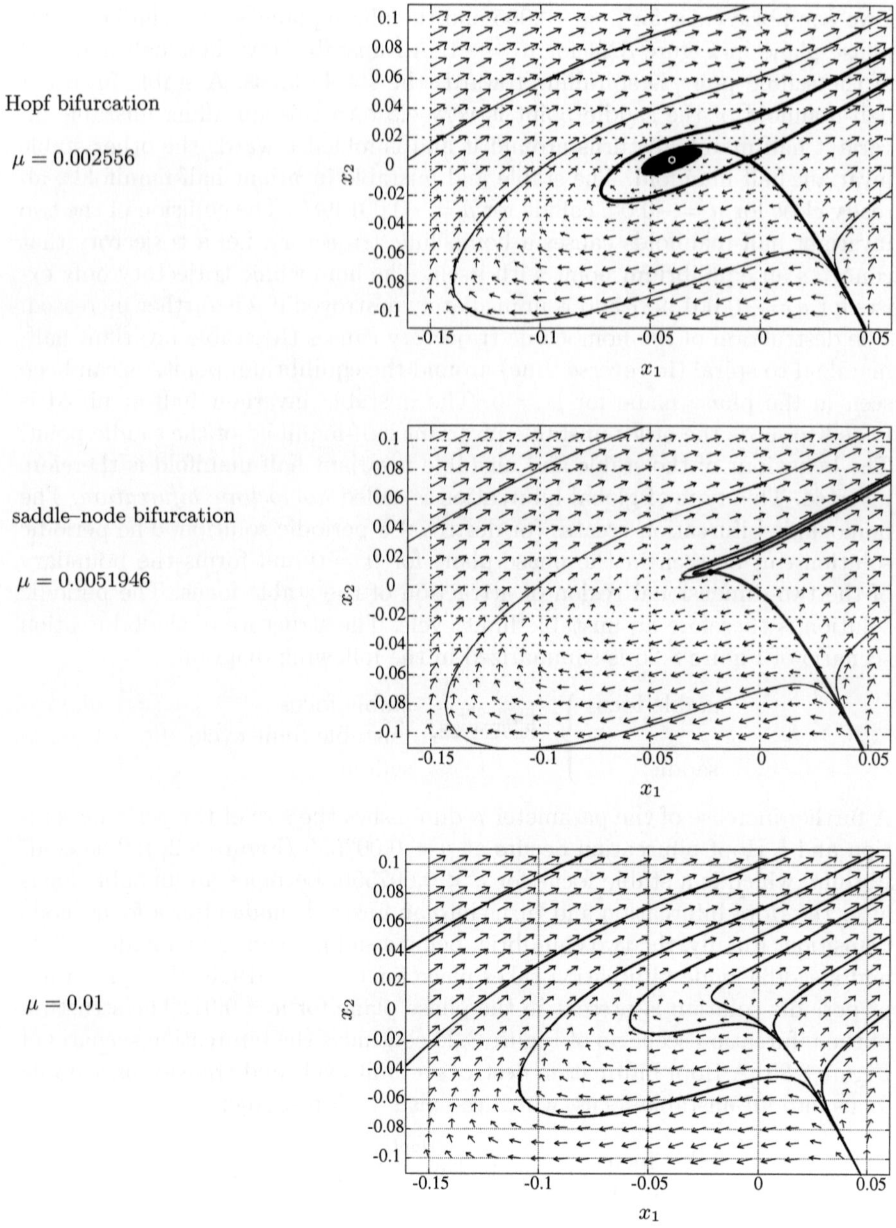

Fig. 8.26. Hopf and saddle–node bifurcation, system (8.99), $\varepsilon = 20$.

Two equilibria exist for $\mu = -0.005$ and the phase plane is very similar to the phase plane in Figure 8.24 for $\mu = -1$. An unstable invariant half-manifold of the saddle point is spiralling towards the stable focus. A stable invariant half-manifold of the saddle point is very close to this spiralling unstable invariant half-manifold, turns around it and is folded towards the other stable invariant half-manifold. The stable and unstable invariant half-manifolds, already close for $\mu = -0.05$, collide for $\mu = -0.0013987$. The collision of the two invariant half-manifolds causes a homoclinic trajectory, i.e. a trajectory that connects an equilibrium point with itself. The homoclinic trajectory only exists for $\mu = -0.0013987$ and is immediately destroyed if μ is further increased. The destruction of the homoclinic trajectory causes the stable invariant half-manifold to spiral (in reverse time) around the equilibrium point, as can been seen in the phase plane for $\mu = 0$. The unstable invariant half-manifold is folded towards the other unstable invariant half-manifold of the saddle point. The behaviour of the stable and unstable invariant half-manifold is therefore inverted. This type of global bifurcation is called *homoclinic bifurcation*. The homoclinic bifurcation creates (or destroys) a periodic solution. The periodic solution can be seen in the phase plane for $\mu = 0$ and forms the boundary of the two-dimensional region of attraction of the stable focus. The periodic solution is therefore an unstable limit cycle. The structure of the bifurcation scenario of Figure 8.25 is summarized in the following diagram:

$$\left.\begin{array}{l}\text{stable focus}\\ \\ \text{saddle}\end{array}\right\}\xrightarrow[\text{bifurcation}]{\text{homoclinic}}\begin{array}{l}\text{stable focus}\\ \text{unstable limit cycle}\\ \text{saddle}\end{array}$$

A further increase of the parameter μ diminishes the size of the periodic solution and a Hopf bifurcation occurs at $\mu = 0.002556$ (Figure 8.26). The equilibrium, which is a stable focus for $\mu < 0.002556$, becomes an unstable focus after the Hopf bifurcation and turns into an unstable node after a focus–node transition. Finally, the two equilibria, being a saddle point and a node, collide and a saddle–node bifurcation takes place for $\mu = 0.0051946$. No equilibrium or periodic solution is present in the phase plane for $\mu = 0.01$. The structure of the bifurcation scenario of Figure 8.26 continues the bifurcation scenario of Figure 8.25. A Hopf bifurcation destroys a limit cycle and transforms a stable focus into an unstable focus, which co-exists with a saddle:

$$\begin{array}{l}\left.\begin{array}{l}\text{stable focus}\\ \text{unstable limit cycle}\end{array}\right\}\xrightarrow[\text{bifurcation}]{\text{Hopf}}\ \text{unstable focus}\\ \text{saddle}\qquad\qquad\qquad\qquad\qquad\qquad\ \ \text{saddle}\end{array}$$

Subsequently, the unstable focus is transformed into an unstable node and is destroyed together with the co-existing saddle by a saddle–node bifurcation:

$$\begin{array}{l}\text{unstable focus}\xrightarrow[\text{transition}]{\text{focus–node}}\left.\begin{array}{l}\text{unstable node}\\ \\ \text{saddle}\end{array}\right\}\xrightarrow[\text{bifurcation}]{\text{saddle–node}}\emptyset\\ \text{saddle}\end{array}$$

The smoothing of the non-smooth terms causes the eigenvalues to be a single-valued function of the parameter μ. The multiple crossing bifurcation of (8.93) is therefore, for this particular choice of the smoothing function, torn apart in two single crossing bifurcations (a Hopf bifurcation and a saddle–node bifurcation) and a global bifurcation (a homoclinic bifurcation). The bifurcation structure of the non-smooth continuous system shows only one discontinuous multiple crossing bifurcation:

$$\left.\begin{array}{r}\text{stable focus}\\ \text{saddle}\end{array}\right\} \xrightarrow[\text{bifurcation}]{\text{multiple crossing}} \emptyset$$

which replaces the complex structure of the smooth approximating system. Again, the question rises how to name this particular multiple crossing bifurcation. The multiple crossing bifurcation is basically a discontinuous turning point bifurcation for which Theorem 8.9 applies. However, the behaviour of the discontinuous bifurcation is much more complex than just the behaviour of a turning point, which is reflected by the complex structure of single crossing bifurcations of the smooth approximating system (8.99). It has become clear that the problem of terminology is becoming extremely difficult when studying more complex bifurcations.

8.8 Discussion and Conclusions

In the previous sections a number of examples of discontinuous bifurcations have been shown. All classical bifurcations of equilibria of smooth systems have been discussed and a parallel has been shown with bifurcations in non-smooth continuous systems. For each of the classical continuous bifurcations a discontinuous single crossing bifurcation has been found in Section 8.6. A continuous bifurcation is in fact a special case of a single crossing bifurcation, for which the set-valuedness of the eigenvalues reduces to a singleton. A number of multiple crossing bifurcations have been discussed in Section 8.7 and have been shown to behave much more complex than single crossing bifurcations.

Some fundamental questions about discontinuous bifurcations have been raised in Section 8.4. With the preceding examples in mind, we will try to answer these questions one by one.

1. *Does a jump of an eigenvalue (or a pair of them) over the imaginary axis under the influence of a parameter imply a bifurcation in the sense of Definition 8.1?* In all examples the conclusion that a bifurcation exists has been taken from *inspection* of the bifurcation diagram. If there is a change in the number of equilibria and (quasi-) periodic solutions for a certain parameter value, then there is a bifurcation at this parameter value according to Definition 8.1. We also observed that a 'jump' of the eigenvalue over the imaginary axis exists. For planar piecewise linear systems with a

single switching boundary, which are of the special form (8.12), it follows from Theorems 8.9 and 8.11 that at a bifurcation point the set of eigenvalues of the generalized Jacobian form a path which crosses the imaginary axis. Although it is intuitively appealing to state that at a bifurcation point the set of eigenvalues of the generalized Jacobian *must* have a crossing with the imaginary axis for a general non-smooth continuous system, we do not have a rigourous proof for this.

2. *Does a continuous bifurcation exist if the non-smooth continuous system, exposing a discontinuous bifurcation, is approximated by a smooth system?* In Subsection 8.6.2 a counter example has been given by taking a non-symmetric smoothing function. The smoothened version of the non-smooth system does not expose a bifurcation. It is noted that if the smoothing parameter is increased, all branches (of equilibria) converge to the branches of the non-smooth system but the branches may not necessarily intersect (implying a bifurcation).
3. *Is the discontinuous bifurcation of the non-smooth system related to the continuous bifurcation of the smoothened system (assuming that it exists)?* In Subsection 8.6.3 a non-smooth system exposes a discontinuous pitchfork bifurcation. A non-symmetric smoothing of the non-smooth system exposes a saddle–node bifurcation. We conclude that the bifurcation(s) occurring in the smoothened system is (are) not always related to the discontinuous bifurcation of the original non-smooth system.
4. *Do discontinuous bifurcations exist that do not have a continuous counterpart? In other words: do discontinuous bifurcations exist that behave qualitatively different from any continuous bifurcation?* Example 8.2 of Section 8.7 shows a discontinuous bifurcation which can be looked upon as the combination of two continuous bifurcations (a Hopf and a pitchfork bifurcation). This bifurcation can therefore not be classified as simply the counterpart of a continuous bifurcation. However, the particular discontinuous bifurcation still behaves in the same way as the two continuous bifurcations. The qualitative behaviour of the bifurcation is (in this particular case) simply the combination of the behaviour of a Hopf and a pitchfork bifurcation. The multiple crossing bifurcation encountered in Example 8.7 shows a bifurcation behaviour that is not simply the combination of behaviour of continuous bifurcations.
5. *Can we classify a discontinuous bifurcation by inspecting the point(s) where the set of eigenvalues of the generalized Jacobian intersects the imaginary axis?* The set-valued eigenvalues of the generalized Jacobian form a path in the complex plane if the bifurcation point is located on a single switching boundary or if the system is one-dimensional, for which the path is a subset of the real axis. The set-valued eigenvalues generally form an area in the complex plane. The discontinuous single crossing bifurcations of the preceding examples have been classified by comparing their nature with a certain type of continuous bifurcation. If at a discontinuous single crossing bifurcation the change of equilibria is the same as

for a certain type of continuous bifurcation, then the discontinuous single crossing bifurcation can be regarded as the discontinuous counterpart of that type of continuous bifurcation. We observe that all discontinuous static bifurcations expose a path of a set-valued eigenvalue intersecting the imaginary axis at the origin, like for the continuous static bifurcations. For the discontinuous Hopf bifurcation, the set of eigenvalues of the generalized Jacobian has a complex conjugated intersection with the imaginary axis, consistent with the continuous Hopf bifurcation. The examples of Section 8.7 have made clear that a classification of multiple crossing bifurcations can not be given. A terminology for multiple crossing bifurcations based on the continuous bifurcations, which arise when smoothing the non-smooth system, is senseless because the resulting sequence/structure of continuous bifurcations is generally not unique. A multiple crossing bifurcation can sometimes be regarded as the collision of two (or more) continuous bifurcations (like for the Hopf–pitchfork bifurcation). However, the behaviour of a multiple crossing bifurcation is generally not the sum of continuous bifurcations (consider Example 8.7). A terminology based on the behaviour of the multiple crossing bifurcation is therefore not possible. At least for *single* crossing bifurcations it seems intuitively correct, that we can make a classification based on the crossing of the eigenvalue(s) with the imaginary axis.

We presume that a discontinuous single crossing bifurcation of an equilibrium of a non-smooth continuous system may be classified by inspecting the point where the path of the set-valued eigenvalue (or a pair of complex conjugated eigenvalues) crosses the imaginary axis. We do not say anything about the possible classification of multiple crossing bifurcations. One could argue whether or not the crossing points of the path of eigenvalues with the imaginary axis of a multiple crossing bifurcation still give some information about the behaviour of the bifurcation. Indeed, some multiple crossing bifurcations show the behaviour of two continuous bifurcations but this is for a general multiple crossing bifurcation not true as it can still show features not covered by two (or more) continuous bifurcations separately. The crossing points of the eigenvalues with the imaginary axis therefore *may* give a hint on the behaviour of the multiple crossing bifurcation.

Periodic solutions can be looked upon as equilibria on a Poincaré map. The generalization of eigenvalues for equilibria are Floquet multipliers for periodic solutions. If a bifurcation of an equilibrium occurs when the set of eigenvalues of the generalized Jacobian crosses the imaginary axis, then we can expect a discontinuous bifurcation of a periodic solution if the set of Floquet multiplier of the generalized fundamental solution matrix crosses the unit circle. Similarly, if one can identify a path of Floquet multipliers, then one can make a distinction between a single and a multiple crossing bifurcation. The type of discontinuous single crossing bifurcation of a periodic solution could be inferred from the crossing point of the path of the Floquet multiplier(s) with the unit circle. We will discuss bifurcations of periodic solutions in the next chapter.

9

Bifurcations of Periodic Solutions

In this chapter we will study bifurcations of periodic solutions in discontinuous systems of Filippov-type. The Poincaré map, which is introduced in Section 9.1, converts the continuous-time system to a discrete map. Fixed points of piecewise linear maps, which represent periodic solutions of Filippov systems, are studied in Section 9.2. Bifurcations of periodic solutions in smooth systems are briefly addressed in Section 9.3. Section 9.4 explains how a discontinuous bifurcation of a periodic solution can be created when a periodic solution touches a non-smooth switching boundary. The relation with discontinuous bifurcations of equilibria in non-smooth continuous systems is discussed. Bifurcations of fixed points in planar maps are discussed in Sections 9.5 and 9.6. Fundamental questions about discontinuous bifurcations of periodic solutions are raised in Section 9.7. Sections 9.8 and 9.9 treat a number of numerical examples which show discontinuous bifurcations. Section 9.10 draws conclusions from the numerical examples and tries to give answers to the questions of Section 9.7.

In this chapter we will study periodic solutions of nonlinear systems with a discontinuous right-hand side, as defined in Chapter 3, under the influence of a single parameter. The right-hand side is discontinuous on one or more switching boundaries and is assumed to be linearly bounded. An example of such a system is

$$\dot{\boldsymbol{x}}(t) = \boldsymbol{f}(t, \boldsymbol{x}(t), \mu) = \begin{cases} \boldsymbol{f}_-(t, \boldsymbol{x}(t), \mu), \; \boldsymbol{x} \in \mathcal{V}_-, \\ \boldsymbol{f}_+(t, \boldsymbol{x}(t), \mu), \; \boldsymbol{x} \in \mathcal{V}_+, \end{cases} \tag{9.1}$$

with

$$\begin{aligned} \mathcal{V}_- &= \{\boldsymbol{x} \in \mathbb{R}^n \mid h(\boldsymbol{x}(t), \mu) < 0\}, \\ \Sigma &= \{\boldsymbol{x} \in \mathbb{R}^n \mid h(\boldsymbol{x}(t), \mu) = 0\}, \\ \mathcal{V}_+ &= \{\boldsymbol{x} \in \mathbb{R}^n \mid h(\boldsymbol{x}(t), \mu) > 0\}. \end{aligned} \tag{9.2}$$

The system depends on a single parameter μ. Also, the switching boundary function h is in general dependent on μ, which implies dependence on μ for the switching boundary Σ and the subspaces $\mathcal{V}_-$ and $\mathcal{V}_+$. Discontinuous systems of this type can be extended to differential inclusions by means of the method proposed by Filippov (Chapter 3). The resulting differential inclusion has a set-valued map $\boldsymbol{F}(t, \boldsymbol{x}, \mu)$ which is upper semi-continuous, convex, closed, non-empty and linearly bounded. Existence of the solution to the initial value problem is therefore guaranteed. Uniqueness of the solution, however, is not a priori guaranteed.

9.1 The Poincaré Map

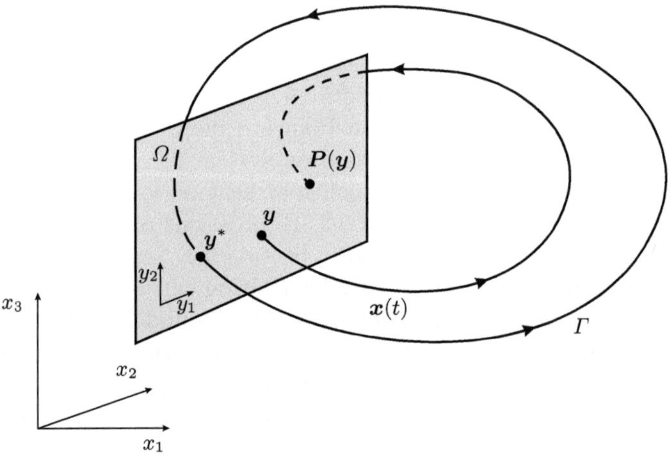

Fig. 9.1. Poincaré map for an autonomous system.

We can elucidate the relation between continuous bifurcations and discontinuous bifurcations by studying the Poincaré map [116]. Consider an n-th order autonomous system

$$\dot{\boldsymbol{x}}(t) = \boldsymbol{f}(\boldsymbol{x}(t)), \tag{9.3}$$

with a periodic solution Γ as shown in Figure 9.1. Let $\boldsymbol{x}^*$ be a point on the periodic solution and let Ω be an $(n-1)$-dimensional hyper-plane transversal to Γ at $\boldsymbol{x}^*$. A trajectory $\boldsymbol{x}(t)$ is said to intersect Ω in a positive direction if

$$\boldsymbol{n}_\Omega^{\mathrm{T}} \boldsymbol{f}(\boldsymbol{x}(t_i)) > 0, \tag{9.4}$$

where $\boldsymbol{n}_\Omega$ is the normal to Ω and with $\boldsymbol{x}(t_i) \in \Omega$. We assume that Ω is chosen such, that $\boldsymbol{n}_\Omega^{\mathrm{T}} \boldsymbol{f}(\boldsymbol{x}) > 0$ holds for all $\boldsymbol{x} \in \Omega$. We will call the hyper-plane Ω a *Poincaré section* and introduce coordinates $\boldsymbol{y} \in \mathbb{R}^{n-1}$ on Ω

$$\boldsymbol{y} = \boldsymbol{v}(\boldsymbol{x}), \qquad \boldsymbol{V} = \left(\frac{\partial \boldsymbol{v}}{\partial \boldsymbol{x}}\right)^{\mathrm{T}}, \tag{9.5}$$

where $\boldsymbol{v} : \Omega \to \Omega$ is a linear function mapping coordinates $\boldsymbol{x} \in \Omega$ to coordinates $\boldsymbol{y}$ on Ω. The matrix $\boldsymbol{V} \in \mathbb{R}^{n \times n-1}$ is constant and $\boldsymbol{V}^{\mathrm{T}}\boldsymbol{V}$ is assumed to be regular. Associated with $\boldsymbol{x}^* \in \Omega$ is the point $\boldsymbol{y}^*$. The trajectory emanating from $\boldsymbol{y}^*$ at $t = 0$ will hit Ω at $\boldsymbol{y}^*$ at $t = T$, where T is the period time of the periodic solution. Trajectories starting on Ω in a sufficiently small neighbourhood of $\boldsymbol{y}^*$ will, in approximately T time units, intersect Ω in the vicinity of $\boldsymbol{y}^*$. Therefore, $\boldsymbol{x}(t)$ and Ω define a mapping $\boldsymbol{P}$ of some neighbourhood $U \subset \Omega$ of $\boldsymbol{y}^*$ onto another neighbourhood $V \subset \Omega$ of $\boldsymbol{y}^*$. The map $\boldsymbol{P} : U \to V$ is defined as

$$\boldsymbol{y}_{i+1} = \boldsymbol{P}(\boldsymbol{y}_i), \tag{9.6}$$

where $\boldsymbol{y}_{i+1} = \boldsymbol{v}(\boldsymbol{x}_{i+1})$, with $\boldsymbol{x}_{i+1} = \boldsymbol{x}(t_{i+1}) \in \Omega$, is the first intersection point of the trajectory $\boldsymbol{x}(t)$ that starts from $\boldsymbol{x}(t_i) = \boldsymbol{x}_i \in \Omega$. The map $\boldsymbol{P}$ is called the *Poincaré map* or *first return map* of the autonomous system. The map can be iterated yielding $\boldsymbol{y}_{i+2} = \boldsymbol{P}(\boldsymbol{P}(\boldsymbol{y}_i))$. The continuous-time system (9.3) thus induces a discrete map $\boldsymbol{P}$ on a section Ω transversal to a periodic solution.

A point $\boldsymbol{y}^*$ on a periodic solution is a fixed point of the mapping $\boldsymbol{P}$

$$\boldsymbol{y}^* = \boldsymbol{P}(\boldsymbol{y}^*). \tag{9.7}$$

The intersection point $\boldsymbol{y}^*$ of a periodic solution Γ with a Poincaré section Ω is therefore a zero of the function

$$\boldsymbol{H}(\boldsymbol{y}) = \boldsymbol{P}(\boldsymbol{y}) - \boldsymbol{y}. \tag{9.8}$$

We can regard $\boldsymbol{H}(\boldsymbol{y})$ as the right-hand side of a discrete dynamical system

$$\Delta \boldsymbol{y} = \boldsymbol{H}(\boldsymbol{y}), \tag{9.9}$$

with an equilibrium point $\boldsymbol{y}^*$, i.e. $\boldsymbol{H}(\boldsymbol{y}^*) = \boldsymbol{0}$. Equilibria are therefore related to periodic solutions through the Poincaré map.

The derivative $\boldsymbol{J}_P \in \mathbb{R}^{n-1 \times n-1}$ of the map $\boldsymbol{P}$ is related to the fundamental solution matrix $\boldsymbol{\Phi}$ (see [116])

$$\begin{aligned} \boldsymbol{J}_P(\boldsymbol{y}_i) &= \left.\frac{\partial \boldsymbol{P}(\boldsymbol{y})}{\partial \boldsymbol{y}}\right|_{\boldsymbol{y}=\boldsymbol{y}_i} \\ &= \boldsymbol{V}^{\mathrm{T}} \frac{\partial \boldsymbol{x}_{i+1}}{\partial \boldsymbol{x}_i} \boldsymbol{V} \left(\boldsymbol{V}^{\mathrm{T}}\boldsymbol{V}\right)^{-1}, \end{aligned} \tag{9.10}$$

with

$$\frac{\partial \boldsymbol{x}_{i+1}}{\partial \boldsymbol{x}_i} = \left[\boldsymbol{I} - \frac{1}{\boldsymbol{n}_\Omega^{\mathrm{T}} \boldsymbol{f}(t_{i+1}, \boldsymbol{x}_{i+1})} \boldsymbol{f}(t_{i+1}, \boldsymbol{x}_{i+1}) \boldsymbol{n}_\Omega^{\mathrm{T}}\right] \boldsymbol{\Phi}(t_{i+1}, t_i), \tag{9.11}$$

where $\boldsymbol{\Phi}(t_{i+1}, t_i)$ is the fundamental solution matrix along the trajectory $\boldsymbol{x}(t)$ of system (9.3) and $\boldsymbol{n}_\Omega \in \mathbb{R}^n$ is the normal to Ω at $\boldsymbol{x}_{i+1}$. The stability of a periodic solution can be inferred from the eigenvalues of the local linearization of the Poincaré map but also from the Floquet multipliers. The $n - 1$

eigenvalues of $\boldsymbol{J}_P(\boldsymbol{y}^*)$ are $\{m_1, \dots, m_{n-1}\}$. The fundamental solution matrix $\boldsymbol{\Phi}_T$ of the periodic solution Γ has n eigenvalues $\{1, m_1, \dots, m_{n-1}\}$, which are also called Floquet multipliers. One Floquet multiplier equals one because the periodic solution of an autonomous system can be shifted in time. The remaining Floquet multipliers are identical to the eigenvalues of the linearization of the Poincaré map.

The Poincaré map can also be defined for non-autonomous time-periodic systems. Consider the system

$$\dot{\boldsymbol{x}}(t) = \boldsymbol{f}(t, \boldsymbol{x}(t)), \tag{9.12}$$

with $\boldsymbol{f}(t, \boldsymbol{x}) = \boldsymbol{f}(t + T, \boldsymbol{x})$ for all $\boldsymbol{x} \in \mathbb{R}^n$. For non-autonomous time-periodic systems the period time T is a priori known. We define an n-dimensional hyper-plane $\Theta \in S^1 \times \mathbb{R}^n$ as

$$\Theta = \{(t, \boldsymbol{x}) \in S^1 \times \mathbb{R}^n \mid t = t_0 + kT,\ k = 0, 1, 2 ...\}. \tag{9.13}$$

Every T time units, the trajectory $\boldsymbol{x}(t)$ of system (9.12) intersects Θ. We can therefore define a mapping $\boldsymbol{P}_N : \Theta \to \Theta$ which maps the state $\boldsymbol{x}(t_0)$ to the state $\boldsymbol{x}(t_0 + T)$ as

$$\boldsymbol{x}(t_0 + T) = \boldsymbol{P}_N(\boldsymbol{x}(t_0)). \tag{9.14}$$

The map $\boldsymbol{P}_N$ is called the Poincaré map for a non-autonomous system or T-advance map. The derivative $\boldsymbol{J}_{P_N}$ of $\boldsymbol{P}_N$ is identical to the fundamental solution matrix $\boldsymbol{\Phi}_T$.

9.2 Piecewise Linear Maps with a Single Switching Boundary

The co-existence of fixed points *for a subclass* of piecewise linear mappings with only one switching boundary can conveniently be analyzed with a method developed by Feigin [30, 32, 33].

Consider a piecewise linear map of the following special form

$$\boldsymbol{x}_{i+1} = \boldsymbol{P}(\boldsymbol{x}_i) = \begin{cases} \boldsymbol{A}_- \boldsymbol{x}_i + \boldsymbol{c}\mu, & \boldsymbol{x}_i \subset \mathcal{V}_{P-} \bigcup \Sigma_P, \\ \boldsymbol{A}_+ \boldsymbol{x}_i + \boldsymbol{c}\mu, & \boldsymbol{x}_i \in \mathcal{V}_{P+}, \end{cases} \tag{9.15}$$

depending on a parameter μ and being continuous on the switching boundary Σ_P. Let the switching function be given by $h(\boldsymbol{x}) = x_n$, not being dependent on μ, which defines the subspaces $\mathcal{V}_{P-}$, Σ_P and $\mathcal{V}_{P+}$. Note that the class of piecewise linear maps of the form (9.15) is very special. The matrices $\boldsymbol{A}_-$ and $\boldsymbol{A}_+$ do not depend on μ and $\boldsymbol{P}$ is continuously differentiable with respect to μ. The map has only a single switching boundary which is the hyper plane $x_n = 0$. A map of the form (9.15) with an arbitrary switching boundary can always be transformed to have $x_n = 0$ as switching boundary. We explicitly state that the non-smooth continuous maps considered in the next sections, which result from Poincaré maps of Filippov systems, are generally not of the form (9.15).

Continuity at the switching boundary of the map (9.15) implies that the first $n-1$ columns of the matrices $\boldsymbol{A}_{\pm}$ agree, i.e. $A_{-_{jk}} = A_{+_{jk}}, \forall j = 1\ldots n, k = 1\ldots n-1$. Assuming that $\boldsymbol{A}_-$ and $\boldsymbol{A}_+$ are non-singular, at most one fixed point can exist in each of the subspaces $\mathcal{V}_{P-}$ and $\mathcal{V}_{P+}$ because the system is linear within these subspaces.

9.2.1 Co-existence of Fixed Points

Let $\boldsymbol{x}_- \in \mathcal{V}_{P-} \bigcup \Sigma$ and $\boldsymbol{x}_+ \in \mathcal{V}_{P+}$ be fixed points of the map (9.15), i.e.

$$\begin{aligned} \boldsymbol{x}_- &= \boldsymbol{A}_-\boldsymbol{x}_- + \boldsymbol{c}\mu, \\ \boldsymbol{x}_+ &= \boldsymbol{A}_+\boldsymbol{x}_+ + \boldsymbol{c}\mu. \end{aligned} \tag{9.16}$$

If $\boldsymbol{A}_-$ and $\boldsymbol{A}_+$ are non-singular, then we can solve for $\boldsymbol{x}_-$ and $\boldsymbol{x}_+$

$$\begin{aligned} \boldsymbol{x}_- &= -(\boldsymbol{A}_- - \boldsymbol{I})^{-1}\boldsymbol{c}\mu = -\frac{\operatorname{adj}(\boldsymbol{A}_- - \boldsymbol{I})}{\det(\boldsymbol{A}_- - \boldsymbol{I})}\boldsymbol{c}\mu, \\ \boldsymbol{x}_+ &= -(\boldsymbol{A}_+ - \boldsymbol{I})^{-1}\boldsymbol{c}\mu = -\frac{\operatorname{adj}(\boldsymbol{A}_+ - \boldsymbol{I})}{\det(\boldsymbol{A}_+ - \boldsymbol{I})}\boldsymbol{c}\mu, \end{aligned} \tag{9.17}$$

where $\operatorname{adj}(\boldsymbol{A}_{\pm} - \boldsymbol{I})$ denotes the adjoint matrix of $\boldsymbol{A}_{\pm} - \boldsymbol{I}$. It follows from (9.17) that the elements of $\boldsymbol{x}_-$ and $\boldsymbol{x}_+$ can be expressed as

$$x_{-_k} = \frac{b_{-_k}}{\det(\boldsymbol{A}_- - \boldsymbol{I})}\mu, \qquad x_{+_k} = \frac{b_{+_k}}{\det(\boldsymbol{A}_+ - \boldsymbol{I})}\mu, \tag{9.18}$$

with $b_{-_k} = [-\operatorname{adj}(\boldsymbol{A}_- - \boldsymbol{I})\boldsymbol{c}]_k$ and $b_{+_k} = [-\operatorname{adj}(\boldsymbol{A}_+ - \boldsymbol{I})\boldsymbol{c}]_k$. It holds that $b_{-_n} = b_{+_n} =: b_n$ because the matrices $\boldsymbol{A}_-$ and $\boldsymbol{A}_+$ differ only in the n-th column (see also [24]). The n-th elements of the fixed points $\boldsymbol{x}_-$ and $\boldsymbol{x}_+$

$$x_{-_n} = \frac{b_n}{\det(\boldsymbol{A}_- - \boldsymbol{I})}\mu, \qquad x_{+_n} = \frac{b_n}{\det(\boldsymbol{A}_+ - \boldsymbol{I})}\mu, \tag{9.19}$$

are therefore only a function of b_n, the determinant of $\boldsymbol{A}_{\pm} - \boldsymbol{I}$ and the parameter μ. We have to require that $x_{-_n} \leq 0$ and $x_{+_n} > 0$ in order to let the fixed points be admissible $\boldsymbol{x}_- \in \mathcal{V}_{P-} \bigcup \Sigma_P$ and $\boldsymbol{x}_+ \in \mathcal{V}_{P+}$. Only one of the fixed points exists for $\mu = 0$, i.e. $\boldsymbol{x}_{-,\mu=0} = \boldsymbol{0}$, being located at the switching boundary. The two fixed points co-exist for $\mu < 0$ *or* $\mu > 0$ if the elements x_{-_n} and x_{+_n} have opposite signs. Co-existence of the fixed points for $\mu < 0$ ($\mu > 0$) implies non-existence of fixed points for $\mu > 0$ ($\mu < 0$). A necessary and sufficient condition for co-existence of fixed points of map (9.15) is

$$\det(\boldsymbol{A}_- - \boldsymbol{I})\det(\boldsymbol{A}_+ - \boldsymbol{I}) < 0. \tag{9.20}$$

The fixed points exist for opposite signs of μ, i.e. $\boldsymbol{x}_-$ exists for $\mu \geq 0$ ($\mu \leq 0$) and $\boldsymbol{x}_+$ exists for $\mu < 0$ ($\mu > 0$), if

$$\det(\boldsymbol{A}_- - \boldsymbol{I})\det(\boldsymbol{A}_+ - \boldsymbol{I}) > 0. \tag{9.21}$$

The co-existence of fixed points of system (9.15) is therefore determined by the signs of the determinants of the matrices $\boldsymbol{A}_\pm - \boldsymbol{I}$. Moreover, the sign of the determinant depends solely on the number of real valued eigenvalues, because $\det(\boldsymbol{A} - \boldsymbol{I}) = (\lambda_1 - 1)(\lambda_2 - 1)\dots(\lambda_n - 1)$ with $\{\lambda_1, \dots, \lambda_n\} = \mathrm{eig}(\boldsymbol{A})$. It holds that $(\lambda - 1)(\bar{\lambda} - 1) \geq 0$. If the number of real valued eigenvalues which are smaller than one is odd, then the determinant $\det(\boldsymbol{A} - \boldsymbol{I})$ is negative. If the number of real valued eigenvalues smaller than one is even, then the determinant is positive.

9.2.2 Period-doubled Map

The fixed points of a piecewise linear map $\boldsymbol{P}(\boldsymbol{x})$ have been studied in the previous subsection. We now consider the fixed points of the map $\boldsymbol{P}^2(\boldsymbol{x}) = \boldsymbol{P}(\boldsymbol{P}(\boldsymbol{x}))$. A fixed point of $\boldsymbol{P}(\boldsymbol{x})$ is naturally also a fixed point of $\boldsymbol{P}^2(\boldsymbol{x})$. Let $\boldsymbol{x}^a$ be a fixed point of $\boldsymbol{P}^2(\boldsymbol{x})$ that is not a fixed point of $\boldsymbol{P}(\boldsymbol{x})$, i.e.

$$\begin{aligned} \boldsymbol{x}^b &= \boldsymbol{P}(\boldsymbol{x}^a), \\ \boldsymbol{x}^a &= \boldsymbol{P}(\boldsymbol{x}^b), \end{aligned} \tag{9.22}$$

with $\boldsymbol{x}^b \neq \boldsymbol{x}^a$. Clearly, also $\boldsymbol{x}_b$ is a fixed point of $\boldsymbol{P}^2(\boldsymbol{x})$. If $\boldsymbol{x}^a$ and $\boldsymbol{x}^b$ exist, and if $\boldsymbol{P}(\boldsymbol{x})$ is piecewise linear with a single switching boundary, then it must hold that $\boldsymbol{x}^a$ and $\boldsymbol{x}^b$ are located in different subspaces $\mathcal{V}_{P-}$ and $\mathcal{V}_{P+}$. Without loss of generality, assume that $\boldsymbol{x}^a \in \mathcal{V}_{P-}$ and $\boldsymbol{x}^b \in \mathcal{V}_{P+}$, i.e. $x_n^a < 0$ and $x_n^b > 0$. Application of the map (9.15) yields

$$\begin{aligned} \boldsymbol{x}^b &= \boldsymbol{A}_- \boldsymbol{x}^a + \boldsymbol{c}\mu, \\ \boldsymbol{x}^a &= \boldsymbol{A}_+ \boldsymbol{x}^b + \boldsymbol{c}\mu. \end{aligned} \tag{9.23}$$

Substitution of the first equation of (9.23) into the second gives two explicit expressions for the fixed points of $\boldsymbol{P}^2(\boldsymbol{x})$

$$\begin{aligned} \boldsymbol{x}^a &= (\boldsymbol{I} - \boldsymbol{A}_+\boldsymbol{A}_-)^{-1}(\boldsymbol{A}_+ + \boldsymbol{I})\boldsymbol{c}\mu, \\ \boldsymbol{x}^b &= (\boldsymbol{I} - \boldsymbol{A}_-\boldsymbol{A}_+)^{-1}(\boldsymbol{A}_- + \boldsymbol{I})\boldsymbol{c}\mu, \end{aligned} \tag{9.24}$$

assuming that $\det(\boldsymbol{I} - \boldsymbol{A}_+\boldsymbol{A}_-) = \det(\boldsymbol{I} - \boldsymbol{A}_-\boldsymbol{A}_+) \neq 0$. Following the derivation in [24, 31], consider the quantity

$$\Delta\boldsymbol{x} = \boldsymbol{x}^b - \boldsymbol{x}^a. \tag{9.25}$$

Substitution of (9.23) in (9.25) gives

$$\Delta\boldsymbol{x} = \boldsymbol{A}_-\boldsymbol{x}^a - \boldsymbol{A}_+\boldsymbol{x}^b. \tag{9.26}$$

By substitution of $\boldsymbol{x}^a = \boldsymbol{x}^b - \Delta\boldsymbol{x}$ and $\boldsymbol{x}^b = \boldsymbol{x}^a + \Delta\boldsymbol{x}$ using (9.25) we obtain the two equations

$$\begin{aligned}(\boldsymbol{A}_+ + \boldsymbol{I})\Delta\boldsymbol{x} &= (\boldsymbol{A}_- - \boldsymbol{A}_+)\boldsymbol{x}^a,\\ (\boldsymbol{A}_- + \boldsymbol{I})\Delta\boldsymbol{x} &= (\boldsymbol{A}_- - \boldsymbol{A}_+)\boldsymbol{x}^b.\end{aligned} \tag{9.27}$$

Assuming that $\det(\boldsymbol{A}_\pm + \boldsymbol{I}) \neq 0$ and using the continuity conditions $A_{-_{jk}} = A_{+_{jk}}, \forall j = 1 \ldots n, k = 1 \ldots n-1$ equations (9.27) simplify into

$$\begin{aligned}\Delta\boldsymbol{x} &= (\boldsymbol{A}_+ + \boldsymbol{I})^{-1}[\boldsymbol{A}_- - \boldsymbol{A}_+]_n x_n^a,\\ \Delta\boldsymbol{x} &= (\boldsymbol{A}_- + \boldsymbol{I})^{-1}[\boldsymbol{A}_- - \boldsymbol{A}_+]_n x_n^b.\end{aligned} \tag{9.28}$$

Equating the two expressions for the last element of $\Delta\boldsymbol{x}$ yields the condition

$$\Delta x_n = \frac{b_n^+}{\det(\boldsymbol{A}_+ + \boldsymbol{I})} x_n^a = \frac{b_n^-}{\det(\boldsymbol{A}_- + \boldsymbol{I})} x_n^b, \tag{9.29}$$

with $b_n^+ = \operatorname{adj}(\boldsymbol{A}_+ + \boldsymbol{I})[\boldsymbol{A}_- - \boldsymbol{A}_+]_n$ and $b_n^- = \operatorname{adj}(\boldsymbol{A}_- + \boldsymbol{I})[\boldsymbol{A}_- - \boldsymbol{A}_+]_n$. Using the fact that $b_n = b_n^+ = b_n^-$ and $x_n^a x_n^b < 0$ we can derive the condition

$$\det(\boldsymbol{A}_+ + \boldsymbol{I}) \det(\boldsymbol{A}_- + \boldsymbol{I}) < 0. \tag{9.30}$$

Furthermore, it must hold that $\Delta x_n > 0$ in order to let the fixed points be admissible. From (9.24) it is immediately clear that the sign of x_n^a and x_n^b is directly dependent on the sign of μ. The two fixed points $\boldsymbol{x}^a$ and $\boldsymbol{x}^b$ of $\boldsymbol{P}^2(\boldsymbol{x})$ therefore co-exist for $\mu > 0$ *or* for $\mu < 0$ if condition (9.30) is met. If the two fixed points co-exist for $\mu > 0$ ($\mu < 0$) then they do not exist for $\mu < 0$ ($\mu > 0$).

9.3 Bifurcations in Smooth Systems

In the preceding chapter, bifurcations of equilibria were discussed of non-smooth continuous systems. The generalized Jacobian matrix of the system around an equilibrium point can be set-valued if the equilibrium is located on a switching boundary. The set-valuedness of the generalized Jacobian matrix causes the eigenvalues to be set-valued. If the set of eigenvalues has an intersection with the imaginary axis under influence of a parameter, we presumed that a discontinuous bifurcation can occur.

Before introducing the basic idea of discontinuous bifurcations of periodic solutions, we briefly address bifurcations of periodic solutions in smooth systems. A periodic solution can be regarded as a fixed point of a Poincaré map $\boldsymbol{P}(\boldsymbol{x})$ on a Poincaré section. Fixed points of mappings are related to equilibria of dynamical systems. The results on bifurcations of equilibria are therefore useful for the investigation of bifurcations of periodic solutions. The stability of a periodic solution is determined by its Floquet multipliers λ_i ($i = 1,\ldots,n$), which are the eigenvalues of the fundamental solution matrix $\boldsymbol{\Phi}(T + t_0, t_0)$. The Floquet multipliers are the generalization of the eigenvalues at an equilibrium. Because solutions of autonomous systems can be shifted in time, one of the Floquet multipliers equals one for such systems. The periodic solution is stable if all Floquet multipliers (not associated with the phase in the autonomous case) lie within the unit circle. If one or more Floquet multipliers lie outside the unit circle, then the periodic solution is unstable. The periodic solution varies (as well as its Floquet multipliers) as a parameter of the system is varied. The periodic solution changes from stable to unstable (or vice versa) when the Floquet multiplier with the largest magnitude passes through the unit circle.

The bifurcations of periodic solutions which occur in smooth systems under variation of one parameter are (see Figure 9.2):

a) *Fold bifurcation*: a branch of periodic solutions folds around. A Floquet multiplier passes the value $+1$.
b) *Symmetry-breaking bifurcation*: branches of asymmetric periodic solutions are born from a branch of symmetric periodic solutions. This bifurcation can only occur in systems with symmetry. A Floquet multiplier passes the value $+1$.
c) *Period-doubling bifurcation* or *Flip bifurcation*: a branch with periodic solutions with period time $2T$ bifurcates from a branch with periodic solutions with period time T. A Floquet multiplier passes the value -1.
d) *Neimark-Sacker bifurcation* or *Secondary Hopf bifurcation*: a branch with quasi-periodic solutions bifurcates from a branch with periodic solutions. A pair of complex conjugated Floquet multipliers crosses the unit circle.

Different definitions for a bifurcation of a periodic solution exist as has been elaborated in Section 8.1. We will take Definition 8.1 as definition for a bifurcation of equilibria and (quasi)-periodic solutions. According to Definition 8.1, a point is a bifurcation point of a periodic solution if the number of periodic solutions changes for a varying system parameter. One can also give a definition of a bifurcation of a periodic solution based on (the loss of) topological equivalence of the phase portrait, like Definition 8.3. We will discuss in Section 9.10 the difference between these definitions when they are applied to periodic solutions of Filippov systems.

A bifurcation of a periodic solution of a smooth system occurs if a Floquet multiplier (or a pair of them) passes through the unit circle under variation of a system parameter. Floquet multipliers of smooth systems are continuous functions of the varied system parameter. This continuous dependence of the Floquet multipliers on the system parameter is lost when Filippov systems are considered, which complicates the bifurcation behaviour of Filippov systems.

9.4 Discontinuous Bifurcation: The Basic Idea

The examples of non-smooth continuous systems studied in Chapter 8 have shown a number of discontinuous bifurcations of equilibria. Loosely speaking, an eigenvalue 'jumps' over the imaginary axis in all the presented examples of discontinuous bifurcations. To be more precise, the set of eigenvalues of the

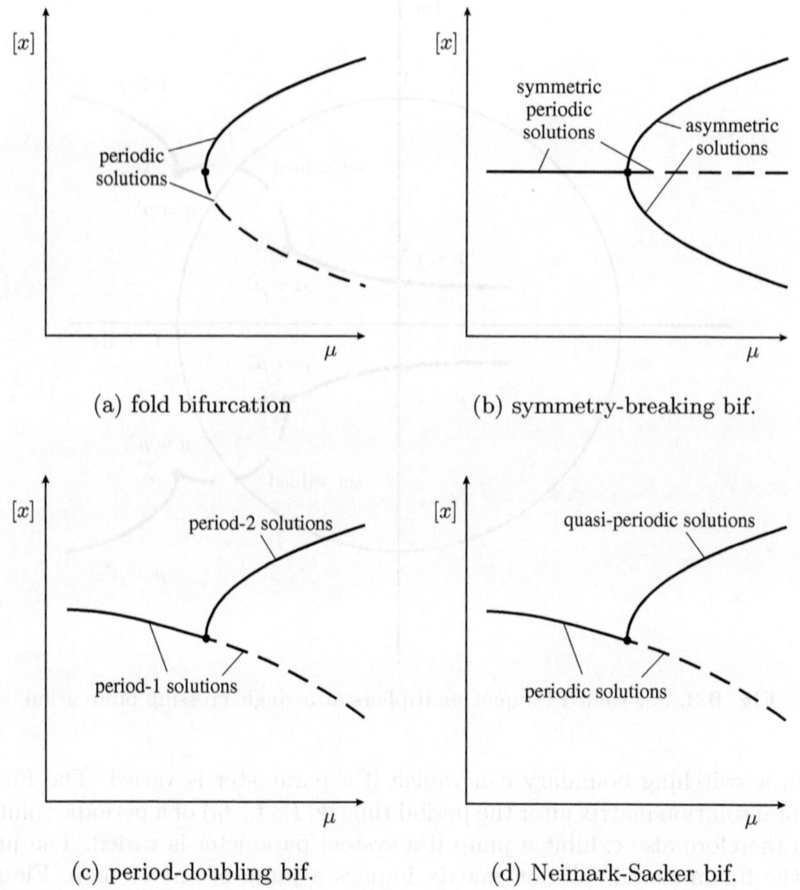

Fig. 9.2. Bifurcations of periodic solutions in smooth systems (solid lines denote stable periodic solutions, dashed lines denote unstable periodic solutions).

generalized Jacobian at the bifurcation point has one or more crossing points with the imaginary axis. We presume that the crossing of the set of eigenvalues with the imaginary axis is a necessary condition for a discontinuous bifurcation of an equilibrium in a non-smooth continuous system. Moreover, we presume that the location of the crossing point determines the type of bifurcation for single crossing bifurcation points. Multiple crossing bifurcation points have been found to be intrinsically more complicated.

In Chapter 7 we have concluded that fundamental solution matrices of discontinuous systems of Filippov-type, $\boldsymbol{\Phi}(t, t_0)$, expose jumps at time instances t on which a trajectory $\boldsymbol{x}(t)$ crosses a switching boundary. Periodic solutions change under influence of a parameter. The crossing of a periodic solution

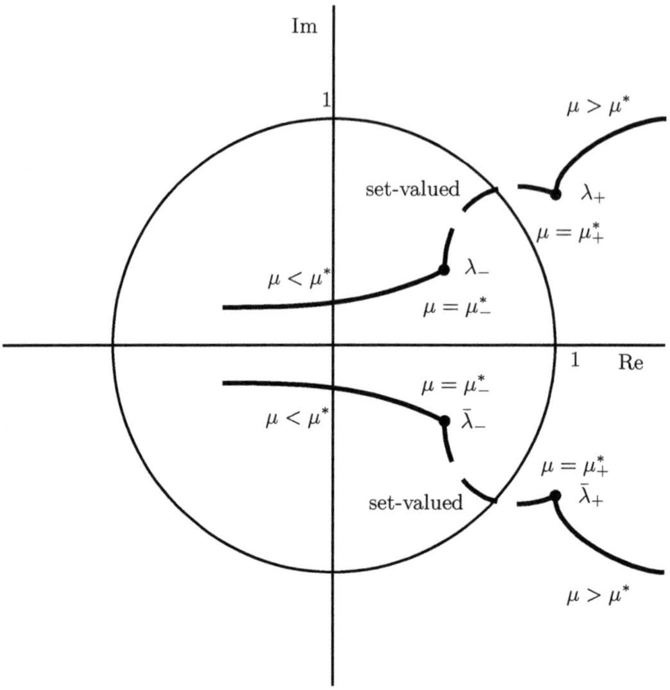

Fig. 9.3. Set-valued Floquet multipliers at a single crossing bifurcation.

with a switching boundary can vanish if a parameter is varied. The fundamental solution matrix after the period time $\boldsymbol{\Phi}(T+t_0, t_0)$ of a periodic solution can therefore also exhibit a jump if a system parameter is varied. The jump of the fundamental solution matrix implies a jump of one or more Floquet multipliers. Instead of regarding the fundamental solution matrix $\boldsymbol{\Phi}(t, t_0)$ as a (matrix) function that is discontinuous, one can also regard it as a derivative (just like the Jacobian matrix is in fact a derivative)

$$\boldsymbol{\Phi}(t, t_0) = \frac{\partial \boldsymbol{\varphi}(t, t_0, \boldsymbol{x}_0)}{\partial \boldsymbol{x}_0}, \tag{9.31}$$

with $\boldsymbol{x}(t) = \boldsymbol{\varphi}(t, t_0, \boldsymbol{x}_0)$. We now replace the classical fundamental solution matrix by a generalized fundamental solution matrix

$$\boldsymbol{\Phi}(t, t_0) = \left(\partial_{\boldsymbol{x}_0} \boldsymbol{\varphi}(t, t_0, \boldsymbol{x}_0)\right)^{\mathrm{T}}, \tag{9.32}$$

by making use of the generalized differential of Clarke (2.10) similar to the generalized Jacobian (2.12). The generalized fundamental solution matrix (9.32) can be set-valued at time instances for which the trajectory crosses a switching boundary.

In a similar way as for bifurcations of equilibria, we can expect a discontinuous bifurcation of a periodic solution when, loosely speaking, a Floquet multiplier 'jumps' over the unit circle under the influence of a parameter. The basic idea is depicted in Figure 9.3. The Floquet multipliers jump at a critical value of the parameter $\mu = \mu^*$ from λ_- and $\bar{\lambda}_-$ to λ_+ and $\bar{\lambda}_+$. To be more precise, the classical fundamental solution matrix is replaced by a generalized fundamental solution matrix which can have set-valued eigenvalues (Floquet multipliers). The set of Floquet multipliers of a set-valued fundamental solution matrix forms generally an area in the complex plane. However, the set of Floquet multipliers forms a one-dimensional path for transitions in which only one switching boundary is touched. The path of Floquet multipliers is obtained from a convex combination of the fundamental solution matrices before and after the transition. The path of the Floquet multipliers can therefore cross the unit circle under influence of a parameter. We presume that this causes a discontinuous bifurcation of a periodic solution.

The numerical examples which will be presented in Sections 9.8 and 9.9 show discontinuous counterparts of various types of continuous bifurcations of periodic solutions, all characterized by a single crossing of the set-valued Floquet multipliers with the unit circle. Like for discontinuous bifurcations of equilibria in non-smooth continuous systems, also multiple crossing bifurcations are possible. A multiple crossing of a real valued Floquet multiplier is shown in Figure 9.4, which extends from a value smaller than -1 to a value larger than $+1$ thereby crossing the unit circle twice. A discontinuous bifurcation of a periodic solution associated with this scenario might reveal the behaviour of a period-doubling bifurcation, because of the crossing at the point -1, and show the behaviour of a fold bifurcation, because of the crossing at $+1$. A discontinuous multiple crossing bifurcation associated with a set of Floquet multipliers as depicted in Figure 9.4 can therefore be a combination of a fold and a period-doubling bifurcation. We have to keep in mind that only some of the examples of multiple crossing bifurcations of equilibria (Section 8.7) can be regarded as a combination of two continuous bifurcations, but that this is not true for general multiple crossing bifurcations. We will discuss an example of a multiple crossing bifurcation of a periodic solution, behaving like a combination of a period-doubling and fold bifurcation, in Section 9.8.2.

9.5 Bifurcations of Planar Maps with a Switching Boundary

In this section we will rigourously prove some results on bifurcations of fixed points of maps in planar continuous piecewise linear maps with a single switching boundary. Consider a planar piecewise linear map of the special form (9.15)

$$\boldsymbol{x}_{i+1} = \boldsymbol{P}(\boldsymbol{x}_i) = \begin{cases} \boldsymbol{A}_-\boldsymbol{x}_i + \boldsymbol{c}\mu, & \boldsymbol{x}_i \in \mathcal{V}_{P-} \bigcup \Sigma_P, \\ \boldsymbol{A}_+\boldsymbol{x}_i + \boldsymbol{c}\mu, & \boldsymbol{x}_i \in \mathcal{V}_{P+}, \end{cases} \tag{9.33}$$

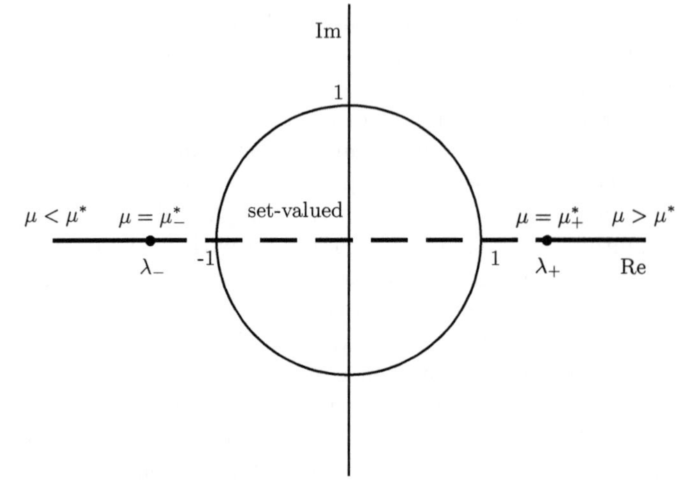

Fig. 9.4. Set-valued Floquet multipliers at a multiple crossing bifurcation.

with $\boldsymbol{x} = \begin{bmatrix} x_1 \, x_2 \end{bmatrix}^{\mathrm{T}}$ and $\boldsymbol{P}(\boldsymbol{x})$ being continuous on the switching boundary $\Sigma_P = \{\boldsymbol{x} \in \mathbb{R}^2 \mid x_2 = 0\}$. The continuity conditions on the switching boundary require that $A_{-11} = A_{+11}$ and $A_{-21} = A_{+21}$. Assuming that the matrices $\boldsymbol{A}_-$ and $\boldsymbol{A}_+$ are non-singular, at most one equilibrium can exist in each of the subspaces $\mathcal{V}_{P-}$ and $\mathcal{V}_{P+}$ because the map is linear and hyperbolic within these subspaces. The map $\boldsymbol{P}$ is non-smooth on Σ_P. We therefore define a generalized derivative of the map in the sense of Clarke

$$\boldsymbol{J}_P(\boldsymbol{x}) = (\partial \boldsymbol{P}(\boldsymbol{x}))^{\mathrm{T}} = \begin{cases} \boldsymbol{A}_-, & \boldsymbol{x} \in \mathcal{V}_{P-}, \\ \overline{\mathrm{co}}(\boldsymbol{A}_-, \boldsymbol{A}_+), & \boldsymbol{x} \in \Sigma_{P-}, \\ \boldsymbol{A}_+, & \boldsymbol{x} \in \mathcal{V}_{P+}. \end{cases} \tag{9.34}$$

In Section 9.2 it has been shown that the fixed points of $\boldsymbol{P}(\boldsymbol{x})$ co-exist for $\mu < 0$ or for $\mu > 0$ if

$$\det(\boldsymbol{A}_- - \boldsymbol{I}) \det(\boldsymbol{A}_+ - \boldsymbol{I}) < 0, \tag{9.35}$$

and one fixed point exists for all μ if

$$\det(\boldsymbol{A}_- - \boldsymbol{I}) \det(\boldsymbol{A}_+ - \boldsymbol{I}) > 0. \tag{9.36}$$

The determinant/trace of a convex combination of two arbitrary matrices $\boldsymbol{A} \in \mathbb{R}^{2\times 2}$ and $\boldsymbol{B} \in \mathbb{R}^{2\times 2}$, with $A_{11} = B_{11}$ and $A_{21} = B_{21}$, is a convex combination of the determinants/traces of $\boldsymbol{A}$ and $\boldsymbol{B}$ as follows from Proposition 8.8. For a planar continuous piecewise linear map of the form (9.33) it therefore holds that

$$\det(q\boldsymbol{A}_+ + (1-q)\boldsymbol{A}_- - \boldsymbol{I}) = q \det(\boldsymbol{A}_+ - \boldsymbol{I}) + (1-q)\det(\boldsymbol{A}_- - \boldsymbol{I}). \tag{9.37}$$

The co-existence criterion (9.35) suggests that a turning point bifurcation can occur in the system (9.33).

Theorem 9.1 (Turning Point Bifurcation Theorem of Maps). *If a planar continuous piecewise linear map with a single switching boundary of the form (9.33) undergoes a turning point bifurcation, then the path of the set-valued eigenvalues crosses the unit circle through the point* $+1$.

Proof: If there exists a turning point bifurcation, then it follows from (9.35) that

$$\det(\boldsymbol{A}_{-} - \boldsymbol{I}) \det(\boldsymbol{A}_{+} - \boldsymbol{I}) < 0.$$

The turning point $\boldsymbol{x}^*$ must be located on the switching boundary Σ_P. The generalized derivative $\boldsymbol{J}_P$ of the fixed point on Σ_P is the set $\boldsymbol{J}_P = \overline{\text{co}}(\boldsymbol{A}_{-}, \boldsymbol{A}_{+}) = \{\boldsymbol{J}_{Pq} \mid \boldsymbol{J}_{Pq} = q\boldsymbol{A}_{+} + (1-q)\boldsymbol{A}_{-},\ q \in [0,1]\}$. It therefore must hold, using (9.37), that there exists a $q \in [0,1]$ for which $\det(\boldsymbol{J}_{Pq} - \boldsymbol{I}) = 0$. Consequently, there exists a set-valued eigenvalue λ_i of the generalized derivative $\boldsymbol{J}_P$ containing the point $+1$, i.e.

$$1 \in \lambda_i, \qquad \boldsymbol{\lambda} = \text{eig}(\boldsymbol{J}_P).$$

The path of the set-valued eigenvalues of the generalized derivative of the map at the discontinuous turning point bifurcation, therefore passes the point $+1$. □

Similarly, the co-existence criterion (9.30) for fixed points of $\boldsymbol{P}^2(\boldsymbol{x})$ suggests that a period-doubling bifurcation can occur in the system (9.33).

Theorem 9.2 (Period-doubling Bifurcation Theorem of Maps). *If a planar continuous piecewise linear map with a single switching boundary of the form (9.33) undergoes a period-doubling bifurcation, then the path of the set-valued eigenvalues crosses the unit circle through the point* -1.

Proof: If there exists a period-doubling bifurcation, then it follows from (9.30) that

$$\det(\boldsymbol{A}_{-} + \boldsymbol{I}) \det(\boldsymbol{A}_{+} + \boldsymbol{I}) < 0.$$

The period-doubling bifurcation point $\boldsymbol{x}^*$ must be located on the switching boundary Σ_P. The generalized derivative $\boldsymbol{J}_P$ of the fixed point on Σ_P is the set $\boldsymbol{J}_P = \overline{\text{co}}(\boldsymbol{A}_{-}, \boldsymbol{A}_{+}) = \{\boldsymbol{J}_{Pq} \mid \boldsymbol{J}_{Pq} = q\boldsymbol{A}_{+} + (1-q)\boldsymbol{A}_{-},\ q \in [0,1]\}$. It therefore must hold, using (9.37), that there exists a $q \in [0,1]$ for which $\det(\boldsymbol{J}_{Pq} + \boldsymbol{I}) = 0$. Consequently, there exists a set-valued eigenvalue λ_i of the generalized derivative $\boldsymbol{J}_P$ containing the point -1, i.e.

$$-1 \in \lambda_i, \qquad \boldsymbol{\lambda} = \text{eig}(\boldsymbol{J}_P).$$

The path of the set-valued eigenvalues of the generalized derivative of the map at the discontinuous period-doubling bifurcation point, therefore passes the point -1. □

Theorems 9.1 and 9.2 have direct consequences for bifurcations of periodic solutions of Filippov systems. A Poincaré map of a three-dimensional Filippov system is a planar continuous map. If the switching boundary is non-smooth, then the Poincaré map is in general a non-smooth continuous map (see Section 9.7). A periodic solution of the Filippov system corresponds to a fixed point of the associated Poincaré map. If the periodic solution is touching a non-smooth point on a switching boundary, then the associated fixed point is located on a non-smooth point of the Poincaré map. The Poincaré map is therefore locally a planar continuous piecewise linear map with a single switching boundary. The eigenvalues of the (generalized) derivative of the Poincaré map at a fixed point of the map are also Floquet multipliers of the associated periodic solution. The set-valued Floquet multipliers of a fold (turning point) bifurcation therefore contain a crossing point with the unit circle at +1. Similarly, set-valued Floquet multipliers of a period-doubling bifurcation contain a crossing point with the unit circle at −1.

9.6 Poincaré Maps at Discontinuous Bifurcations

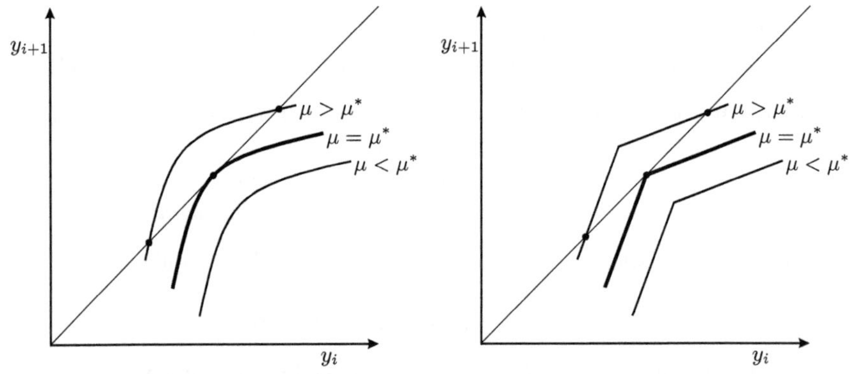

(a) Continuous (b) Discontinuous

Fig. 9.5. Poincaré map at a fold bifurcation.

A Poincaré map $\boldsymbol{P}(\boldsymbol{y})$ of a smooth dynamical system is a (locally) smooth function in $\boldsymbol{y}$. Discontinuities in a Poincaré map of a smooth system can arise due to the presence of saddle points. Two solutions on either side of an incoming invariant half-manifold of a saddle point first evolve towards the saddle point, but then bend off, each solution along a different outgoing invariant half-manifold. To be more precise, the Poincaré map of a smooth system is discontinuous on the intersection of invariant manifolds of saddle points with

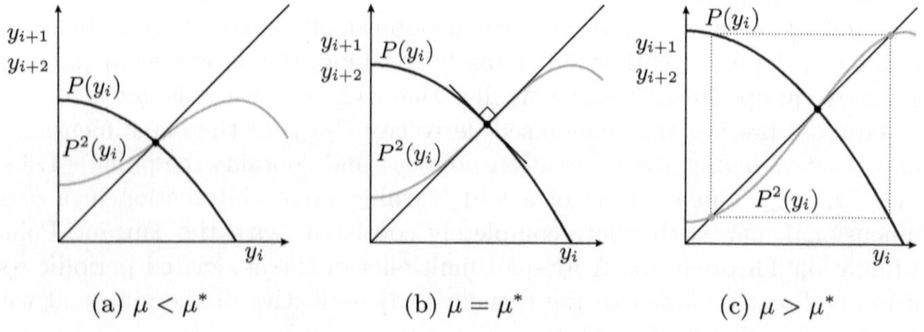

Fig. 9.6. Poincaré map at a continuous period-doubling bifurcation.

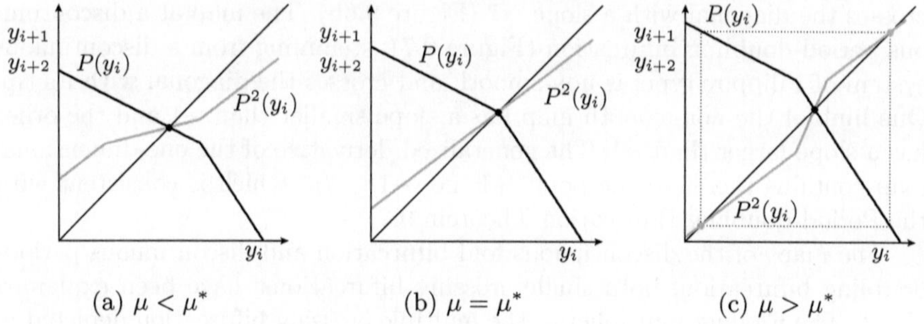

Fig. 9.7. Poincaré map at a discontinuous period-doubling bifurcation.

the Poincaré section. It was explained in Section 7.2 that the fundamental solution matrix of a Filippov system is discontinuous when a switching boundary is crossed. The derivative of the Poincaré map is directly related to the fundamental solution matrix. The Poincaré map $\boldsymbol{P}(\boldsymbol{y})$ of a Filippov system is therefore a (locally) non-smooth continuous function in $\boldsymbol{x}$. We will elucidate the difference between various continuous and discontinuous bifurcations of periodic solutions using the Poincaré map. The Poincaré map is generally a mapping from $\mathbb{R}^{n-1}$ to $\mathbb{R}^{n-1}$, which can not easily be visualized if $n > 2$. For the sake of illustration, consider a one-dimensional Poincaré map $P(y)$ as depicted in Figure 9.5. The map has no intersection points with the diagonal $y_{i+1} = y_i$ for $\mu < \mu^*$. The map is tangent to the diagonal for $\mu = \mu^*$. For $\mu > \mu^*$ the map has two intersection points, which correspond to periodic solutions of the dynamical system. Two periodic solutions are therefore created/destroyed at $\mu = \mu^*$, which is consequently a fold bifurcation. The map at a continuous fold bifurcation, stemming from a smooth system, is (locally) smooth and is tangent to the diagonal $y_{i+1} = y_i$, i.e. the slope is $+1$. The map at a discontinuous fold bifurcation, stemming from a discontinuous system of Filippov-type, is non-smooth and touches the diagonal with its tip. One limb of the non-smooth map has a slope smaller than +1 and the other limb has a slope larger than +1. Loosely speaking, the derivative of the map therefore 'jumps' from a value smaller than +1 to a value larger than +1. To be more precise, the generalized derivative $J_P(y)$ of the one-dimensional map is set-valued at the bifurcation point y^* and contains the point +1, i.e. $+1 \in J_P(y^*)$. This scenario of a fold (turning point) bifurcation in a one-dimensional map is therefore completely consistent with the Turning Point Bifurcation Theorem 9.1. A Floquet multiplier of the associated periodic solution is directly related to the (generalized) derivative of the map and will therefore also 'jump' through the point +1.

The Poincaré map P at a period-doubling bifurcation (or flip bifurcation) is depicted in Figure 9.6 together with the second iterate P^2. The map at a continuous period-doubling bifurcation, stemming from a smooth system, crosses the diagonal with a slope −1 (Figure 9.6b). The map at a discontinuous period-doubling bifurcation (Figure 9.7), stemming from a discontinuous system of Filippov-type, is non-smooth and crosses the diagonal with its tip. One limb of the non-smooth map has a slope smaller than −1 and the other has a slope larger than −1. The generalized derivative of the one-dimensional map contains therefore the point −1, i.e. $-1 \in J_P$, which is consistent with the Period-doubling Bifurcation Theorem 9.2.

The maps of the discontinuous fold bifurcation and discontinuous period-doubling bifurcation, both single crossing bifurcations, have been explained above. The Floquet multiplier of the multiple crossing bifurcation depicted in Figure 9.4 'jumps' from $\lambda < -1$ to $\lambda > +1$. As the slope of the Poincaré map is directly related to the Floquet multiplier, the map should be non-smooth having two limbs where one has a slope larger than +1 and the other has a slope smaller than −1. This map is depicted in Figure 9.8 and appears to be the tent map. This type of bifurcation, with an underlying tent map, will be encountered in Section 9.8.2.

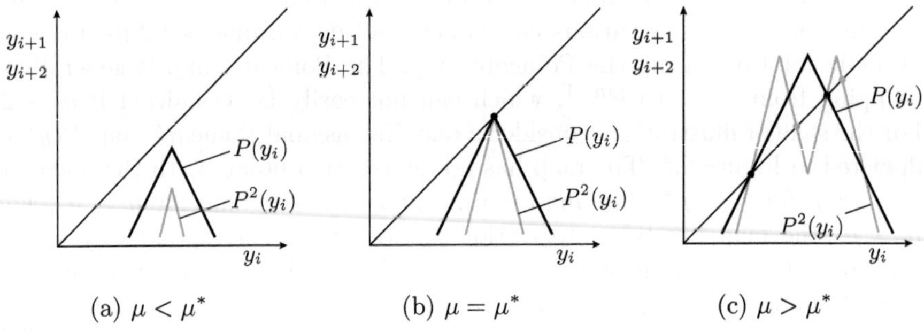

Fig. 9.8. Tent map P and second iterate P^2.

We close this section with an example of a one-degree-of-freedom mechanical system with a one-dimensional Poincaré map showing a discontinuous fold bifurcation.

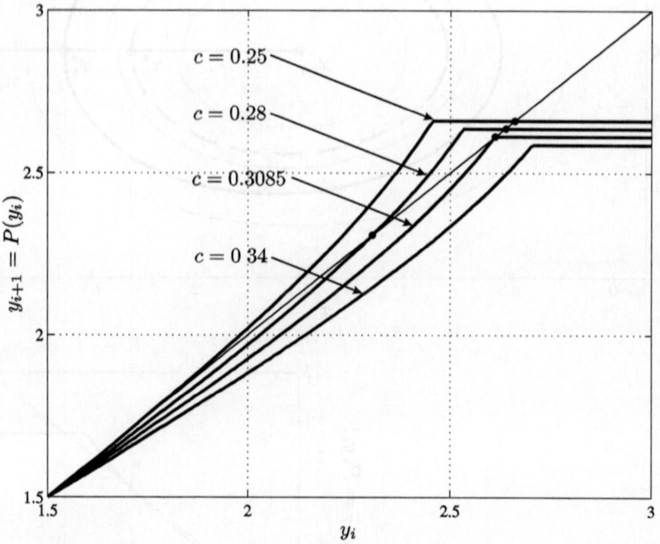

Fig. 9.9. Poincaré map of the stick-slip system (9.38) for different values of c.

Example 9.1
Consider the stick-slip system depicted in Figure 6.8, consisting of a mass on a driving belt which is supported by a spring and a dashpot. The equation of motion reads as

$$m\ddot{x} + c\dot{x} + kx = F_T(v_{\mathrm{rel}}), \tag{9.38}$$

with $v_{\mathrm{rel}} = \dot{x} - v_{\mathrm{dr}}$. The friction characteristic of F_T is of type (b) in Figure 4.2

$$F_T(v) \in -\left(\frac{F_{\mathrm{s}} - F_{\mathrm{d}}}{1 + \delta|v|} + F_{\mathrm{d}}\right) \mathrm{Sign}(v). \tag{9.39}$$

The system (9.38) is therefore a differential inclusion of Filippov-type. The equilibrium point of the system is

$$x_{\mathrm{eq}} = \frac{1}{k}\left(\frac{F_{\mathrm{s}} - F_{\mathrm{d}}}{1 + \delta|v_{\mathrm{dr}}|} + F_{\mathrm{d}}\right). \tag{9.40}$$

The parameter values are $m = 1$ kg, $k = 1$ N/m, $v_{\mathrm{dr}} = 1$ m/s, $\delta = 1$, $F_{\mathrm{s}} = 2$ N and $F_{\mathrm{d}} = 1$ N. The system exhibits limit cycles for certain parameter sets. The dynamics of (9.38) is studied for different values of the damping constant c Ns/m. We define the Poincaré section

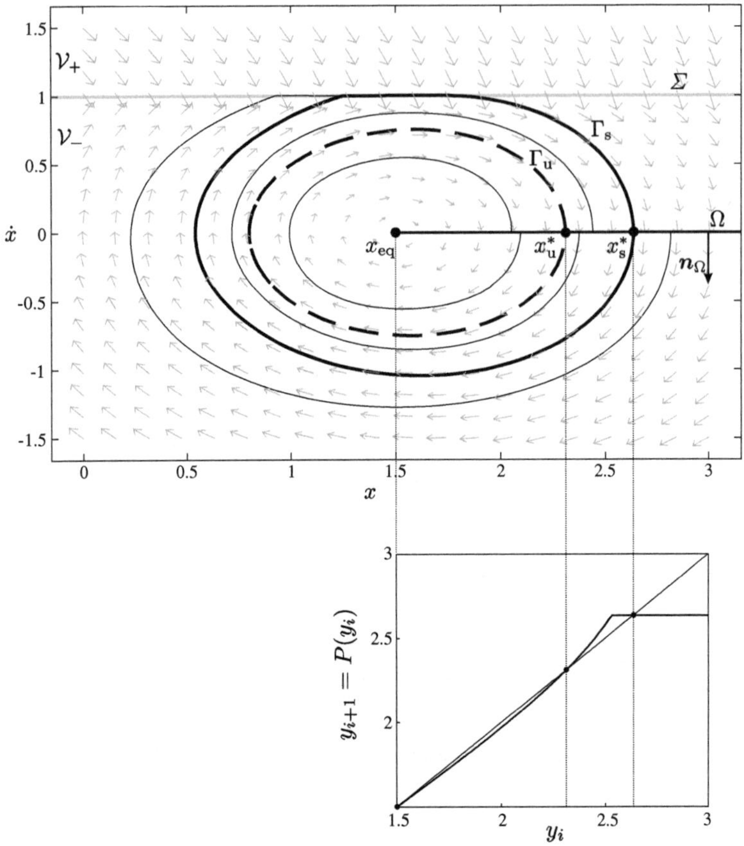

Fig. 9.10. Phase plane of the stick-slip system (9.38) for $c = 0.28$.

$$\Omega = \{(x, \dot{x}) \in \mathbb{R}^2 \mid \dot{x} = 0,\, x \geq x_{\text{eq}}\}, \tag{9.41}$$

with the coordinate $y = x$ on Ω and the normal $\boldsymbol{n}_\Omega = \begin{bmatrix} 0 & -1 \end{bmatrix}^{\mathrm{T}}$. If the system (9.38) has a periodic solution, then it is clearly transversal to Ω, because periodic motion a one-degree-of-freedom system implies sign reversals of the velocity. Consider the one-dimensional Poincaré map

$$y_{i+1} = P(y_i), \tag{9.42}$$

which maps a position $x(t_i) = y_i \in \Omega$ to the point $x(t_{i+1}) = y_{i+1} \in \Omega$ where $(x(t), \dot{x}(t))$ intersects Ω in a positive direction (see Equation (9.4)) at $t_i, t_{i+1}, \ldots$. The equilibrium point $x_{\text{eq}} = 1.5$ is a fixed point of the Poincaré map P. Every intersection point $x^* > x_{\text{eq}}$ of a periodic solution (or limit cycle) with Ω is also a fixed point of P. The one-dimensional Poincaré map can be obtained by numerical integration and is depicted in Figure 9.9 for four different values of the damping constant c. If $c = 0.34$, then the dissipation of the linear damping annihilates the destabilizing Stribeck effect and the equilibrium is the only fixed point of P and appears to be globally asymptotically stable. Note that the slope of the Poincaré map at the fixed point x_{eq} is smaller than one, which corresponds to a stable equilibrium. However, if c is lowered to 0.28, then the equilibrium is still locally asymptotically stable, but two additional fixed points of P exist, x_{s}^* and x_{u}^*, which are intersection points of a stable an unstable limit cycle with Ω. The phase plane of the stick-slip system (9.38) for $c = 0.28$ is depicted in Figure 9.10. The fixed point x_{u}^*, at which the slope of P is larger than one, corresponds to the unstable limit cycle Γ_{u}, while the fixed point x_{s}^*, with a slope smaller than one, corresponds to the stable limit cycle Γ_{s}. The stable limit cycle Γ_{s} is a stick-slip limit cycle, which contains both a slip and a stick phase. The unstable limit cycle Γ_{u} remains within the subspace $\mathcal{V}_- = \{\dot{x} - v_{\mathrm{dr}} < 0\}$ during its entire period. The slope of the Poincaré map near x_{s}^* is flat, because neighbouring solutions to the stable limit cycle Γ_{s} are completely attracted to Γ_{s} in finite time. The two limit cycles Γ_{s} and Γ_{u} are created by a fold bifurcation, which occurs for $c = 0.3085$, at which the map P becomes tangent to the diagonal $y_{i+1} = y_i$ (see Figure 9.9). The tangent point, however, is a non-smooth point of P and the fold bifurcation point is therefore a discontinuous bifurcation, because the generalized derivative of P is set-valued at this point and contains the value $+1$ in its set. The Poincaré maps in Figure 9.9 show locally the behaviour depicted in Figure 9.5b. If we decrease the value of the damping c even more, then the fixed point x_{u}^* moves towards $x_{\mathrm{eq}} = 1.5$, i.e. the unstable limit cycle shrinks and approaches the equilibrium. The unstable limit cycle disappears through a continuous Hopf bifurcation, which changes the stability of the equilibrium. The Poincaré map for $c = 0.25$ shows the unstable fixed point of the equilibrium together with the stable fixed point of the stable limit cycle.

The preceding example shows a discontinuous fold bifurcation in a one-dimensional Poincaré map. An example of a period-doubling bifurcation in a one-dimensional *Poincaré* map can not be given, because a continuous or discontinuous period-doubling bifurcation can only occur in a three- or higher-dimensional system. However, if a higher-dimensional system contains many stick-slip transitions or impact events, then it is sometimes possible to construct a one-dimensional map which still captures the dynamics of the system. This approach, which is not generally applicable, is called the event map method [43, 81].

9.7 Intersection of Switching Boundaries

In Chapter 7 we elaborated how fundamental solution matrices of Filippov systems can 'jump', or become set-valued, as the trajectory crosses a switching boundary. A set-valued eigenvalue of the generalized Jacobian, which contains a point on the imaginary axis, is by definition (see Definition 8.5) essential for a discontinuous bifurcation of an equilibrium in a non-smooth continuous system, as has been discussed in the preceding chapter. Similarly, a set-valued Floquet multiplier of a generalized fundamental solution matrix is essential for a discontinuous bifurcation of a periodic solution in a Filippov system. The conditions under which a periodic solution can have set-valued Floquet multipliers are studied in this section.

We consider the following scenario (Figure 9.11). The switching boundary Σ divides the state-space in the two subspaces $\mathcal{V}_+$ and $\mathcal{V}_-$. It holds that $\boldsymbol{f}_-(\boldsymbol{x}) \neq \boldsymbol{f}_+(\boldsymbol{x})$ for $\boldsymbol{x} \in \Sigma$, but Σ itself is a smooth hyper-surface. Assume that a periodic solution exists which changes under influence of a parameter μ. For a value $\mu < \mu^*$ the periodic solution is denoted by periodic solution I and assume that periodic solution I does not cross Σ. If we increase the parameter μ to $\mu = \mu^*$, then the periodic solution changes to periodic solution II. Let periodic solution II be tangent at one point to Σ. If we increase μ even more to $\mu > \mu^*$, then the periodic solution becomes periodic solution III, which crosses Σ twice at points A and B. Assume that periodic solution I comes infinitely close to Σ (without crossing it) and that periodic solution III stays an infinitely small time in $\mathcal{V}_+$ but crosses Σ twice. The periodic solutions I and III are therefore (almost) identical, but the fundamental solution matrix of periodic solution III will jump twice with saltation matrices $\boldsymbol{S}_A$ and $\boldsymbol{S}_B$. The crossings occur at $t_A = t_B = t_\Sigma$ as periodic solution III stays an infinitely small time in $\mathcal{V}_+$. We can now express the fundamental solution matrix $\boldsymbol{\Phi}_{\mathrm{III}}$ in $\boldsymbol{\Phi}_{\mathrm{I}}$ and the saltation matrices $\boldsymbol{S}_A$ and $\boldsymbol{S}_B$

$$\boldsymbol{\Phi}_{\mathrm{III}}(T + t_0, t_0) = \boldsymbol{\Phi}_{\mathrm{I}}(T + t_0, t_\Sigma + t_0)\boldsymbol{S}_B\boldsymbol{S}_A\boldsymbol{\Phi}_{\mathrm{I}}(t_\Sigma + t_0, t_0). \tag{9.43}$$

Assuming that $\boldsymbol{S}_A$ and $\boldsymbol{S}_B$ are non-singular, we conclude from (7.64) and (7.65) that $\boldsymbol{S}_B = \boldsymbol{S}_A^{-1}$. The fundamental solution matrix of periodic solution III is therefore identical to the one of periodic solution I, $\boldsymbol{\Phi}_{\mathrm{III}}(T + t_0, t_0) = \boldsymbol{\Phi}_{\mathrm{I}}(T + t_0, t_0)$. This scenario, in which a single smooth switching boundary is crossed twice, can consequently not lead to a discontinuous bifurcation of a periodic solution if $\boldsymbol{S}_A$ is non-singular. The singular case arises in sliding mode problems (see Subsection 9.9.1).

The preceding scenario does not lead to set-valued Floquet multipliers, because the saltation matrix over a smooth switching boundary is equal to the inverse (if it exists) of the saltation matrix in opposite direction over the same switching boundary at that point. We will study a second scenario which is depicted in Figure 9.12. The switching boundary Σ is now non-smooth and consists of two parts Σ_A and Σ_B. The periodic solution III enters $\mathcal{V}_+$ by crossing Σ_A at point A and leaves $\mathcal{V}_+$ by crossing Σ_B at point B. The saltation matrix $\boldsymbol{S}_A$ is (in general) not equal to $\boldsymbol{S}_B^{-1}$. Consequently, the fundamental solution matrix of periodic solution III is not identical to the one of periodic solution I. Therefore, at $\mu = \mu^*$, the fundamental solution matrix over the period time $\boldsymbol{\Phi}(T + t_0, t_0)$ will jump from $\boldsymbol{\Phi}_{\mathrm{I}}$ to $\boldsymbol{\Phi}_{\mathrm{III}}$. In Section 9.4, the generalized fundamental solution matrix (9.32) was introduced. The generalized

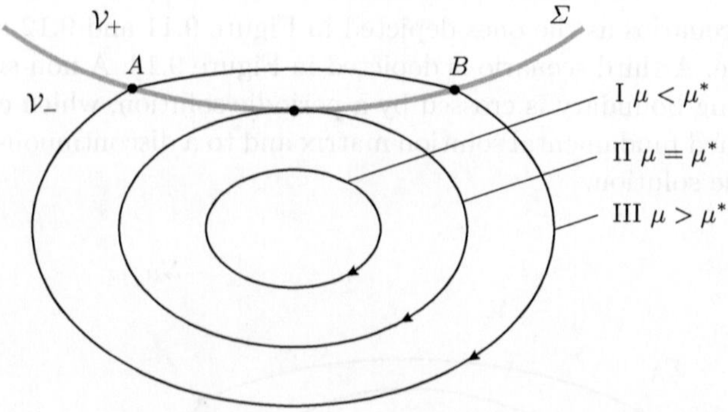

Fig. 9.11. Double intersection of a smooth switching boundary.

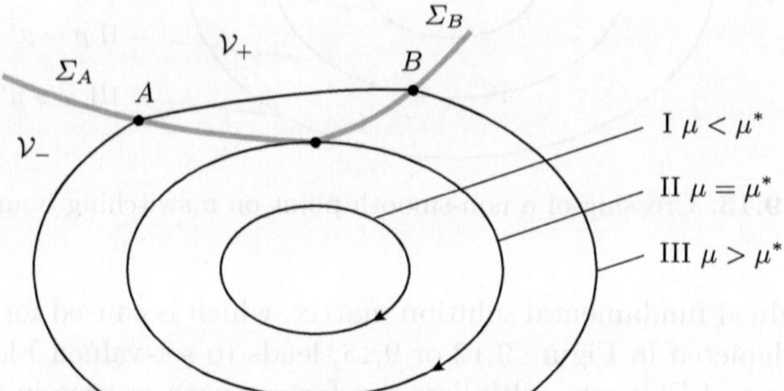

Fig. 9.12. Double intersection of a non-smooth switching boundary.

fundamental solution matrix $\boldsymbol{\Phi}_{\mathrm{II}}$ at $\mu = \mu^*$ is the closed convex hull of $\boldsymbol{\Phi}_{\mathrm{I}}$ and $\boldsymbol{\Phi}_{\mathrm{III}}$

$$\begin{aligned}\boldsymbol{\Phi}_{\mathrm{II}} &= \overline{\mathrm{co}}\{\boldsymbol{\Phi}_{\mathrm{I}}(T+t_0,t_0), \boldsymbol{\Phi}_{\mathrm{III}}(T+t_0,t_0)\}\\ &= \{\boldsymbol{\Phi}_q \mid \boldsymbol{\Phi}_q = q\boldsymbol{\Phi}_{\mathrm{III}}(T+t_0,t_0) - (1-q)\boldsymbol{\Phi}_{\mathrm{I}}(T+t_0,t_0)), \forall 0 \le q \le 1\}\\ &= \boldsymbol{\Phi}_{\mathrm{I}}(T+t_0, t_\Sigma + t_0)\boldsymbol{S}_{BA}\boldsymbol{\Phi}_{\mathrm{I}}(t_\Sigma + t_0, t_0),\end{aligned} \tag{9.44}$$

with the set-valued saltation matrix

$$\boldsymbol{S}_{BA} = \{\boldsymbol{S}_q \mid \boldsymbol{S}_q = q\boldsymbol{S}_B\boldsymbol{S}_A + (1-q)\boldsymbol{I}, \forall 0 \le q \le 1\}. \tag{9.45}$$

In fact, (9.44) defines *how* the fundamental solution matrix of the periodic solution 'jumps' from $\boldsymbol{\Phi}_{\mathrm{I}}$ to $\boldsymbol{\Phi}_{\mathrm{III}}$ if μ is increased from $\mu < \mu^*$ to $\mu > \mu^*$. From the set-valued generalized fundamental solution matrix $\boldsymbol{\Phi}_{\mathrm{II}}$ we can obtain the set-valued Floquet multipliers. We can look at $\mathrm{eig}(\boldsymbol{\Phi}_{\mathrm{II}})$ together with (9.44)) as if it gives a path of Floquet multipliers 'during' the jump as q is varied from 0 to 1.

Other scenarios as the ones depicted in Figure 9.11 and 9.12 are of course also possible. A third scenario is depicted in Figure 9.13. A non-smooth point of a switching boundary is crossed by a periodic solution, which can also lead to a set-valued fundamental solution matrix and to a discontinuous bifurcation of a periodic solution.

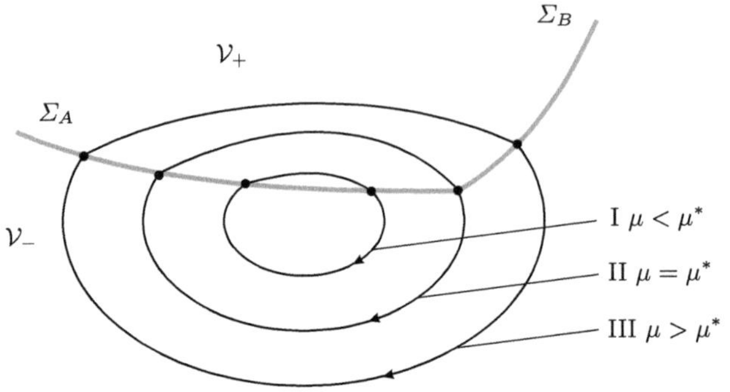

Fig. 9.13. Crossing of a non-smooth point on a switching boundary.

A set-valued fundamental solution matrix, which is caused for instance by a scenario depicted in Figure 9.12 or 9.13, leads to set-valued Floquet multipliers. The set of Floquet multipliers can form a path or area in the complex plane, which can cross the unit circle. Similar to a discontinuous bifurcation of an equilibrium, we define a discontinuous bifurcation of a periodic solution as a bifurcation for which one or more set-valued Floquet multipliers cross the unit circle. If the generalized Jacobian determines the existence and type of discontinuous bifurcation of equilibria in planar non-smooth continuous systems, then the generalized fundamental solution matrix determines the existence and type of bifurcation of periodic solutions in Filippov systems for which a one- or two-dimensional Poincaré map exists.

A Filippov system may also be approximated by a smooth approximating system. We have concluded in the previous chapter that the discontinuous bifurcation of an equilibrium may vanish if the vector field is approximated by a smooth vector field. This problem probably also exists for bifurcations of periodic solutions in Filippov systems. Floquet multipliers of smooth systems do not 'jump' but move 'fast' when the solution is passing through a 'stiff' part of the vector field. A smooth approximation, which preserves the existence of the bifurcation, will yield a continuous path of the Floquet multipliers through the unit circle. A smooth approximation, which does not preserve the existence of the bifurcation, will yield two unconnected branches that come close to each other. To each branch belongs a continuous path of Floquet multipliers which does not necessarily cross the unit circle.

The above reflections can be summarized by the following fundamental questions:

1. Do discontinuous bifurcations of periodic solutions of Filippov systems exist?
2. Does a discontinuous bifurcation, caused when a periodic solution touches a switching boundary, exist?
3. Does a jump of a Floquet multiplier over the unit circle under the influence of a parameter lead to a bifurcation of a periodic solution?
4. Can we classify the bifurcation by inspecting the point where the path of the set-valued Floquet multiplier(s) crosses the unit circle?
5. Do discontinuous bifurcations of periodic solutions exist that do not have a continuous counterpart? In other words: do discontinuous bifurcations exist that behave qualitatively different from any continuous bifurcation?
6. Can the geometric definition of bifurcation and the definition of bifurcation through topological equivalence be inconsistent when they are applied to periodic solutions of Filippov systems?

These questions are related to each other and are also related to the questions raised in Section 8.4. Some numerical examples in the following sections show discontinuous bifurcations of periodic solutions which illustrate these questions. We will return to the questions in the last section.

9.8 Bifurcations of Periodic Solutions in Filippov Systems

Filippov systems, or discontinuous systems that become Filippov systems when using Filippov's convex method, form an important class of non-smooth systems. In this section we will study the bifurcation behaviour of a number of mechanical systems that belong to the class of Filippov systems. The examples of the next subsections will reveal continuous as well as discontinuous bifurcations of periodic solutions. The discontinuous bifurcations are either single crossing bifurcations or multiple crossing bifurcations. The bifurcation diagrams in this chapter are made with a path-following technique based on the shooting method [35,116,144]. The fundamental solution matrix $\boldsymbol{\Phi}_T$, used in the shooting method, is obtained with a sensitivity analysis [86].

9.8.1 Fold Bifurcation; Trilinear System

In this subsection we study a Filippov system that exposes a discontinuous fold bifurcation of a periodic solution. It is well known that the forced oscillation of a mass–spring–damper–system with a cubic spring stiffness leads to the Duffing equation [52, 54, 103, 104]

$$m\ddot{x} + c\dot{x} + kx + k_3x^3 = f_0 \cos(\omega t). \tag{9.46}$$

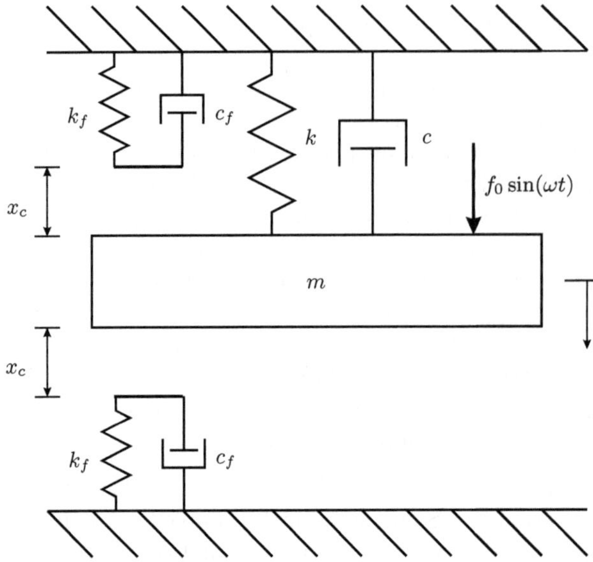

Fig. 9.14. Trilinear system.

The Duffing equation (9.46) is a classical example of a nonlinear system with a harmonic resonance peak, which is bent to one side in the frequency response diagram. The bent in the harmonic peak gives rise to two fold bifurcations (also called turning point bifurcations), which causes a hysteresis phenomenon. In this subsection, a similar mass-spring-damper system, where the cubic spring is replaced by a trilinear spring, will be considered. Additionally, trilinear damping is added to the model (see Figure 9.14). The trilinear damping will turn out to be essential for the existence of a *discontinuous* fold bifurcation in this particular system.

The model is very similar to the model of Natsiavas [101, 102], but the transitions from contact with the support to no contact are different from those in the model of Natsiavas. The model of Natsiavas switches as the position of the mass passes the contact distance (in both transition directions). In the model studied in this subsection, contact is made when the position of the mass passes the contact distance (for growing $|x|$), and contact is lost when the contact force becomes zero. It is remarked that the modification of the Natsiavas model is not introduced to obtain an improved physical representation, but in order to illustrate the theory developed in this book.

We consider the system depicted in Figure 9.14. The model is similar to the discontinuous support of Example II in Section 7.4, but now has two supports on equal contact distances x_c. The supports are first-order systems, which relax to their unstressed state if there is no contact with the mass. If we assume that the relaxation time of the supports is much smaller than the time interval between two contact events, then we can neglect the free motion of the supports. It is therefore assumed that the supports are at rest at the moment that contact is made. This is not an essential assumption but simplifies our treatment as the system reduces to a second-order differential equation.

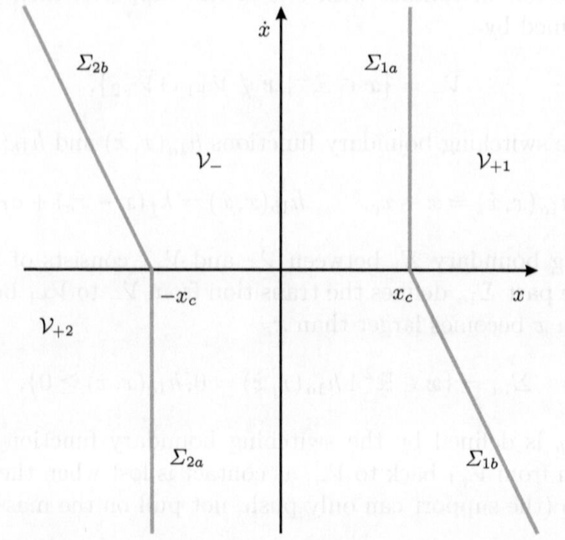

Fig. 9.15. Subspaces of the trilinear system (9.47).

The *discontinuous* second-order differential equation of this system is

$$m\ddot{x} + C(\dot{x}) + K(x) = f_0 \sin(\omega t), \tag{9.47}$$

in which

$$K(x) = \begin{cases} kx, & \boldsymbol{x} \in \mathcal{V}_-, \\ kx + k_f(x - x_c), & \boldsymbol{x} \in \mathcal{V}_{+1}, \\ kx + k_f(x + x_c), & \boldsymbol{x} \in \mathcal{V}_{+2}, \end{cases} \tag{9.48}$$

is the trilinear restoring spring force and

$$C(\dot{x}) = \begin{cases} c\dot{x}, & \boldsymbol{x} \in \mathcal{V}_-, \\ (c + c_f)\dot{x}, & \boldsymbol{x} \in \mathcal{V}_{+1} \cup \mathcal{V}_{+2}, \end{cases} \tag{9.49}$$

is the trilinear damping force with $\boldsymbol{x} = \begin{bmatrix} x \ \dot{x} \end{bmatrix}^{\mathrm{T}}$.

The state-space is divided into three subspaces $\mathcal{V}_-$, $\mathcal{V}_{+1}$ and $\mathcal{V}_{+2}$ (Figure 9.15). If the mass is in contact with the lower support, then the state is in space $\mathcal{V}_{+1}$

$$\mathcal{V}_{+1} = \{\boldsymbol{x} \in \mathbb{R}^2 \mid x > x_c, k_f(x - x_c) + c_f\dot{x} > 0\},$$

whereas if the mass is in contact with the upper support, then the state is in space $\mathcal{V}_{+2}$

$$\mathcal{V}_{+2} = \{\boldsymbol{x} \in \mathbb{R}^2 \mid x < -x_c, k_f(x + x_c) + c_f\dot{x} < 0\}.$$

If the mass is not in contact with one of the supports, then the state is in space $\mathcal{V}_{-}$ defined by

$$\mathcal{V}_{-} = \{\boldsymbol{x} \in \mathbb{R}^2 \mid \boldsymbol{x} \notin \overline{\mathcal{V}}_{+1} \cup \overline{\mathcal{V}}_{+2}\}.$$

We define the switching boundary functions $h_{1a}(x, \dot{x})$ and $h_{1b}(x, \dot{x})$ as

$$h_{1a}(x, \dot{x}) = x - x_c, \qquad h_{1b}(x, \dot{x}) = k_f(x - x_c) + c_f\dot{x}. \tag{9.50}$$

The switching boundary Σ_1 between $\mathcal{V}_{-}$ and $\mathcal{V}_{+1}$ consists of two parts Σ_{1a} and Σ_{1b}. The part Σ_{1a} defines the transition from $\mathcal{V}_{-}$ to $\mathcal{V}_{+1}$ because contact is made when x becomes larger than x_c

$$\Sigma_{1a} = \{\boldsymbol{x} \in \mathbb{R}^2 \mid h_{1a}(x, \dot{x}) = 0, h_{1b}(x, \dot{x}) \geq 0\}. \tag{9.51}$$

The part Σ_{1b} is defined by the switching boundary function which defines the transition from $\mathcal{V}_{+1}$ back to $\mathcal{V}_{-}$, as contact is lost when the support-force becomes zero (the support can only push, not pull on the mass)

$$\Sigma_{1b} = \{\boldsymbol{x} \in \mathbb{R}^2 \mid h_{1a}(x, \dot{x}) \geq 0, h_{1b}(x, \dot{x}) = 0\}. \tag{9.52}$$

Similarly, the switching boundary Σ_2 between $\mathcal{V}_{-}$ and $\mathcal{V}_{+2}$ consists of two parts Σ_{2a} and Σ_{2b} defined by the switching boundary functions

$$h_{2a}(x, \dot{x}) = x + x_c, \qquad h_{2b}(x, \dot{x}) = k_f(x + x_c) + c_f\dot{x}. \tag{9.53}$$

We use Filippov's solution concept for the discontinuous second-order system (9.47) and we therefore replace it through Filippov's convex method by the differential inclusion

$$m\ddot{x} + c\dot{x} + kx = -F_{s1}(\boldsymbol{x}) - F_{s2}(\boldsymbol{x}) + f_0 \sin(\omega t), \tag{9.54}$$

with the support forces

$$\begin{aligned} F_{s1}(\boldsymbol{x}) &\in \begin{cases} 0, & \boldsymbol{x} \in \mathcal{V}_{-} \bigcup \Sigma_{1b}, \\ [0, c_f\dot{x}], & \boldsymbol{x} \in \Sigma_{1a}, \\ k_f(x - x_c) + c_f\dot{x}, & \boldsymbol{x} \in \mathcal{V}_{+1}, \end{cases} \\ F_{s2}(\boldsymbol{x}) &\in \begin{cases} 0, & \boldsymbol{x} \in \mathcal{V}_{-} \bigcup \Sigma_{2b}, \\ [0, c_f\dot{x}], & \boldsymbol{x} \in \Sigma_{2a}, \\ k_f(x + x_c) + c_f\dot{x}, & \boldsymbol{x} \in \mathcal{V}_{+2}. \end{cases} \end{aligned} \tag{9.55}$$

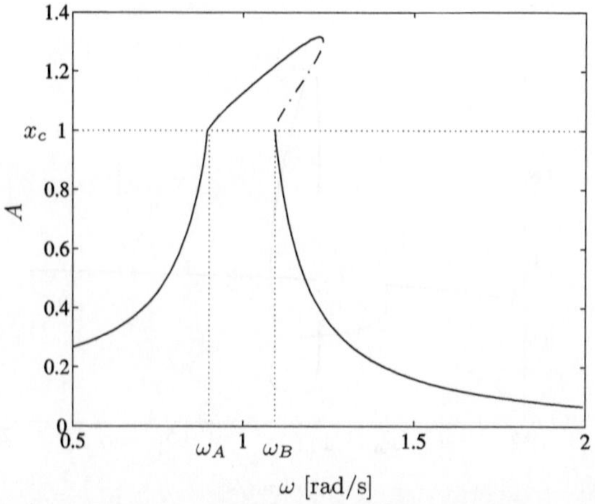

Fig. 9.16. Response diagram of trilinear system (9.47).

Note that the support forces F_{s1} and F_{s2} are an element of a set on Σ_{1a} and Σ_{2a} respectively. The image of the support forces become single-valued for $c_f = 0$. It is easily verified that only transversal intersections can take place on the switching boundaries Σ_{1a} and Σ_{2a}.

The saltation matrices on the four parts of the switching boundaries $\Sigma_{1a,b}$ and $\Sigma_{2a,b}$ are of course similar to those of Subsection 7.4

$$\boldsymbol{S}_{1a} = \boldsymbol{S}_{2a} = \begin{bmatrix} 1 & 0 \\ -\frac{c_f}{m} & 1 \end{bmatrix}, \qquad \boldsymbol{S}_{1b} = \boldsymbol{S}_{2b} = \boldsymbol{I}. \tag{9.56}$$

The switching boundaries Σ_1 and Σ_2 are non-smooth at the corner points $(\pm x_c, 0)$, and the saltation matrices on each side of the corner points are not each others inverse, i.e. $\boldsymbol{S}_{1a} \neq \boldsymbol{S}_{1b}^{-1}$ and $\boldsymbol{S}_{2a} \neq \boldsymbol{S}_{2b}^{-1}$. According to Section 9.4 we now have all the ingredients for the existence of a discontinuous bifurcation.

The response diagram of the trilinear system is shown in Figure 9.16 with the forcing frequency ω as bifurcation parameter and with the 'amplitude' $A = \max(x)$ on the vertical axis. Stable branches of periodic solutions are indicated by solid lines and unstable branches by dashed-dotted lines. The parameter values used in the simulations are $m = 1$ kg, $c = 0.05$ N/(ms), $k = 1$ N/m, $x_c = 1$ m, $k_f = 4$ N/m, $c_f = 0.5$ N/(ms), $f_0 = 0.2$ N.

A periodic solution with an amplitude smaller than x_c has no contact with the supports during its entire period time and is located in subspace $\mathcal{V}_-$. The system is locally linear in subspace $\mathcal{V}_-$. The lower part of the response curve for $A < x_c$ is therefore just the linear harmonic peak. For amplitudes above x_c, periodic solutions will be for some time in contact with the supports during the period time, which will cause a hardening behaviour of the response curve. The resonance peak of the response curve bends to the right like for the Duffing system with a hardening spring. The amplitude becomes equal to x_c at the two frequencies $\omega = \omega_A$ and $\omega = \omega_B$, on both sides of the peak.

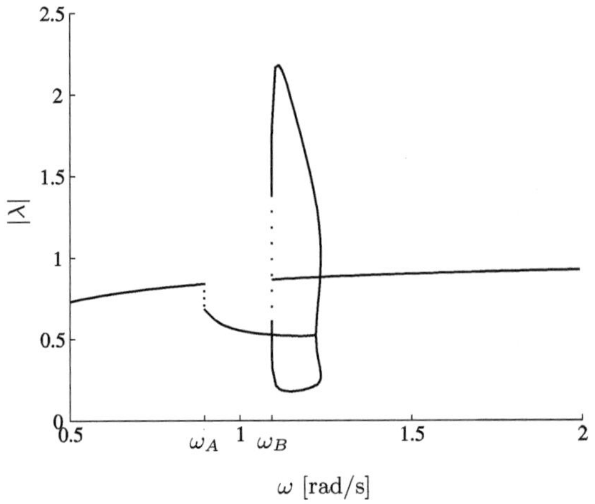

Fig. 9.17. Magnitude of the Floquet multipliers of the trilinear system (9.47).

The response curve is non-smooth at $\omega = \omega_A$ and $\omega = \omega_B$. The periodic solution touches the corners of Σ_1 and Σ_2 for $A = x_c$, similar to solution II in Figure 9.12. The Floquet multipliers can therefore be set-valued for the periodic solutions with frequencies $\omega = \omega_A$ and $\omega = \omega_B$. The magnitude of the Floquet multipliers is shown in Figure 9.17. The two Floquet multipliers form a complex conjugated pair for $A < x_c$ (and therefore have the same magnitude). The Floquet multipliers are set-valued at $\omega = \omega_A$, but the set does not intersect the unit circle. It follows from the numerical calculations depicted in Figure 9.16, that no bifurcation exists (in the sense of Definition 8.1) at $\omega = \omega_A$. However, at $\omega = \omega_B$ the complex pair 'jumps' to two distinct real Floquet multipliers, one with a magnitude larger than one. One of those Floquet multipliers therefore 'jumps' through the unit circle, i.e. a set-valued Floquet multiplier contains a value on the unit circle. A bifurcation is observed in Figure 9.16 at $\omega = \omega_B$ (one periodic solution exists for $\omega < \omega_B$ and three periodic solutions co-exist for $\omega > \omega_B$). We conclude that this bifurcation is caused by the crossing of the set-valued Floquet multiplier(s) with the unit circle. The path of the set-valued Floquet multipliers is obtained from the generalized fundamental solution matrix (9.44). The observed bifurcation resembles a continuous fold bifurcation of a smooth system. The bifurcation at $\omega = \omega_B$, at which a set-valued Floquet multiplier crosses the point $+1$, is therefore called a *discontinuous fold bifurcation.*

Damping of the support is essential for the existence of this discontinuous fold bifurcation. For $c_f = 0$, all saltation matrices would be equal to the identity matrix and the corner between Σ_{1a} and Σ_{1b} would disappear (and also between Σ_{2a} and Σ_{2b}); therefore no discontinuous bifurcation could take place and the fold bifurcation would be continuous (see also Example 7.4). The model of Natsiavas [101,102] does not contain a *discontinuous* fold bifurcation because the transitions are modelled such that $\boldsymbol{S}_{1a} = \boldsymbol{S}_{1b}^{-1}$ and $\boldsymbol{S}_{2a} = \boldsymbol{S}_{2b}^{-1}$.

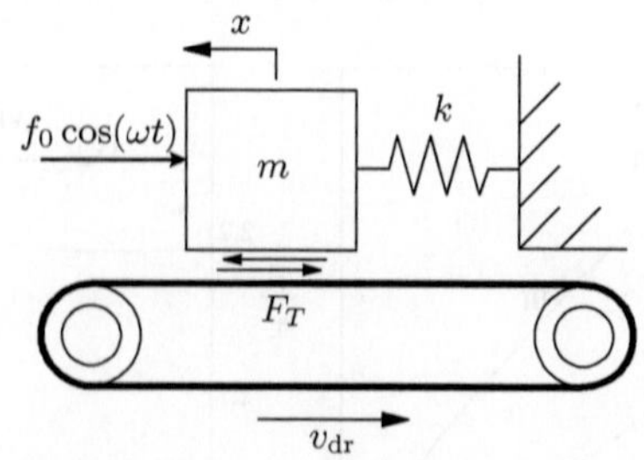

Fig. 9.18. The forced stick-slip system (9.57).

9.8.2 Multiple Crossing Bifurcations

In this subsection a multiple crossing bifurcation is studied which occurs in a forced stick-slip system (see also [152]). The system depicted Figure 9.18 is similar to the stick-slip system of Section 6.5.1 (Figure 6.4) without linear damping and with a sinusoidal excitation. The friction model for the contact between block and belt is of the type depicted in Figure 4.2b, which is set-valued with a minimal closed upper semi-continuous convex set. The system can be expressed by the second-order differential inclusion

$$m\ddot{x} + kx = F_T(v_{\text{rel}}) + f_0 \cos(\omega t), \tag{9.57}$$

with $v_{\text{rel}} = \dot{x} - v_{\text{dr}}$. The set-valued friction model reads as

$$-F_T(v_{\text{rel}}) \in \alpha_0 \operatorname{Sign}(v_{\text{rel}}) - \alpha_1 v_{\text{rel}} + \alpha_3 v_{\text{rel}}^3. \tag{9.58}$$

The bifurcation diagrams depicted in Figures 9.19 to 9.21 were obtained numerically using the parameter values $m = 1$ kg, $k = 1$ N/m, $v_{\text{dr}} = 1$ m/s, $\alpha_0 = 1.5$ N, $\alpha_1 = 1.5$ Ns/m, $\alpha_3 = 0.45$ Ns^3/m^3 and $f_0 = 0.1$ N.

The resonance curve of system (9.57) has been published in [152] for $0.2 \leq \omega \leq 4$. The 1/2-subharmonic closed resonance curve is of special interest and depicted in Figure 9.19a and an enlargement in Figure 9.19b. The real part of the largest Floquet multiplier (in magnitude) is depicted in Figure 9.20. All Floquet multipliers are real except on a part of branch III. Stable branches of periodic solutions are denoted by solid lines and unstable branches by dashed-dotted lines. The set-valuedness of Floquet multipliers is denoted by dotted lines.

The 1/2-subharmonic closed resonance curve possesses several discontinuous and continuous bifurcations. Branches I-V are period-2 solutions, i.e. periodic solutions with period time $2T$. Branches VI and VII are period-4, and branch VIII is period-8. A discontinuous fold bifurcation at point A connects the stable branch I to the unstable branch II and its largest Floquet multiplier 'jumps' through +1 (similar to the discontinuous fold bifurcation in Subsection 9.8.1). The stable branch I smoothly folds into branch V and stability is exchanged. At point B, the unstable branch V is folded into the

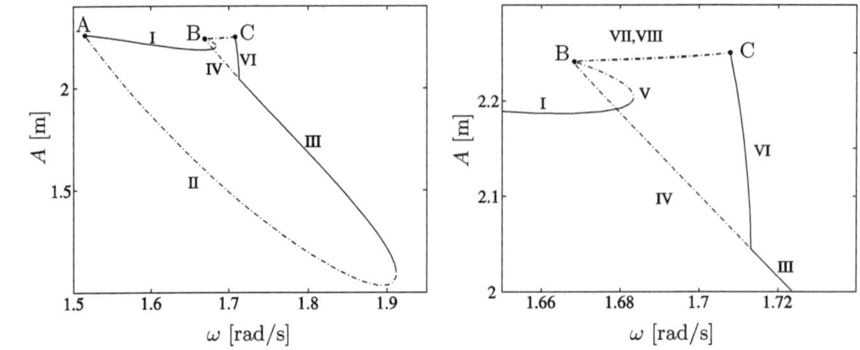

(a) (b) zoom-in of a)

Fig. 9.19. Bifurcation diagram of the forced stick-slip system (9.57).

unstable branch IV *without exchanging stability*. The set-valued Floquet multiplier at point B crosses the unit circle twice as it 'jumps' from $\lambda > 1$ on branch V to $\lambda < -1$ on branch IV, i.e. the Floquet multiplier contains the points $+1$ and -1 on the unit circle in its set (Figure 9.4). The associated bifurcation is therefore a multiple crossing bifurcation of a periodic solution. In Section 9.4 it has been presumed that such a multiple crossing bifurcation *might* show a behaviour which is a combination of a fold and a period-doubling bifurcation. The analysis of multiple crossing bifurcations of equilibria in non-smooth continuous systems makes it clear that a multiple crossing bifurcation can generally *not* be interpreted as the combination of continuous bifurcations. The fold action of the bifurcation point B in Figure 9.20 is clear as the branch is folded. A conventional continuous period-doubling bifurcation causes a period-doubled branch to bifurcate from the main branch. Branches IV and V are period-2 and branch VII emanates indeed from point B and is period-4. The bifurcation at point B therefore also shows a period-doubling action. In addition, branch VIII also bifurcates from bifurcation point B and is period-8. Bifurcation point B shows the behaviour of a continuous fold and period-doubling bifurcation but also shows the creation of many more branches (chaos). Consequently, this particular multiple crossing bifurcation is not simply the combination of two single crossing bifurcations.

A better understanding of the phenomenon can be obtained by looking at the Poincaré map depicted in Figure 9.21. Note that the map is indeed similar to the tent map of Figure 9.8. In fact, the 'full' Poincaré map is a mapping from $\mathbb{R}^2$ to $\mathbb{R}^2$, because the system is three-dimensional, and the Poincaré map can therefore not easily be visualized. Instead, a section of this map is depicted with the displacement $x_n = x(nT)$, where $T = 2\pi/\omega$, on the abscissa and the displacement after two periods x_{n+2} on the ordinate (because

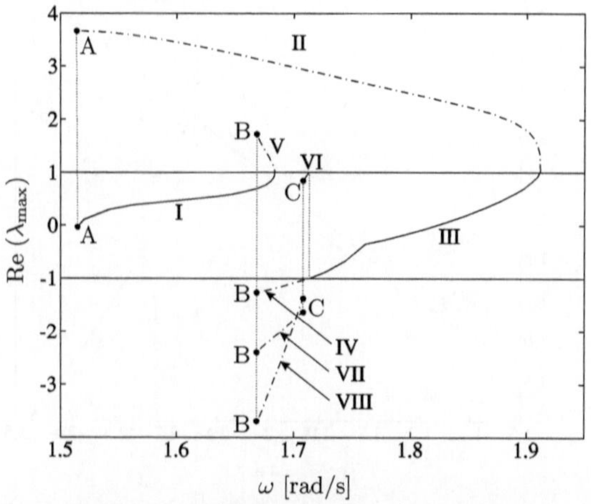

Fig. 9.20. Floquet multipliers of system (9.57).

we study period-2 oscillations). The section of the map is such, that $\dot{x}_n$ is equal to $\dot{x}_{n+2}$. Fixed points of this reduced map are periodic solutions of period-2 (or period-1) as $x(nT) = x((n+2)T)$ holds and likewise $\dot{x}(nT) = \dot{x}((n+2)T)$. The map is calculated for $\omega = 1.67587$ [rad/s], which is just to the right of the bifurcation point B. It can be seen that there are three fixed points, which correspond to the periodic solutions at the branches I, IV and V. The map exposes a peak between the fixed points IV and V. Although this map is a section of a higher-dimensional map, the 'full' map will also be similar to the tent map.

The one-dimensional tent map has been studied thoroughly in literature [21, 45, 90, 126]. The tent map is a non-smooth piecewise linear version of the logistic map (both non-invertible). The logistic map is smooth and leads to a cascade of period-doublings, which is a well known route to chaos. The distance between two succeeding period-doublings is finite for the logistic map. An infinite number of period-doublings occur at the same bifurcation value for the tent map, which leads directly to chaos.

The results on the tent map could explain the behaviour at the bifurcation point B. The similarity between the tent map and the Poincaré map suggests that there are infinitely many period-doublings. This would result in an infinite number of other unstable branches starting from point B (period-8, 16, 32...). A period-8 branch (VIII) starting from point B has indeed been found besides the 'expected' period-4 branch (VII). The infinitely many other branches become more unstable as their period-doubling number increases and the branches become closely located to each other, which makes it difficult to find them numerically. These facts agree with the analytical results on the tent map. Similar to the tent map, the system will (presumably) behave chaotically for ω-values just to the right of point B.

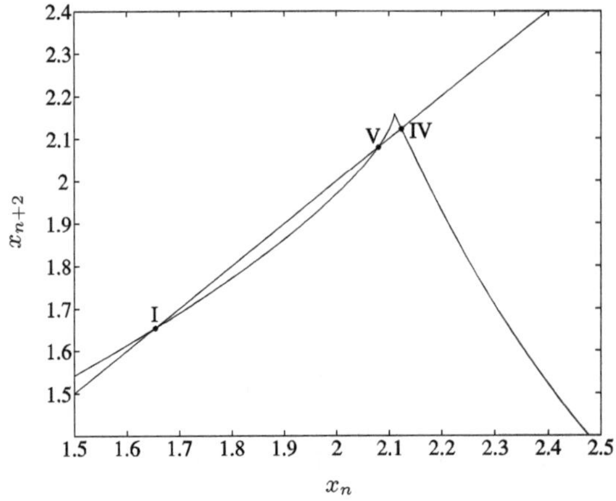

Fig. 9.21. Poincaré map of (9.57) ($\omega = 1.67587$ [rad/s]).

In [152] the system (9.57) is studied carefully and it is shown that the underlying Poincaré map is similar to the tent map but the branches VII and VIII (or higher period-doublings) are not presented. Discontinuous fold and period-doubling bifurcations, for which the Floquet multipliers 'jump' at the bifurcation point, are presented in [152] but the relation with set-valued Floquet multipliers, saltation matrices and generalized fundamental solution matrices, as elaborated here, is not given in [152]. The region between B and C is classified as chaotic in [152] and the similarity with the bifurcations found by [106] is mentioned. Nusse and York [106] have presented a study of low-dimensional non-smooth mappings similar to a tent map and denote discontinuous bifurcations by 'border-collision bifurcations'. The numerical calculations in [106] only show stable fixed points. Nusse and York [106] do not give a method to classify discontinuous bifurcations but conclude that this is still an open question.

A multiple crossing bifurcation of a periodic solution has been discussed in this subsection and it has been shown that it is related to the one-dimensional tent map. It is suggested that infinitely many branches meet at the same bifurcation point and these branches are all period-doublings of the branch under bifurcation.

9.9 Bifurcations in Systems with Stiction Friction Model

Mechanical systems with a stiction friction model can not be regarded as differential inclusions and do therefore not belong to the class of Filippov systems as explained in Section 4.2. We do therefore not immediately have a solution concept for systems with a stiction friction model. Instead, we will use the Switch Model for a stiction friction model. The Switch Model has been presented in Section 6.5.2 as a *numerical* algorithm, but is used in this section as a *theoretical* solution concept. The thickness parameter $\eta > 0$ of the band around the switching boundary is considered to be infinitely small for the theoretical solution concept. A small but finite value is chosen as a numerical approximation in the following examples. Although a mechanical system with a stiction friction model can not be regarded as a differential inclusion and is therefore not a Filippov system, the (convexified version of) the Switch Model of a system with a stiction friction model is discontinuous and of Filippov-type. We can therefore still use the developed theory on sliding modes and saltation matrices on switching boundaries, not on the original switching boundary $\Sigma = \{\boldsymbol{x} \in \mathbb{R}^n \mid h(\boldsymbol{x}) = v_{\mathrm{rel}} = 0\}$ but on the switching boundaries between the subspaces $\mathcal{U}$, $\mathcal{V}_-$, $\mathcal{V}_-$, $\mathcal{T}_-$ and $\mathcal{T}_+$.

9.9.1 Infinitely Unstable Periodic Solutions

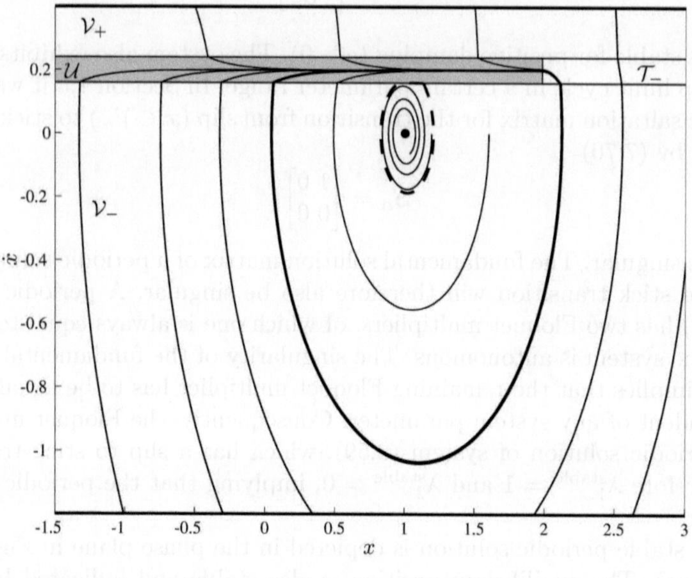

Fig. 9.22. Phase plane of the stick-slip system (9.59) with stiction friction model.

A discontinuous fold bifurcation has been studied in Subsection 9.8.1, at which (loosely speaking) a Floquet multiplier 'jumped' over the unit circle, through +1, to a finite value. In this subsection we will study a discontinuous fold bifurcation in which a Floquet multiplier 'jumps' to infinity (crossing +1). To be more precise, we will study a bifurcation for which a set-valued Floquet multiplier is an unbounded subset of the complex plane. This results in periodic solutions with a Floquet multiplier at infinity, which we will call *infinitely unstable periodic solutions.*

We consider again the stick-slip system with stiction friction model of Section 6.5.2 depicted in Figure 6.8. The equation of motion of the system is given by

$$\dot{\boldsymbol{x}} = \boldsymbol{f}(\boldsymbol{x}) = \begin{bmatrix} \dot{x} \\ -\frac{k}{m}x - \frac{c}{m}\dot{x} + \frac{F(v_{\mathrm{rel}})}{m} \end{bmatrix}, \tag{9.59}$$

with $\boldsymbol{x} = \begin{bmatrix} x & \dot{x} \end{bmatrix}^{\mathrm{T}}$, $v_{\mathrm{rel}} = \dot{x} - v_{\mathrm{dr}}$ and the friction force $F(v_{\mathrm{rel}})$ is given by a stiction friction model as depicted in Figure 4.2c with maximal static friction force F_{s} and dynamic friction force F_{d}. The same parameter values are chosen as in Section 6.5.2. A small but finite value $\eta = 10^{-6}$ m/s is chosen as a numerical approximation in the Switch Model. We assume that the thickness η is small enough to approximate the qualitative behaviour of the theoretically infinitely thin band to a sufficient extent.

The system (9.59) has one equilibrium

$$\boldsymbol{x}_{eq} = \begin{bmatrix} \frac{F_{\mathrm{slip}}}{k} \\ 0 \end{bmatrix}, \tag{9.60}$$

which is stable for positive damping ($c > 0$). The system also exhibits a stable stick-slip limit cycle in a certain parameter range. In Section 7.3 it was shown that the saltation matrix for the transition from slip ($\boldsymbol{x} \in \mathcal{V}_-$) to stick ($\boldsymbol{x} \in \mathcal{U}$) is given by (7.70)

$$\boldsymbol{S}_{\alpha} = \begin{bmatrix} 1 & 0 \\ 0 & 0 \end{bmatrix},$$

which is singular. The fundamental solution matrix of a periodic solution with a slip to stick transition will therefore also be singular. A periodic solution of (9.59) has two Floquet multipliers, of which one is always equal to one because the system is autonomous. The singularity of the fundamental solution matrix implies that the remaining Floquet multiplier has to be equal to zero, independent of any system parameter. Consequently, the Floquet multipliers of a periodic solution of system (9.59), which has a slip to stick transition, are therefore $\lambda_1^{\mathrm{stable}} = 1$ and $\lambda_2^{\mathrm{stable}} = 0$, implying that the periodic solution is stable.

The stable periodic solution is depicted in the phase plane in Figure 9.22 (bold line). The equilibrium position is also stable and indicated by a dot. The thickness parameter η is taken fairly large in Figure 9.22 to make the subspaces $\mathcal{U}$ and $\mathcal{T}_-$ visible but should be taken very small to obtain an accurate numerical simulation of (9.59).

A trajectory outside the stable periodic solution depicted in Figure 9.22, will spiral inwards to the stable periodic solution and reach the stick-phase $\mathcal{U}$. A trajectory in the stick-phase will rapidly converge towards the middle of the band, i.e. to the switching boundary Σ, and therefore also to the stable periodic solution. A trajectory in $\mathcal{U}$ will converge exactly to the stable periodic solution in finite time if η is taken infinitely small. Every point in $\mathcal{U}$ is therefore part of the basin of attraction of the stable periodic solution.

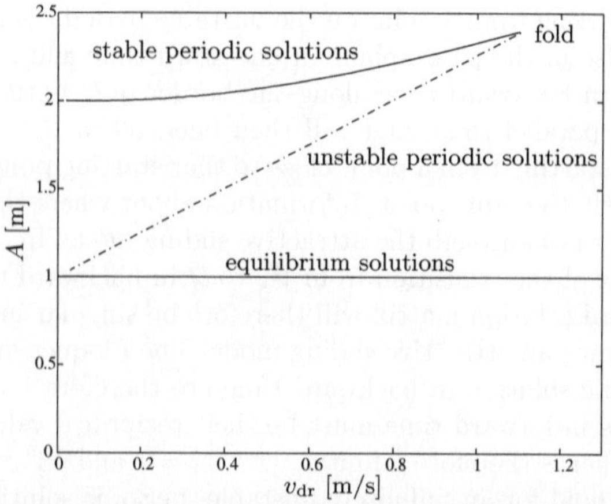

Fig. 9.23. Bifurcation diagram of the stick-slip system.

A trajectory which starts in the interior of the stable periodic solution, might spiral around the equilibrium position and hit $\mathcal{U}$ whereupon it converges rapidly to the stable periodic solution. But a trajectory inside the stable periodic solution might also spiral around the equilibrium position and not reach the stick phase $\mathcal{U}$. Such a trajectory will then be attracted to the equilibrium position. Consequently, there must exist a boundary of attraction between the two attracting limit sets (the equilibrium and the stable periodic solution). This boundary is the unstable periodic solution depicted by a dashed line in Figure 9.22. The boundary of attraction is partly along the border between $\mathcal{U}$ and $\mathcal{V}_-$ because trajectories in $\mathcal{U}$ will attract to the stable periodic solution and just outside $\mathcal{U}$ to the equilibrium position. The unstable periodic solution is therefore defined by the trajectory in $\mathcal{V}_-$ which hits the border of $\mathcal{U}$ tangentially and by a part along the border of $\mathcal{U}$ and $\mathcal{V}_-$. The part of the unstable periodic solution along the border of $\mathcal{U}$ is therefore a sliding mode along a switching boundary as discussed in Chapter 3. The solution on either side of the sliding mode is repulsing from the switching boundary and is therefore a repulsive sliding mode. The solution starting from a point on a repulsive sliding mode is non-unique in forward time as was discussed in Chapter 3. Non-uniqueness of solutions in forward time causes one Floquet multiplier of the unstable periodic solution to be at infinity because a solution may drift away from the periodic solution without any disturbance from the periodic solution. The unstable periodic solution can be called 'infinitely unstable' meaning that at least one Floquet multiplier has an unbounded magnitude. As the periodic solution is infinitely unstable, it is not possible to calculate it in forward time. However, calculation of the periodic solution in backward time is possible. The vector field in backward time is identical to the one in forward time but opposite in direction. The repulsive sliding mode in forward time will turn into an attractive sliding mode in backward time.

The solution starting from a point on the unstable periodic solution will move counter-clockwise in the phase-plane in backward time and hit the border of $\mathcal{U}$. It will slide in backward time along the border of $\mathcal{U}$ until the vector field in $\mathcal{V}_-$ becomes parallel to $\mathcal{U}$, and will then bend off in $\mathcal{V}_-$. Any solution in backward time starting from a point close to that starting point will hit $\mathcal{U}$ and leave $\mathcal{U}$ at exactly the same point. Information about where the solution came from is therefore lost through the attractive sliding mode. In other words: the saltation matrix of the transition from $\mathcal{V}_+$ to $\mathcal{U}$ in backward time is singular. The fundamental solution matrix will therefore be singular in backward time because it contains an attractive sliding mode. The Floquet multipliers of the unstable periodic solution in backward time are therefore 1 and 0. The Floquet multipliers in forward time must be their reciprocal values. The second Floquet multiplier is therefore infinity, $\lambda_1^{\text{unstable}} = 1$ and $\lambda_2^{\text{unstable}} = \infty$, which of course must hold for an 'infinitely unstable' periodic solution.

The bifurcation diagram of the system is shown in Figure 9.23 with the velocity of the belt v_{dr} as parameter and the 'amplitude' $A = \max(x)$ of the periodic solution on the vertical axis. The equilibrium branch and the stable and unstable periodic branches are depicted. The unstable periodic branch is of course located between the stable periodic branch and the equilibrium branch, as can be inferred from Figure 9.22. The stable and unstable periodic branches are connected through a fold bifurcation point. The second Floquet multiplier 'jumps' from $\lambda_2 = 0$ to $\lambda_2 = \infty$ at the bifurcation point, and therefore through $+1$ on the unit circle. More precisely, the second Floquet multiplier of the periodic solution at the bifurcation point is set-valued $\lambda_2 = [0, \infty)$ and contains the point $+1$ in its set. We will call this bifurcation therefore a discontinuous fold bifurcation. The fold bifurcation occurs when v_{dr} is such that a trajectory which leaves the stick phase $\mathcal{U}$, traverses $\mathcal{V}_-$, and hits $\mathcal{U}$ tangentially (like the unstable periodic solution). The stable and unstable periodic solutions coincide for the bifurcation point $v_{\text{dr}} = v_{\text{dr}}^*$. The periodic solution for the critical value $v_{\text{dr}} = v_{\text{dr}}^*$ traverses the lower right corner of the boundary of $\mathcal{U}$ (the point at which $\mathcal{V}_-$, $\mathcal{U}$ and $\mathcal{T}_-$ meet). Note that this scenario is similar to Figure 9.13, in which a periodic solution is depicted that crosses a corner point on a non-smooth switching boundary. The saltation matrices are not each others inverse, $\boldsymbol{S}_\alpha \boldsymbol{S}_\beta \neq \boldsymbol{I}$, which is essential for the existence of a discontinuous bifurcation. Moreover, $\boldsymbol{S}_\alpha$ is singular.

Poincaré maps of the system are depicted in Figure 9.24 for three different values of v_{dr}; before, at and after the bifurcation point $v_{\text{dr}} = v_{\text{dr}}^* = 1.1$ m/s. The Poincaré section is chosen as $\Omega = \{\boldsymbol{x} \in \mathbb{R}^2 \mid x \geq F_{\text{d}}/k, \dot{x} = 0\}$. The three intersection points of the Poincaré map with the diagonal in Figure 9.24a indicate the equilibrium position at $x = 1$ and the unstable and stable periodic solutions. The slope of the Poincaré map at the intersection points of the periodic solutions with the diagonal is equal to the second Floquet multiplier, which is consistent with $\lambda_2^{\text{stable}} = 0$ and $\lambda_2^{\text{unstable}} = \infty$. The Poincaré map of Figure 9.24b touches the diagonal with its tip similar to the Poincaré map of the discontinuous fold bifurcation in Figure 9.5. The stable and unstable

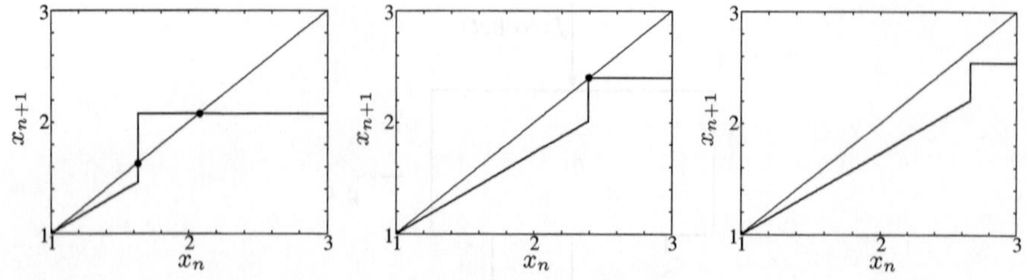

(a) $v_{\mathrm{dr}} = 0.5$ m/s (b) $v_{\mathrm{dr}} = 1.1$ m/s (c) $v_{\mathrm{dr}} = 1.3$ m/s

Fig. 9.24. Poincaré maps of the stick-slip system (9.59) with stiction friction model.

periodic solutions have disappeared in Figure 9.24c as is shown in Figure 9.23 for $v_{\mathrm{dr}} > v_{\mathrm{dr}}^*$.

A similar model is studied in [145] with a very accurately smoothed friction curve of the stiction friction model. The stable branch is followed for increasing values of v_{dr}, but the fold bifurcation could not be rounded to proceed on the unstable branch in [145]. As the unstable branch is infinitely unstable for the discontinuous model, it is extremely unstable for the smoothed approximating model. The branch can therefore not be followed numerically in forward time if the stiction friction model is approximated accurately.

The stable branch in Figure 9.23 has been followed in forward time up to the bifurcation point. The path-following algorithm is stopped and restarted in backward time to follow the unstable branch.

This subsection shows that infinitely unstable periodic solutions come into being through repulsive sliding modes. Filippov theory turns out to be essential for the understanding of infinitely unstable periodic solutions. Infinitely unstable periodic solutions and their branches can be found through backward integration. Smoothing of a discontinuous model is not sufficient to obtain a complete bifurcation diagram of a mechanical system with a stiction friction model, as infinitely unstable branches of periodic solutions can not be found.

9.9.2 Symmetry-Breaking Bifurcation

Another type of bifurcation of a periodic solution which will be studied is the *symmetry-breaking bifurcation.* Suppose a non-autonomous time-periodic smooth system has the following symmetry property (also called 'inversion symmetry')

$$\boldsymbol{f}(t, \boldsymbol{x}) = -\boldsymbol{f}(t + \frac{1}{2}T, -\boldsymbol{x}), \tag{9.61}$$

where T is the period. If $\boldsymbol{x}_1(t) = \boldsymbol{x}(t)$ is a periodic solution of the system, then also $\boldsymbol{x}_2(t) = -\boldsymbol{x}(t + \frac{1}{2}T)$ must be a periodic solution. The periodic

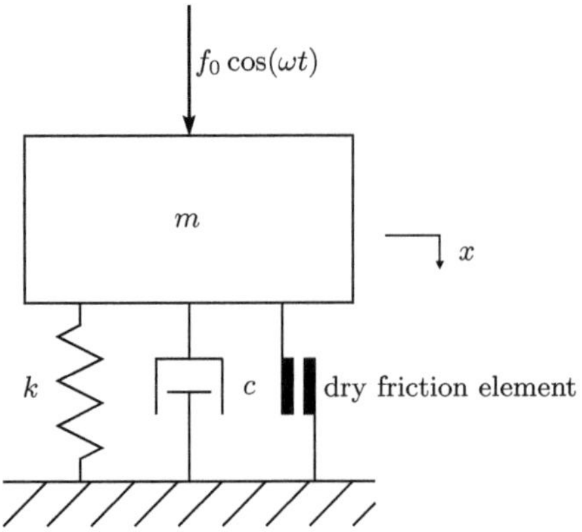

Fig. 9.25. Forced vibration with dry friction of system (9.62).

solution is called *symmetric* if $\boldsymbol{x}_1(t) = \boldsymbol{x}_2(t)$ and asymmetric if $\boldsymbol{x}_1(t) \neq \boldsymbol{x}_2(t)$. When a Floquet multiplier passes through the unit circle at $+1$, then the associated bifurcation depends on the nature of the periodic solution prior to the bifurcation. Suppose that the periodic solution prior to the bifurcation is a symmetric periodic solution. Then, if the bifurcation breaks the symmetry of the periodic solution, it is called a symmetry-breaking bifurcation [103, 133].

We will show in this subsection that the continuous symmetry-breaking bifurcation has a discontinuous counterpart. Consider the forced vibration of the one-degree-of-freedom system depicted in Figure 9.25. The mass is supported by a spring, damper and dry friction element. The equation of motion reads as

$$\dot{\boldsymbol{x}} = \boldsymbol{f}(\boldsymbol{x}) = \begin{bmatrix} \dot{x} \\ -\frac{k}{m}x - \frac{c}{m}\dot{x} + \frac{F(\dot{x})}{m} \end{bmatrix}, \tag{9.62}$$

with $\boldsymbol{x} = \begin{bmatrix} x \ \dot{x} \end{bmatrix}^{\mathrm{T}}$ and the friction force $F(\dot{x})$ is given by a stiction friction model as depicted in Figure 4.2c with maximal static friction force F_{s} and dynamic friction force F_{d}. It can be verified that the system (9.62) has the symmetry property (9.61). The parameter values used for the numerical analysis are $m = 1$ kg, $c = 0.01$ N/(ms), $k = 1$ N/m, $f_0 = 2.5$ N, $F_{\mathrm{d}} = 1$ N, $F_{\mathrm{s}} = 2$ N and $\eta = 10^{-4}$ m/s. The bifurcation diagram of (9.62), for varying values of the forcing frequency ω, is depicted in Figure 9.26 and shows two branches with periodic solutions. Branch I is unstable between the points A and B. Branch II bifurcates from branch I at points A and B. For large amplitudes, the dry friction force will be much less than the mean forces in the linear spring and dashpot. Near the resonance frequency, $\omega_{\mathrm{res}} = \sqrt{k/m} = 1$ [rad/s], branch I will therefore be close to the harmonic resonance peak of a linear one-degree-of-freedom system. If the velocity of the mass $\dot{x}$ becomes zero for a non-zero time-interval, then the system is said to 'stick' and we speak of a 'stick event'. A periodic solution on branch I for $\omega > \omega_B$ does not contain a 'stick event' in which the periodic solution passes the stick phase. The number of stick

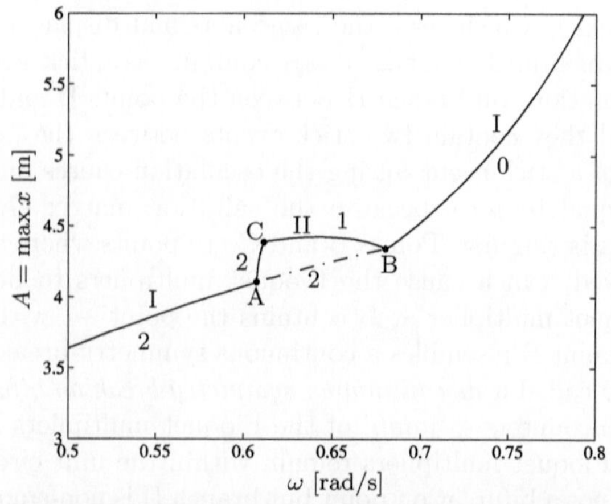

Fig. 9.26. Bifurcation diagram of system (9.62).

events on a part of a branch is indicated by numbers (0,1,2) in Figure 9.26. The Floquet multipliers on branch I for $\omega > \omega_B$ are complex conjugated (Figure 9.27). The system behaves much like a linear oscillator for $\omega > \omega_B$ (for frequencies near resonance). The periodic solutions on branch I are all symmetric.

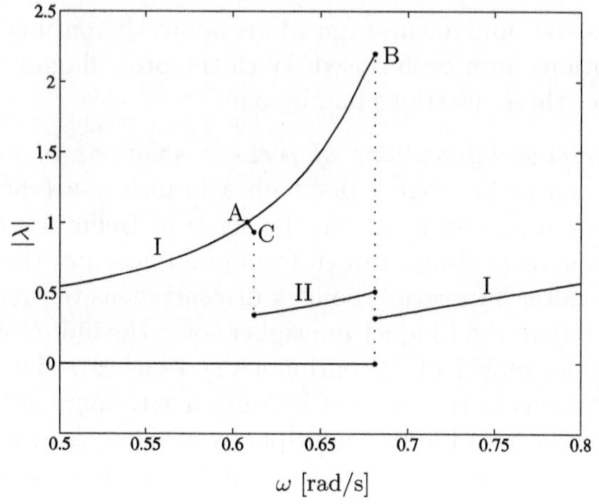

Fig. 9.27. Floquet multipliers of system (9.62).

If branch I for $\omega > \omega_B$, with almost sinusoidal symmetric periodic solutions, is followed for decreasing values of ω, then bifurcation point B is met. At bifurcation point B (for $\omega = \omega_B$), the symmetric branch I becomes unstable and a second branch II with asymmetric periodic solutions is created. In fact, on the bifurcated asymmetric branch two distinct periodic solutions

$x_1(t) \neq x_2(t)$ exist, which have the same maximal displacement. The periodic solutions on branch I for $\omega < \omega_B$ contain two stick events per cycle. The periodic solutions on branch II between the points B and C contain one stick event, and they contain two stick events between the points A and C. The existence of a stick event during the oscillation causes one Floquet multiplier to be equal to zero, because the saltation matrix (7.70) for slip to stick transitions is singular. Points B and C are points where stick events are created/destroyed, which cause the Floquet multipliers to be set-valued. A set-valued Floquet multiplier at B contains the point +1 within its set. The bifurcation at point B resembles a continuous symmetry-breaking bifurcation and is therefore called a *discontinuous symmetry-breaking bifurcation.*

Branch II encounters a 'jump' of the Floquet multipliers at point C but the set-valued Floquet multipliers remain within the unit circle. We observe that point C is not a bifurcation point but branch II is non-smooth at point C.

The asymmetric branch meets the symmetric branch again at point A. The Floquet multipliers pass +1 continuously and point A is therefore a continuous symmetry-breaking bifurcation. No new stick events are created at point A because all branches have two stick events per cycle. Remark that branch I behaves smooth at bifurcation point A and non-smooth at bifurcation B.

9.10 Discussion and Conclusions

In Section 9.7 some fundamental questions about discontinuous bifurcations of periodic solutions have been raised. With the preceding examples in mind, we try to answer these questions one by one.

1. *Do discontinuous bifurcations of periodic solutions of Filippov systems exist?* The conclusion that a point on a branch is a (continuous or discontinuous) bifurcation point (in the sense of Definition 8.1) was in the preceding examples always drawn from observation of the bifurcation diagram. We call a bifurcation point a discontinuous bifurcation point if it exposes a jump of the Floquet multipliers over the unit circle. Sections 9.8 an 9.9 show a number of discontinuous systems exposing bifurcations of periodic solutions in the sense of Definition 8.1. Some of the bifurcations are accompanied by a Floquet multiplier which passes in a continuous way through the unit circle. The branch of periodic solutions at such bifurcation points remains smooth. We classify those bifurcations as continuous bifurcations. Other bifurcation points are accompanied by a Floquet multiplier which crosses the unit circle discontinuously. The branch on such bifurcation points is continuous but non-smooth. We classify those bifurcations as discontinuous bifurcations of periodic solutions because of the discontinuous behaviour of the Floquet multipliers.
2. *Does a discontinuous bifurcation, caused when a periodic solution touches a switching boundary, exist?* A discontinuous bifurcation can only occur if

the periodic solution touches the tip of a non-smooth switching boundary as has been explained in Section 9.7. Figure 9.16 shows a bifurcation diagram in which there are two points where a periodic solution touches the tip of a non-smooth switching boundary as depicted in Figure 9.12. The point at $\omega = \omega_A$ is not a bifurcation point and is accompanied by a jump of the Floquet multipliers within the unit circle. The point at $\omega = \omega_B$ is a discontinuous fold bifurcation point and is accompanied by a jump of the Floquet multipliers through unit circle at $+1$. The stable periodic solution depicted in Figure 9.22 undergoes a discontinuous fold bifurcation depicted in Figure 9.23 if the periodic solution touches the lower-right corner of the stick region $\mathcal{U}$ (Figure 9.22). The discontinuous bifurcation is accompanied by a jump through the unit circle of the second Floquet multiplier.

3. *Does a jump of a Floquet multiplier through the unit circle under the influence of a parameter lead to a bifurcation of a periodic solution?* In all examples the conclusion that a bifurcation exists is taken from *inspection* of the bifurcation diagram. If there is a change in the number of periodic solutions for a certain parameter value, then there is a bifurcation of a periodic solution at this parameter value according to Definition 8.1. We also observe that a 'jump' of the Floquet multipliers over the unit circle exists. For systems with planar piecewise linear Poincaré maps with a single switching boundary, which are of the special form (9.33), it follows from Theorems 9.1 and 9.2 that at a bifurcation point the set of Floquet multipliers of the generalized fundamental solution matrix form a path which crosses the unit circle. Although it is intuitively appealing to state that at a bifurcation point the set of Floquet multipliers of the generalized fundamental solution matrix *must* have a crossing with the unit circle for a general Filippov system, we do not have a rigourous proof for this.
4. *Can we classify the bifurcation by inspecting the point(s) where the path of the set-valued Floquet multiplier(s) crosses the unit circle?* The discontinuous single crossing bifurcations of the preceding examples were 'classified' by comparing their nature with a certain type of continuous bifurcation. If at a discontinuous single crossing bifurcation the change of the number and type of periodic solutions is the same as for a certain type of continuous bifurcation, then the discontinuous single crossing bifurcation can be regarded as the discontinuous counterpart of that type of continuous bifurcation. For instance, Sections 9.8.1 and 9.9.1 show bifurcations which behave qualitatively like fold bifurcations. They are called discontinuous fold bifurcations because they resemble a conventional fold bifurcation and because of the discontinuous behaviour of the Floquet multipliers at the bifurcation point. The 'classification' of those bifurcations as fold bifurcations seems consistent with the fact that they are accompanied by a jump of a Floquet multiplier through $+1$. It has been made clear in Chapter 8 on bifurcations of equilibria in non-smooth continuous systems, that a classification of multiple crossing bifurcations can not be given, as the

terminology itself is already arbitrary. The same argument holds for bifurcations of periodic solutions in Filippov systems, because a periodic solution of a Filippov system can be regarded as a fixed point on a non-smooth (locally) continuous Poincaré map. A terminology for multiple crossing bifurcations based on the continuous bifurcations, which arise when smoothing the non-smooth system, is useless because the resulting sequence/structure of continuous bifurcations is generally not unique. A multiple crossing bifurcation can sometimes be regarded as the collision of two (or more) continuous bifurcations. However, the behaviour of a multiple crossing bifurcation is generally not the sum of continuous bifurcations. Consider for instance bifurcation point B in Section 9.8.2 exposing a double crossing through the unit circle at -1 and $+1$. A classification of this bifurcation as a discontinuous fold–period-doubling bifurcation because it resembles both bifurcations, is not completely satisfactory because branch VIII and the possibly infinitely many branches that are created at the discontinuous bifurcation point resemble neither a continuous fold nor a continuous period-doubling bifurcation. A terminology based on the behaviour of the multiple crossing bifurcation is therefore not possible. At least for *single* crossing bifurcations it seems intuitively correct, that we can make a kind of 'classification', or a useful terminology, based on the crossing of the Floquet multiplier(s) with the unit circle.

5. *Do discontinuous bifurcations of periodic solutions exist that do not have a continuous counterpart? In other words: do discontinuous bifurcations exist that behave qualitatively different from any continuous bifurcation?* Clearly, bifurcation point B in Figure 9.19 is a bifurcation point that does not have a direct continuous counterpart. This bifurcation behaves qualitatively different from any continuous bifurcation.
6. *Can the geometric definition of bifurcation and the definition of bifurcation through topological equivalence be inconsistent when they are applied to periodic solutions of Filippov systems?* The geometric definition of bifurcation (Definition 8.1) is based on a change of the number of equilibria and (quasi)-periodic solutions at a critical value of a parameter of the system. Definition 8.3 is based on a topological change of the phase portrait under variation of a parameter. These two definitions can be inconsistent. If we compare a periodic solution I which is not along an attractive sliding mode with a periodic solution II which is partly along an attractive sliding mode, then we can map every trajectory in the phase plane of periodic solution I to a trajectory in the phase plane of periodic solution II. However, the inverse map (mapping trajectories from II to I) does not exist due to the non-uniqueness in backward time of attractive sliding modes (see Chapter 3). Consequently, there is no homeomorphism between the two phase planes. A periodic solution with an attractive sliding mode is therefore topologically different from a periodic solution without a sliding mode. Consider point C in Figure 9.26. The point is not a bifurcation point according to Definition 8.1. The number of stick intervals changes

at this point from 2 to 1. Each stick interval is an attractive sliding mode. Consequently, point C is a bifurcation according to Definition 8.3. In [26] the term 'sliding bifurcation' is introduced for a change of a periodic solution with a sliding mode to a periodic solution without a sliding mode under influence of a parameter. The term 'multi-sliding bifurcation' is introduced for a change in the number of sliding modes. Point C in Figure 9.26 would according to this definition be a multi-sliding bifurcation. Although not explicitly stated in [26], it seems that a 'sliding bifurcation' is a bifurcation in the sense of Definition 8.3.

The different definitions for a bifurcation can lead to confusion when they are applied to non-smooth systems. This urges for a consensus about what is to be understood by the term 'bifurcation'. Definition 8.1 seems to be a good candidate, as it is very clear from the bifurcation diagram whether or not a point on a branch is a bifurcation point.

Equilibria in non-smooth continuous systems are related to periodic solutions in Filippov systems through the Poincaré map. We close this chapter by stating that the results on bifurcations of periodic solutions of Filippov systems seem to be consistent with the results on equilibria of non-smooth continuous systems.

10

Concluding Remarks

The preceding chapters present to some extent a theoretical framework as well as numerical methods for the analysis of non-smooth systems with an emphasis on mechanical systems with dry friction. Theory and methods are presented from Non-smooth Mechanics *and* Nonlinear Dynamics, which is the main objective of the book.

In the first part of the book, the reader has been made familiar with topics from Convex Analysis, Non-smooth Analysis and differential inclusions. The convexification method of Filippov transforms first-order discontinuous differential equations into first-order differential inclusions and is commonly used in Nonlinear Dynamics. From the mechanical point of view, one can regard non-smooth constitutive equations such as Coulomb friction as set-valued force laws, which leads to a description of the dynamics in terms of second-order differential inclusions (with possible state re-initializations at collisions). Set-valued force laws can be regarded as subdifferentials from non-smooth potentials. The theory of non-smooth potentials together with the principle of Gauß opens the way to optimization theory. The augmented Lagrangian method leads to an unconstrained optimization problem for the description of the dynamics of mechanical systems with set-valued force laws. The stationarity conditions of the augmented Lagrangian method result in a set of differential algebraic equations.

Numerical methods from the field of Non-smooth Mechanics and Nonlinear Dynamics have been presented for the numerical integration of non-smooth systems. The integration methods from Non-smooth Mechanics, such as the event-driven integration method and the time-stepping method, are able to deal with mechanical systems with many unilateral and/or frictional contact points. These integration methods require that a contact problem is solved in some or every time-step. The augmented Lagrangian method provides a general approach to solve the contact problem of complicated set-valued force laws.

The first part of the book deals with the formulation and numerical simulation of mechanical systems with set-valued force laws. Most methods and concepts of the first part of the book are developed in the Non-smooth Mechanics community. The second part of the book discusses the stability and bifurcation of periodic solutions, which is classically a topic of Nonlinear Dynamics. The saltations (or 'jumps') of the fundamental solution matrix have been addressed, which is of crucial importance when studying the stability of periodic solutions in Filippov systems.

The last two chapters are devoted to the bifurcation behaviour of non-smooth systems. Numerous examples are studied revealing a variety of bifurcation phenomena that are not encountered in smooth systems and are due to the non-smoothness of the systems. The considered systems are almost always mechanical, but similar bifurcation phenomena occur in for instance non-smooth electrical networks or non-smooth economical models. Loosely speaking, the term 'discontinuous bifurcation' is launched to denote bifurcations that are typical for non-smooth systems. A number of concepts and analysis techniques for discontinuous bifurcations are presented in the last two chapters. However, a complete theoretical framework for bifurcations in non-smooth systems is currently not available.

Such a theoretical framework should firstly include an analysis of possible bifurcations in low-dimensional systems. There might exist infinitely many types of bifurcations in non-smooth systems (even in the low-dimensional case), which would make it senseless to speak of a 'classification'. We therefore deliberately refrain from the word 'classification' and only speak of an 'analysis of possible bifurcations', in order to be as cautious as possible.

Secondly, the theoretical framework should contain a kind of reduction method of bifurcations in higher-dimensional non-smooth systems to 'normal forms' of discontinuous bifurcations, such as the Center Manifold Reduction for smooth dynamical systems. It remains dubious to speak of 'normal forms' as long as such a reduction method for non-smooth systems does not exist.

The construction of bifurcation diagrams is becoming a common analysis technique to study the behaviour of dynamical systems. All currently available software packages for the continuation of branches of equilibria or periodic solutions are restricted to the analysis of smooth dynamical systems. However, applying the existing continuation methods to non-smooth systems leads to theoretical as well as numerical problems at discontinuous bifurcation points. The continuation method is often not able to continue after the discontinuous bifurcation, or, if continuation succeeds, then it is not clear if there might be other branches emanating from the same bifurcation point. A future analysis of the discontinuous bifurcations may provide continuation methods for non-smooth systems with the necessary information of the number and type of branches which are connected by the bifurcation point. Moreover, an analysis of discontinuous bifurcations forms the starting point for the development of specialized continuation algorithms for non-smooth systems.

As the need for an analysis of discontinuous bifurcations is rising, attempts are being made to classify the bifurcations in subclasses of non-smooth systems. Bifurcations of periodic solutions in Filippov systems are nowadays receiving considerable attention as well as the related bifurcations of equilibria in non-smooth continuous systems. Some bifurcation theorems for planar systems can be proven (Chapter 8 and 9), but more general results seem currently not available. Surprisingly, the bifurcation behaviour of equilibria in Filippov systems has received very little attention in literature. Filippov systems generally contain sets of equilibria which are located on switching boundaries. The bifurcation analysis of equilibrium sets is hampered by the lack of theory on stability properties of equilibrium sets on switching boundaries. Moreover, the bifurcation behaviour of non-smooth systems with non-uniqueness (or even non-existence) of solutions in part of the state-space, as for instance occurs in the Painlevé problem, has barely been studied in literature. The framework of (non)linear complementarity problems and related constrained optimization problems appears to be a very useful tool to understand non-uniqueness/existence problems and might provide a means to analyze (if not to say classify) the related bifurcations.

References

1. Aizerman, M. A., and Gantmakher, F. R. On the stability of periodic motions. *Journal of Applied Mathematics and Mechanics (translated from Russian) 1* (1958), 1065–1078.
2. Alart, P., and Curnier, A. A mixed formulation for frictional contact problems prone to Newton like solution methods. *Computer Methods in Applied Mechanics and Engineering 92* (1991), 353–375.
3. Andersson, I. *Stick-slip motion in one-dimensional continuous systems and systems with several degrees-of-freedom.* Ph.D. thesis, Chalmers University of Technology, Göteborg, Sweden, 1980.
4. Andronov, A. A., Vitt, A. A., and Khaikin, S. E. *Theory of Oscillators (translated from Russian).* Pergamon Press, Oxford, 1966.
5. Anitescu, M., and Potra, F. A. Formulating dynamic multi-rigid-body contact problems with friction as solvable linear complementarity problems. *Nonlinear Dynamics 14*, 3 (1997), 231–247.
6. Armstrong-Hélouvry, B., Dupont, P., and Canudas de Wit, C. A survey of models, analysis tools and compensation methods for the control of machines with friction. *Automatica 30*, 7 (1994), 1083–1138.
7. Aubin, J.-P., and Cellina, A. *Differential Inclusions*, vol. 264 of *Grundlehren der mathematischen Wissenschaften.* Springer-Verlag, Berlin, 1984.
8. Begley, C. J., and Virgin, L. N. A detailed study of the low-frequency periodic behavior of a dry friction oscillator. *Journal of Dynamic Systems, Measurement, and Control 119* (1997), 491–487.
9. Begley, C. J., and Virgin, L. N. Impact response under the influence of friction. *Journal of Sound and Vibration 211*, 5 (1998), 801–818.
10. Blazejczyk-Okolewska, B., and Kapitaniak, T. Dynamics of impact oscillator with dry friction. *Chaos, Solitons & Fractals 7*, 9 (1996), 1455–1459.
11. Bockman, S. F. Lyapunov exponents for systems described by differential equations with discontinuous right-hand sides. *Proceedings of the American Control Conference* (1991), 1673–1678.
12. Brogliato, B. *Nonsmooth Mechanics*, 2 ed. Springer, London, 1999.
13. Brogliato, B., Ten Dam, A. A., Paoli, L., Génot, F., and Abadie, M. Numerical simulation of finite dimensional multibody nonsmooth mechanical systems. *ASME Applied Mechanics Reviews 55*, 2 (2002), 107–150.

14. Clarke, F. H. *Optimization and Nonsmooth Analysis.* Wiley, New York, 1983.
15. Clarke, F. H., Ledyaev, Y. S., Stern, R. J., and Wolenski, P. R. *Nonsmooth Analysis and Control Theory*, vol. 178 of *Graduate Texts in Mathematics.* Springer, New York, 1998.
16. Contensou, P. Couplage entre frottement de glissement et frottement de pivotement dans la théorie de la toupie. In *Kreiselprobleme; Gyrodynamics* (Berlin, 1963), H. Ziegler, Ed., Springer-Verlag, pp. 201–216. IUTAM Symposium Celerina 1962.
17. Cottle, R. W., and Dantzig, G. B. Complementary pivot theory of mathematical programming. *Linear Algebra and its Applications 1* (1968), 103–125.
18. Curnier, A., Ed. *Proc. Contact Mechanics International Symposium, Lausanne, Switzerland, October 7-9, 1992* (Lausanne, 1992), Presses Polytechniques et Universitaires Romandes.
19. Curnier, A. Unilateral contact. In Pfeiffer and Glocker [120], pp. 1–68.
20. Dankowicz, H., and Nordmark, A. B. On the origin and bifurcations of stick-slip oscillations. *Physica D 136*, 3-4 (2000), 280–302.
21. de Melo, W., and van Strien, S. *One dimensional dynamics*, vol. 25 of *A Series of Modern Surveys in Mathematics.* Springer Verlag, Berlin, 1993.
22. Deimling, K. *Multivalued differential equations.* De Gruyter, Berlin, 1992.
23. Deimling, K., and Szilágyi, P. Periodic solutions of dry friction problems. *Zeitschrift für angewandte Mathematik und Physik 45* (1994), 53–60.
24. di Bernardo, M., Feigin, M. I., Hogan, S. J., and Homer, M. E. Local analysis of C-bifurcations in n-dimensional piecewise-smooth dynamical systems. *Chaos, Solitons & Fractals 10*, 11 (1999), 1881–1908.
25. di Bernardo, M., Garafalo, F., Ianelli, L., and Vasca, F. Bifurcations in piece-wise smooth feedback systems. *International Journal of Control 75*, 16 (2002), 1243–1259.
26. di Bernardo, M., Johansson, K., and Vasca, F. Sliding orbits and their bifurcations in relay feedback systems. In *Proceedings of the 38th Conference on Decision & Control* (December 1999), pp. 708–713. CD-ROM publication.
27. Eich-Soellner, E., and Führer, C. *Numerical Methods in Multibody Dynamics.* Teubner, Stuttgart, 1998.
28. Elmer, F.-H. Nonlinear dynamics of dry friction. *Journal of Physica A: Mat. Gen. 30*, 17 (1997), 6057–6063.
29. Feeny, B. F., Guran, A., Hinrichs, N., and Popp, K. A historical review of dry friction and stick-slip phenomena,. *Applied Mechanics Reviews 51*, 5 (1998), 321–341.
30. Feigin, M. I. On the generation of sets of subharmonic modes in a piecewise-continuous system. *PMM 38*, 5 (1974), 810–818.
31. Feigin, M. I. On the the behavior of dynamic systems in the vicinity of existence boundaries of periodic motions. *PMM 41*, 4 (1977), 628–636.
32. Feigin, M. I. On the structure of C-bifurcation boundaries of piecewise-continuous systems. *PMM 42*, 5 (1978), 820–829.
33. Feigin, M. I. The increasingly complex structure of the bifurcation tree of a piecewise-smooth system. *Journal of Applied Mathematics and Mechanics 59*, 6 (1995), 853–863.
34. Fečkan, M. Bifurcations of periodic and chaotic solutions in discontinuous systems. *Archivum Mathematicum 34* (1998), 73–82.

35. Fey, R. H. B. *Steady-state behaviour of reduced dynamic systems with local nonlinearities.* Ph.D. thesis, Eindhoven University of Technology, The Netherlands, 1992.
36. Filippov, A. F. Differential equations with discontinuous right-hand side. *American Mathematical Society Translations, Series 2 42* (1964), 199–123.
37. Filippov, A. F. *Differential equations with discontinuous right-hand sides.* Kluwer Academic, Dordrecht, 1988.
38. Fletcher, R., Ed. *Optimization.* Academic Press, London, 1969.
39. Foale, S., and Bishop, R. Bifurcations in impacting oscillations. *Nonlinear Dynamics 6* (1994), 285–299.
40. Fortin, M. Minimization of some non-differentiable functionals by the augmented Lagrangian method of Hestenes and Powell. *Appl. Math. Optim. 2*, 4 (1976), 236–250.
41. Génot, F., and Brogliato, B. New results on Painlevé Paradoxes. *European Journal of Mechanics – A/Solids 18* (1999), 653–677.
42. Galvanetto, U., Bishop, S. R., and Briseghella, L. Mechanical stick–slip vibrations. *International Journal of Bifurcation and Chaos 5*, 3 (1995), 637–651.
43. Galvanetto, U., and Knudsen, C. Event maps in a stick-slip system. *Nonlinear Dynamics 13*, 2 (1997), 99–115.
44. Gamkrelidze, R. V. *Analysis II, Convex Analysis and Approximation Theory*, vol. 14 of *Encyclopaedia of Mathematical Sciences.* Springer-Verlag, New York, 1989.
45. Glendinning, P. *Stability, instablility and chaos: an introduction to the theory of nonlinear differential equations.* Cambridge University Press, Cambridge, 1994.
46. Glocker, Ch. *Dynamik von Starrkörpersystemen mit Reibung und Stößen*, vol. 18, no. 182 of *Fortschr.-Ber. VDI.* VDI Verlag, Düsseldorf, 1995.
47. Glocker, Ch. The Principles of d'Alembert, Jourdain and Gauß in Nonsmooth Mechanics, Part I: Scleronomic Multibody Systems. *Zeitschrift für angewandte Mathematik und Mechanik 78*, 1 (1998), 21–37.
48. Glocker, Ch. Discussion of d'Alembert's Principle for Non-smooth Unilateral Constraints. *Zeitschrift für angewandte Mathematik und Mechanik 79*, 1 (1999), S91–S94.
49. Glocker, Ch. Formulation of spatial contact situations in rigid multibody systems. *Computer Methods in Applied Mechanics and Engineering 177* (1999), 199–214.
50. Glocker, Ch. Scalar force potentials in rigid multibody systems. In Pfeiffer and Glocker [120], pp. 69–146.
51. Glocker, Ch. *Set-Valued Force Laws, Dynamics of Non-Smooth Systems*, vol. 1 of *Lecture Notes in Applied Mechanics.* Springer-Verlag, Berlin, 2001.
52. Guckenheimer, J., and Holmes, P. *Nonlinear Oscillations, Dynamical Systems, and Bifurcations of Vector Fields*, vol. 42 of *Applied Mathematical Sciences.* Springer-Verlag, New York, 1983.
53. Haessig, J. D. A., and Friedland, B. On the modeling and simulation of friction. *ASME Journal of Dynamic Systems, Measurement and Control 113* (1991), 354–362.
54. Hagedorn, P. *Non-Linear Oscillations.* Clarendon Press, Oxford, 1988.

55. Hairer, E., and Wanner, G. *Solving Ordinary Differential Equations II; Stiff and Differential-Algebraic Problems*, 2 ed., vol. 14 of *Springer Series in Computational Mathematics*. Springer, Berlin, 2002.
56. Heemels, M. *Linear complementarity systems: a study in hybrid dynamics*. Ph.D. thesis, Eindhoven University of Technology, The Netherlands, 1999.
57. Hestenes, M. Multiplier and gradient methods. *J. Opt. Th. Appl. 4* (1969), 303–320.
58. Hinrichs, N. *Reibungsschwingungen mit Selbst- und Fremderregung: Experiment, Modellierung und Berechnung*, vol. 11, no. 240 of *Fortschr.-Ber. VDI*. VDI Verlag, Düsseldorf, 1997.
59. Hinrichs, N., Oestreich, M., and Popp, K. On the modelling of friction oscillators. *Journal of Sound and Vibration 216*, 3 (1998), 435–459.
60. Hiriart-Urruty, J.-P., and Lemaréchal, C. *Convex Analysis and Minimization Algorithms I*, vol. 305 of *Grundlehren der mathematischen Wissenschaften*. Springer-Verlag, Berlin, 1993.
61. Hiriart-Urruty, J.-P., and Lemaréchal, C. *Convex Analysis and Minimization Algorithms II*, vol. 306 of *Grundlehren der mathematischen Wissenschaften*. Springer-Verlag, Berlin, 1993.
62. Ibrahim, R. A. Friction-induced vibration, chatter, squeal, and chaos; Part I: Mechanics of contact and friction. *ASME Applied Mechanics Reviews 47*, 7 (1994), 209–226.
63. Ibrahim, R. A. Friction-induced vibration, chatter, squeal, and chaos; Part II: Dynamics and modeling. *ASME Applied Mechanics Reviews 47*, 7 (1994), 227–253.
64. Ibrahim, R. A., and Soom, A., Eds. *ASME WAM, Proc. Symp. on Friction Induced Vibration, Chatter, Squeal, and Chaos* (New York, 1992), vol. 49, ASME.
65. Ivanov, A. P. Analytical methods in the theory of vibro-impact systems. *Journal of Applied Mathematics and Mechanics 57*, 2 (1993), 221–236.
66. Ivanov, A. P. Bifurcations in impact systems. *Chaos, Solitons & Fractals 7*, 10 (1996), 1615–1634.
67. Jean, M., and Moreau, J. J. Unilaterality and dry friction in the dynamics of rigid body collections. In Curnier [18], pp. 31–48.
68. Jellet, J. H. *Treatise on the Theory of Friction*. Hodges, Foster and Co, 1872.
69. Johnson, K. L. *Contact Mechanics*. Cambridge University Press, Cambridge, 1985.
70. Karnopp, D. Computer simulation of stick-slip friction in mechanical dynamic systems. *ASME Journal of Dynamic Systems, Measurement and Control 107* (1985), 100–103.
71. Khalil, H. K. *Nonlinear systems*. Prentice-Hall, New Jersey, 1996.
72. Klarbring, A. Mathematical Programming and Augmented Lagrangian Methods for Frictional Contact Problems. In Curnier [18], pp. 409–422.
73. Kunze, M. *Non-Smooth Dynamical Systems*. Springer, Berlin, 2000.
74. Kunze, M., and Küpper, T. Qualitative bifurcation analysis of a non-smooth friction oscillator model. *Zeitschrift für angewandte Mathematik und Physik 48*, 1 (1997), 87–101.
75. Kuznetsov, Y. A. *Elements of Applied Bifurcation Theory*. Springer-Verlag, New York, 1995.

76. LAURSEN, T. A., AND SIMO, J. C. Algorithmic symmetrization of Coulomb frictional problems using augmented Lagrangians. *Computer Methods in Applied Mechanics and Engineering 108* (1993), 133–146.

77. LECORNU, L. Sur le frottement de glissement. *Comptes Rendu des Séances de l'Academie des Sciences 140*, 6 (March 1905), 635–637,847–848.

78. LEINE, R. I. *Bifurcations in Discontinuous Mechanical Systems of Filippov-Type.* Ph.D. thesis, Eindhoven University of Technology, The Netherlands, 2000.

79. LEINE, R. I., BROGIATO, B., AND NIJMEIJER, H. Periodic motion and bifurcations induced by the Painlevé paradox. *European Journal of Mechanics – A/Solids 21*, 5 (2002), 869–896.

80. LEINE, R. I., AND GLOCKER, CH. A set-valued force law for spatial Coulomb–Contensou friction. *European Journal of Mechanics – A/Solids 22* (2003), 193–216.

81. LEINE, R. I., GLOCKER, CH., AND VAN CAMPEN, D. H. Nonlinear dynamics and modeling of various wooden toys with impact and friction. *Journal of Vibration and Control 9*, 1 (2003), 25–78.

82. LEINE, R. I., AND VAN CAMPEN, D. H. Fold bifurcations in discontinuous systems. *Proceedings of DETC'99 ASME Design Engineering Technical Conferences* (September 12-15 1999). Las Vegas, CD-ROM, DETC99/VIB-8034.

83. LEINE, R. I., AND VAN CAMPEN, D. H. Discontinuous bifurcations of periodic solutions. *Mathematical and Computer Modelling 36*, 3 (2002), 259–273.

84. LEINE, R. I., AND VAN CAMPEN, D. H. Discontinuous fold bifurcations in mechanical systems. *Archive of Applied Mechanics 72* (2002), 138–146.

85. LEINE, R. I., AND VAN CAMPEN, D. H. Discontinuous fold bifurcations. *Systems Analysis Modelling Simulation 43* (2003), 321–332.

86. LEINE, R. I., VAN CAMPEN, D. H., DE KRAKER, A., AND VAN DEN STEEN, L. Stick-slip vibrations induced by alternate friction models. *Nonlinear Dynamics 16*, 1 (1998), 41–54.

87. LEINE, R. I., VAN CAMPEN, D. H., AND KEULTJES, W. J. G. Stick-slip whirl interaction in drillstring dynamics. *ASME Journal of Vibration and Acoustics 124*, 2 (2002), 209–220.

88. LEINE, R. I., VAN CAMPEN, D. H., AND VAN DE VRANDE, B. L. Bifurcations in nonlinear discontinuous systems. *Nonlinear Dynamics 23*, 2 (2000), 105–164.

89. LÖTSTEDT, P. Coulomb friction in two-dimensional rigid-body systems. *Zeitschrift für angewandte Mathematik und Mechanik 61* (1981), 605–615.

90. MARTELLI, M. *Discrete dynamical systems and chaos.* Longman Scientific & Technical, New York, 1992.

91. MASON, M. T., AND WANG, Y. On the inconsistency of rigid-body frictional planar mechanics. *Proc. IEEE Int. Conf. Robotics and Automation* (1988), 524–528.

92. MEIJAARD, J. P. A mechanism for the onset of chaos in mechanical systems with motion-limiting stops. *Chaos, Solitons & Fractals 7*, 10 (1996), 1649–1658.

93. MEIJAARD, J. P. Efficient numerical integration of the equations of motion of non-smooth mechanical systems. *Zeitschrift für angewandte Mathematik und Mechanik 76*, 6 (1997), 419–427.

94. MIHAJLOVIĆ, N., VAN VEGGEL, A. A., VAN DE WOUW, N., AND NIJMEIJER, H. Analysis of friction-induced limit cycling in an experimental drill-string system. *ASME Journal of Dynamic Systems, Measurements and Control* (2004). in press.

95. MOLENAAR, J., DE WEGER, J. G., AND VAN DE WATER, W. Mappings of grazing-impact oscillators. *Nonlinearity 14* (2001), 301–321.
96. MOREAU, J. J. Dynamique de systèmes à liaisons unilatérales avec frottement secéventuel; essais numériques. Tech. Rep. 85-1, LMGC, Montpellier, France, 1986.
97. MOREAU, J. J. Unilateral contact and dry friction in finite freedom dynamics. In Moreau and Panagiotopoulos [98], pp. 1–82.
98. MOREAU, J. J., AND PANAGIOTOPOULOS, P. D., Eds. *Non-Smooth Mechanics and Applications*, vol. 302 of *CISM Courses and Lectures.* Springer, Wien, 1988.
99. MÜLLER, P. C. Calculation of Lyapunov exponents for dynamic systems with discontinuities. *Chaos, Solitons and Fractals 5*, 9 (1995), 1671–1681.
100. MURTY, K. G. *Linear Complementarity, Linear and Nonlinear Programming*, vol. 3 of *Sigma Series in Applied Mathematics.* Heldermann Verlag, Berlin, 1988.
101. NATSIAVAS, S. Periodic response and stability of oscillators with symmetric trilinear restoring force. *Journal of Sound and Vibration 134*, 2 (1989), 315–331.
102. NATSIAVAS, S., AND GONZALEZ, H. Vibration of harmonically excited oscillators with asymmetric constraints. *ASME Journal of Applied Mechanics 59* (1992), 284–290.
103. NAYFEH, A. H., AND BALACHANDRAN, B. *Applied Nonlinear Dynamics; Analytical, Computational, and Experimental Methods.* John Wiley & Sons, New York, 1995.
104. NAYFEH, A. H., AND MOOK, D. T. *Nonlinear Oscillations.* Wiley-Interscience, Chichester, 1979.
105. NORDMARK, A. B. Universal limit mapping in grazing bifurcations. *Physical review E 55*, 1 (1997), 266–270.
106. NUSSE, H. E., AND YORK, J. A. Border-collision bifurcations including "period two to period three" for piecewise smooth systems. *Physica D 57* (1992), 39–57.
107. OANCEA, V. G., AND LAURSEN, T. A. Investigations of low frequency stick-slip motion: experiments and numerical modelling. *Journal of Sound and Vibration 213*, 4 (1998), 577–600.
108. OESTREICH, M. *Untersuchung von Schwingern mit nichtglatten Kennlinien*, vol. 11, no. 258 of *Fortschr.-Ber. VDI.* VDI Verlag, Düsseldorf, 1998.
109. PAINLEVÉ, P. *Leçon sur le frottement.* Hermann, Paris, 1895.
110. PAINLEVÉ, P. Sur les lois du frottement de glissement. *Comptes Rendu des Séances de l'Academie des Sciences 121* (1895), 112–115.
111. PAINLEVÉ, P. Sur les lois du frottement de glissement. *Comptes Rendu des Séances de l'Academie des Sciences 141* (August 1905), 401–405 and 546–552.
112. PANAGIOTOPOULOS, P. D. A nonlinear programming approach to the unilateral contact-, and friction-boundary value problem in the theory of elasticity. *Ingenieur Archiv 44* (1975), 421–432.
113. PANAGIOTOPOULOS, P. D. *Hemivariational Inequalities.* Springer, Berlin, 1993.
114. PAOLI, L., AND SCHATZMAN, M. A numerical scheme for impact problems I: the one-dimensional case. *SIAM Journal on Numerical Analysis 40*, 2 (2002), 702–733.

115. Paoli, L., and Schatzman, M. A numerical scheme for impact problems II: the multidimensional case. *SIAM Journal on Numerical Analysis 40*, 2 (2002), 734–768.
116. Parker, T. S., and Chua, L. O. *Practical Numerical Algorithms for Chaotic Systems.* Springer-Verlag, New York, 1989.
117. Peterka, F. Bifurcations and transition phenomena in an impact oscillator. *Chaos, Solitons & Fractals 7*, 10 (1996), 1635–1647.
118. Pfeiffer, F., and Glocker, Ch. *Multibody dynamics with unilateral contacts.* Wiley, New York, 1996.
119. Pfeiffer, F., and Glocker, Ch., Eds. *Proc. IUTAM Symposium on Unilateral Multibody Contacts, Munich, Germany, August 3–7, 1998* (Dordrecht, 1998), Kluwer Academic Pulishers.
120. Pfeiffer, F., and Glocker, Ch., Eds. *Multibody Dynamics with Unilateral Contacts*, vol. 421 of *CISM Courses and Lectures.* Springer, Wien, 2000.
121. Pfeiffer, F., and Hajek, M. Stick-slip motion of turbine blade dampers. *Philosophical Transactions of the Royal Society A338* (1992), 503–517.
122. Poincaré. *Les Méthodes Nouvelles de la Méchanique Céleste.* Gautier-Villars, Paris, 1899.
123. Popp, K. Some model problems showing stick-slip motion and chaos. In Ibrahim and Soom [64], pp. 1–12.
124. Popp, K., Hinrichs, N., and Oestreich, M. Dynamical behaviour of a friction oscillator with simultaneous self and external excitation. *Sādhāna, Academy Proceedings in Engineering Sciences 20, Parts 2-4* (1995), 627–654.
125. Powell, B. T. A method for nonlinear constraints in optimization problems. In Fletcher [38].
126. Rasband, S. N. *Chaotic Dynamics of Nonlinear Systems.* Wiley, Chichester, 1990.
127. Reithmeier, E. *Periodic solutions of nonlinear dynamical systems.* Springer-Verlag, Berlin, 1991.
128. Rockafellar, R. T. *Convex Analysis.* Princeton Landmarks in Mathematics. Princeton University Press, Princeton, New Jersey, 1970.
129. Rockafellar, R. T. Augmented Lagrangians and applications of the proximal point algorithm in convex programming. *Mathematics of Operations Research 1*, 2 (1976), 97–116.
130. Rockafellar, R. T., and Wets, R.-B. *Variational Analysis.* Springer, Berlin, 1998.
131. Sastry, S. *Nonlinear Systems: Analysis, Stability and Control*, vol. 10 of *Interdisciplinary Applied Mathematics.* Springer, New York, 1999.
132. Sepehri, N., Sassani, F., Lawrence, P. D., and Ghasempoor, A. Simulation and experimental studies of gear backlash and stick-slip friction in hydraulic excavator swing motion. *ASME Journal of Dynamic Systems, Measurement and Control 118* (1996), 463–467.
133. Seydel, R. *Practical Bifurcation and Stability Analysis; From Equilibrium to Chaos.* Springer-Verlag, New York, 1994.
134. Simo, J. C., and Laursen, T. A. An Augmented Lagrangian treatment of contact problems involving friction. *Computers & Structures 42*, 1 (1992), 97–116.
135. Stelter, P. Nonlinear vibrations of structures induced by dry friction. *Nonlinear Dynamics 3* (1992), 329–345.

136. STELTER, P., AND SEXTRO, W. Bifurcations in dynamical systems with dry friction. *International Series of Numerical Mathematics 97* (1991), 343–347.
137. STEWART, D. E., AND TRINKLE, J. C. An implicit time-stepping scheme for rigid body dynamics with inelastic collisions and Coulomb friction. *International Journal for Numerical Methods in Engineering 39*, 15 (1996), 2673–2691.
138. STIEGELMEYR, A. Chimney dampers. In Pfeiffer and Glocker [119], pp. 299–308.
139. STIEGELMEYR, A. *Zur numerischen Berechnung strukturvarianter Mehrkörpersysteme*, vol. 18, no. 271 of *Fortschr.-Ber. VDI.* VDI Verlag, Düsseldorf, 2001.
140. UTKIN, V. I. Equations of slipping regimes in discontinuous systems I. *Automation and Remote Control 32* (1971), 1897–1907.
141. UTKIN, V. I. Equations of slipping regimes in discontinuous systems II. *Automation and Remote Control 33* (1972), 211–219.
142. UTKIN, V. I. *Sliding modes in control and optimization.* Springer-Verlag, Berlin, 1992.
143. UTKIN, V. I., GULDNER, J., AND SHI, J. *Sliding Mode Control in Electromechanical Systems.* Taylor & Francis, London, 1999.
144. VAN DE VORST, E. L. B. *Long Term Dynamics and Stabilization of Nonlinear Mechanical Systems.* Ph.D. thesis, Eindhoven University of Technology, The Netherlands, 1996.
145. VAN DE VRANDE, B. L., VAN CAMPEN, D. H., AND DE KRAKER, A. An approximate analysis of dry-friction-induced stick-slip vibrations by a smoothing procedure. *Nonlinear Dynamics 19*, 2 (1999), 157–169.
146. VAN DEN STEEN, L. *Suppressing stick-slip-induced drillstring oscillations: a hyperstability approach.* Ph.D. thesis, University of Twente, The Netherlands, 1997.
147. VAN DER SCHAFT, A. J., AND SCHUMACHER, J. M. *An Introduction to Hybrid Dynamical Systems*, vol. 251 of *Lecture Notes in Control and Information Sciences, vol. 251.* Springer, London, 2000.
148. WIERCIGROCH, M. On modelling discontinuities in dynamic systems. *Machine Vibration 5* (1996), 112–119.
149. WIERCIGROCH, M., AND DE KRAKER, A., Eds. *Applied Nonlinear Dynamics and Chaos of Mechanical Systems with Discontinuities*, vol. 28 of *World Scientific series on nonlinear science, Series A.* World Scientific, Singapore, 2000.
150. WIGGINS, S. *Global Bifurcations and Chaos.* Springer Verlag, Berlin, 1988.
151. WOLFSTEINER, P. *Dynamik von Vibrationsfördern*, vol. 2, no. 511 of *Fortschr.-Ber. VDI.* VDI Verlag, Düsseldorf, 1999.
152. YOSHITAKE, Y., AND SUEOKA, A. Forced self-excited vibration with dry friction. In Wiercigroch and de Kraker [149], ch. 10.

Index

Printing: Strauss GmbH, Mörlenbach
Binding: Schäffer, Grünstadt

MIX
Papier aus verantwortungsvollen Quellen
Paper from responsible sources
FSC® C105338

If you have any concerns about our products,
you can contact us on
ProductSafety@springernature.com

In case Publisher is established outside the EU,
the EU authorized representative is:
Springer Nature Customer Service Center GmbH
Europaplatz 3, 69115 Heidelberg, Germany

Printed by Libri Plureos GmbH
in Hamburg, Germany